THE
RACHEL LAMBERT MELLON
COLLECTION

S. Le Clerc in. et f.   Goyton ex.

AN OAK SPRING

HERBARIA

HERBS AND HERBALS FROM THE FOURTEENTH TO THE NINETEENTH CENTURIES

A Selection of the Rare Books, Manuscripts and Works of Art in the Collection of Rachel Lambert Mellon

LUCIA TONGIORGI TOMASI &
TONY WILLIS

EDITED WITH A DESCRIPTION
OF THE AMERICAN HERBALS BY
MARK ARGETSINGER

UPPERVILLE VIRGINIA
OAK SPRING GARDEN LIBRARY

2009

PRINTED IN THE UNITED STATES OF AMERICA

LIBRARY OF CONGRESS CONTROL NUMBER
2009926555

ISBN 978-0-9654508-1-2

★

*Translated from
the Italian by Lisa Chien*

FRONTISPIECE:

DENIS DODART, *Mémoires pour servir à l'Histoire des Plantes*, 1676, frontispiece depicting Louis XIV on an official visit to the Académie Royale des Sciences

Dedicated to

my dearest friend & daughter

ELIZA LLOYD MOORE

(1942–2008)

★

With all my love

RACHEL LAMBERT MELLON

CONTENTS

LIST OF ILLUSTRATIONS *page* xi
FOREWORD *by Rachel Lambert Mellon* xxi
ACKNOWLEGEMENTS xxiii
INTRODUCTION *by Lucia Tongiorgi Tomasi* xxvii
DESCRIPTIVE METHOD xlvii
BIBLIOGRAPHY xlix

I · LATE MEDIEVAL HERBALS

1. Konrad von Megenberg, *Buch der Natur* (manuscript), *c.* 1350 3
2. Italian school, Herbal manuscript, *c.* 1425 9
3. *Hortus Sanitatis*, Paris, *c.* 1500, and Venice, 1511 14, 19
4. *The Grete Herball*, London, 1526 23

II · THE GREAT AGE OF RENAISSANCE BOTANY

5. Otto Brunfels, *Herbarum vivæ eicones*, 1530 31
6. Leonhart Fuchs, *De historia stirpium*, 1542 37
7. Pietro Andrea Mattioli, *Commentarii*, 1565, plus original woodblocks 45, 52
8. William Turner, *A New Herball*, 1568 56
9. Carolus Clusius, *Rariorum aliquot stirpium*, 1576 62
10. Rembert Dodoens, *A Niewe Herball*, 1578 67
11. Fabio Colonna, *Phytobasanos*, 1592 72
12. John Gerard, *The Herball, or Generall Historie of Plantes*, enlarged by Thomas Johnson, 1636 79
13. Adriaan van de Spiegel, *Isagoges in rem herbariam*, 1633 85

III · HERBALS AND PLANTS FROM DISTANT LANDS

14. Carolus Clusius, *Aliquot notæ in Garciæ Aromatum Historiam*, 1582 93
15. Prospero Alpini, *De plantis Aegypti liber* and *De balsamo dialogus*, 1592 95
16. Nicolas Monardes, *Joyfull Newes Out of the New-found Worlde*, 1596 99
17. Garcia da Orta, *Dell' Historia dei Semplici Aromati*, 1616 104
18. Jean Robin, *L'Histoire des Plantes Aromatiques*, 1619, and *Histoire des Plantes*, 1620 108
19. Georg Wolfgang Wedel, *Opiologia*, 1682 111
20. Nicolas de Blégny, *Le bon usage du Thé du Caffé et du Chocolat*, 1687 115
21. Thomas Malie, Eight plant studies (drawings), 1728–1741 121
22. Chinese school, Drawing of Chinese plants (John Bradby Blake), *c.* 1770–1774 131
23. Giovanni Domenico Civinini, *Della Storia e Natura del Caffè*, 1731 138
24. Niccolò Gavelli, *Storia Distinta, e Curiosa del Tabacco*, 1758 141
25. John Ellis, 'Dionæa Muscipula or Venus's Flytrap' & 'Directions for bringing over Seeds and Plants' (drawings), 1769, and *An Historical Account of Coffee*, 1774 145
26. Baldassare Cattrani, *Exoticarum atque indigenarum plantarum* (manuscript), 1776–1800 152
27. Ambroise-Marie-François-Joseph Palisot de Beauvois, *Flore d'Oware et de Benin*, 1804–1807 159

IV · HERBALS BY HERBALISTS, PHARMACISTS, AND PHYSICIANS

28. Antoine Du Pinet de Noroy, *Historia Plantarum*, 1561 167
29. Castore Durante, *Herbario Nuovo*, 1585 169
30. Paul de Reneaulme, *Specimen Historiæ Plantarum*, 1611 178

31. Jacques and Paul Contant, *Les Oeuvres*, 1628 180
32. French school, Eight medicinal plant paintings, *c.* 1630 186
33. Antonio Donati, *Trattato de Semplici*, 1631 198
34. William Coles, *Adam in Eden*, 1657 202
35. John Hill, *The British Herball*, 1756 206
36. Johann Wilhelm Weinmann, *Phytanthoza-iconographia*, 1737–1745 210
37. François Alexandre Pierre De Garsault, *Description, Vertus et Usages de … Plantes*, 1767 217
38. John Edwards, *The British Herbal*, 1770 223
39. Andreas Friedrich Happe, *Herbarium pictum* (manuscript), *c.* 1780–1790 226
40. Joshua Webster, *Webster's Distribution of English Medicinal Plants* (manuscript), *c.* 1780 230
41. Nicholas Culpeper, *Culpeper's Complete Herbal*, 1828 235

V · HERBALS OF THE BOTANICAL GARDENS AND PRIVATE GARDENS OF EUROPE

42. Joachim Camerarius the Younger, *Hortus medicus et philosophicus*, 1588 245
43. Guy de la Brosse, *Description du Jardin Royal des Plantes Médecinales*, 1636 251
44. Denis Dodart, *Mémoires pour servir à l'Histoire des Plantes*, 1676 257
45. Jan and Caspar Commelin, *Horti Medici Amstelodamensis Rariorum … Plantarum*, 1697–1701 260
46. Carolus Linnaeus (Carl von Linné), *Hortus Cliffortianus*, 1737 266
47. Giovanni Battista Morandi, *Collectio Plantarum* (manuscript), 1737 270
48. Giorgio Bonelli, *Hortus Romanus*, 1772–1793 277
49. Gaetano Savi, *Materia Medica Vegetabile Toscana*, 1805 286
50. John Lindley, *Flora Medica*, 1838 290

VI · CURIOUS AND STRANGE HERBALS

51. Giovanbattista della Porta, *Phytognomonica*, 1588 297
52. Joachim Camerarius the Younger, *Symbolorum & Emblematum*, 1590–1604 302
53. Claude Duret, *Histoire Admirable des Plantes et Herbes*, 1605 311
54. Pierre Pomet, *A Compleat History of Druggs*, 1712 315
55. Jean Pierre Rambosson, *Histoire et Légendes des Plantes*, 1869 319

VII · DRIED SPECIMENS AND NATURE PRINTING

56. Johannes Harder, *Historia stirpium* (manuscript), *c.* 1595 327
57. English school, Nature prints of plants and trees from England (manuscript), early 1700s 329
58. Carlo Sembertini [attributed], *Pharmaco Diversarum Plantarum* (manuscript), *c.* 1720 334
59. Christian Gottlieb Ludwig, *Ectypa Vegetabilium*, 1760–1764 339

VIII · AMERICAN HERBALS

60. Samuel Henry, *A New and Complete American Medical Family Herbal*, 1814 347
61. Jacob Bigelow, *American Medical Botany*, 1817–1821 355
62. Constantine Samuel Rafinesque, *Medical Flora*, 1828–1830 365
63. Peter Peyto Good, *The Family Flora and Materia Medica Botanica*, 1845 (1854) 376

INDEX 387

VI · CURIOUS AND STRANGE HERBALS

51. Giovanbattista della Porta, *Phytognomonica*, 1588 297

LIST OF ILLUSTRATIONS

DENIS DODART, *Mémoires pour servir à l'Histoire des Plantes*, 1676.
Frontispiece depicting Louis XIV on an official visit to the Académie Royale des Sciences *frontispiece*

H. L. DUHAMEL DU MONCEAU, *Traité des arbres fruitiers*, 1768.
An oak twig tail-piece, volume 1, page 147 *title-page*

JOSHUA WEBSTER, *Webster's Distribution of English Medicinal Plants*, *c.* 1780.
Frontispiece: an angel coming to the aid of a wounded shepherd xxvi

KONRAD VON MEGENBERG, *Buch der Natur*, *c.* 1350.
Plant secretions: *balsam*, extracted into two small bottles, folio 241[r] xxxv

[*H*]*ortus Sanitatis*, Venice, 1511. Herbalists discussing a plant, a1[v] xxxvi

LEONHART FUCHS, *De historia stirpium*, 1542.
Oak tree (*Quercus robur*), page 229 xxxvii

PROSPERO ALPINI, *De Balsamo dialogus*, 1592.
'Balsamum,' Balm of gilead (*Commiphora opobalsamum*), page 78 xxxix

JOACHIM CAMERARIUS, *Hortus medicus et philosophicus*, 1588.
Sunflower (*Helianthus annuus*), volume 1, page 51 xli

DENIS DODART, *Mémoires pour servir à l'Histoire des Plantes*, 1676.
Vignette on page 1 showing a group of academics watching a chemistry experiment xlii

GIOVANNI BATTISTA MORANDI, *Collectio plantarum*, 1737.
Various parts of *Tradescantia virginiana*, 'Ephemerum virginiana,' fol. 5 xliii
Bulb and leaves, possibly of a Squill (*Scilla* sp.), with inflorescence of a Yucca (*Yucca* sp.), folios 92 and 93 xliv

BRADFORD MEDICAL INSTITUTE, *A Brief Treatise on Various Ailments and Their Treatment by Nature's Remedies*, London, *c.* 1890.
Elder or Elderberry (*Sambucus nigra*), page 12 xlvi

KONRAD VON MEGENBERG, *Buch der Natur*, 14th century.
Binding 2
Five small trees: apple, pear, hazel, cherry, and one in center unidentified, folio 194[v] 5
Crocus sp., folio 246[v] 6

KONRAD VON MEGENBERG, *Buch der Natur,* 14th century (*cont.*)
Cucubits (*Cucurbita pepo*) growing on poles, folio 247 v 7

ITALIAN SCHOOL, *An herbal,* circa 1455.
Viper's bugloss (*Echium vulgare*), folio 65 r 10
Flowers of *Viola* sp., folio 79 v 12
Details of various plants: top left, Lady's mantle (*Alchemilla vulgaris*); top right, Grape vine (*Vitis vinifera*); center, Belbine (*Calystegia sepium*); below, Monkshood (*Aconitum* sp.), folio 118 v 13

[*H*]*ortus Sanitatis,* Paris, *c.* 1500 and Venice, 1511.
Paris edition, colophon, volume II, B8 v 14
Paris edition, title-page, volume II 16
Paris edition, verso of title-page, skeleton with the bone names in Latin, volume II 17
Paris edition, Basil (*Ocimum basilicum*) growing in a pot, volume I, f3 v 18
Venice edition, fore-edge painting by Titian's nephew, Cesare Vecellio, of a Numidian crane, roses with stem and foliage, and a lion 20
Venice edition, two woodcuts: 'Capitulum.ccliij,' a member of the Bladderwort family (*Lentibulariaceae*), and 'Capitulum.ccliiij,' possibly a Lentil (*Ervum lens*), P6 v 21
Venice edition, detail, woodcut of Adam and Eve with serpent around an Apple tree (*Malus domestica*), T6 v 22

The Grete Herball, London, 1526.
Title-page 25
Colophon, 2E6 v 26

OTTO BRUNFELS, *Herbarum vivæ eicones,* 1530.
Title-page 30
Stinging nettle (*Urtica dioica*), page 151 33
Liverleaf (*Hepatica*), page 190 35

LEONHART FUCHS, *De historia stirpium,* 1542.
Portrait of Fuchs handling a species of *Veronica,* verso of title-page 38
Herb Paris (*Paris quadrifolia*), page 87 41
The Wild Basil or Hedge Basil (*Clinopodium vulgare*), page 896 43

PIETRO ANDREA MATTIOLI, *Commentarii,* 1565.
A species of Lavender (*Lavandula*), page 32 47
Crocus sp., possibly *Crocus biflorus,* page 69 49

Woodblock: Sea lavender (*Limonium* sp.), image on page 980 53
Woodblock: Rush (*Juncus* sp.), image on page 1,036 54
Woodblock: Wood cranesbill (*Aconitum*) possibly *Geranium sylvaticum*, image on page 1,088 55

WILLIAM TURNER, *A New Herball*, 1568.
Yellow Star of Bethlehem, a species of field *Gagea*, page 97 59
'Autumn crocus,' Wild saffron (*Colchicum autumnale*), left, in flower and right, with fruit, page 156 61

CAROLUS CLUSIUS, *Rariorum aliquot stirpium*, 1576.
'Scammonea Valentina,' Swallow wort (*Cynanchum acutum*), page 226 65

REMBERT DODOENS, *A Niewe Herball*, 1578.
Title-page 69
European common twayblade (*Listera ovata*) and the Bird's nest orchid (*Neottia nidus-avis*), page 223 71

FABIO COLONNA, *Phytobasanos*, 1592.
Two Primeroses (*Primula acaulis* and *Primula officinalis*), page 21 75
A member of the *Umbelliferae* family, possibly a species of *Pimpinella* or of *Oenanthe*, page 76 76

JOHN GERARD, *The Herball, or Generall Historie of Plantes*, 1636.
Two plantains: Great Plantain (*Plantago major*) and Hoary Plantain (*Plantago media*), page 419 81
Virginian Potatoes (*Solanum tuberosum*), page 927 83

ADRIAAN VAN DE SPIEGEL, *Isagoges*, 1633. Title-page 87

CARLOUS CLUSIUS, *Aliquot notæ in Garciæ*, 1582.
Coconut palm (*Cocos nucifera*), page 10 92
Cacao seeds (*Theobroma cacao*), page 29 94

PROSPERO ALPINI, *De plantis Aegypti liber*, 1592.
'Papyrus Burdi' (*Cyperus papyrus*), page 43 97

NICOLAS MONARDES, *Joyfull Newes Out of the New-Found Worlde*, 1596.
Tobacco (*Nicotiana rustica*), folio 34 101
A member of the family Armadillo (*Dasypodidae*), folio 73 102

GARCIA DE ORTA, *Dell'Historia dei Semplici Aromati*, 1616.
Cashew (*Anacardium occidentale*), page 143 105
Sassafras (possibly *Sassafras albidum*), page 407 106

JEAN ROBIN, *L'Histoire des Plantes*, 1619. *Histoire des Plantes*, 1620.
L'Histoire des Plantes, 1619. Title-page 108
L'Histoire des Plantes, 1619. Fruit of the 'Sang de Dragon' (*Calamus draco*), page 652 110
Histoire des Plantes, 1620. 'Canna Indica flore rubro,' Indian shot (*Canna indica*), page 15 110

GEORG WOLFGANG WEDEL, *Opiologia*, 1682.
Title-page (*Papaver somniferum*) 113

NICOLAS DE BLÉGNY, *Le bon Uusage du Thé du Caffé et du Chocolat*, 1687.
Frontispiece and title-page, *Camellia thea* 117
Coffeepot with spirit burner, page 149 119

THOMAS MALIE, Eight Plant Studies, 1728–1741.
'The American Plum' (*Spondias purpurea*) 122
'Anona Maxima' (*Annona muricata*) 123
'The Bichy Tree,' Kola nut (*Cola acuminata*) 124
'Cacao' (*Theobroma cacao*) 125
'The Citron' (*Citrus medica*) with insects 127
'Mammaia Sapota,' Sapote (*Pouteria sapota*) 128
'Mammone,' Cherimoya (*Annona cherimolia*) 129
'Seaside or Mangrove grape tree,' Seagrape (*Cocoloba uvifera*) and 'Cedar tree' (*Cedrela*) 130

CHINESE SCHOOL, Drawings of Chinese Plants, circa 1770–1774.
Binding of silk brocade boards 133
A variety of 'Teene' (*Leguminosae*), volume I, folio 40 134
'Gardenia,' possibly a, Common Gardenia (*Gardenia jasminoides*), volume II, folio 3 135
Two leaves from the Chinese dictionary of various plants, insects, and fowl, 42 verso and 43 recto 137

GIOVANNI DOMENICO CIVININI, *Della Storia e Natura del Caffè*, 1731.
Coffee plant, possibly *Coffea arabica*, between pages 6 and 7 139

NICCOLÒ GAVELLI, *Storia Distinta, et Curiosa de Tobacco*, 1758.
'Nicotiana Tabacco' (*Nicotiana tabacum*), A2 143

JOHN ELLIS, The Venus Fly Trap & Plant Collecting Boxes—*An Historical Account of Coffee*, 1774.
Venus's-flytrap (*Dionaea muscipula*) 148

Traveling containers for seeds and plants 150
Title-page and frontispiece of coffee plant (*Coffea arabica*) 152

BALDASSARE CATTRANI, *Exoticarum*, 1776–1800.
Manuscript title leaf 155
'Convallaria Polygonatum,' Angular Solomon's seal (*Polygonatum odoratum*) 156
'Hypericum Androsæmum,' St.-John's Wort or Tutsan (*Hypericum and rosaemum*) 157

AMBROISE MARIE FRANÇOIS JOSEPH PALISOT DE BEAUVOIS, *Flore d'Oware et de Benin*, 1804–1807.
Napoleona imperialis, a cauliflory, showing the blue flowers on the limbs 163

ANTOINE DU PINET, *Historia Plantarum*, 1561.
Title-page 166
Olive (*Olea europaea*), page 81 168

CASTORE DURANTE, *Herbario Nuovo*, 1585.
'Chameleon Bianco,' Silver thistle (*Carlina acaulis*), and 'Chameleon Nero,' (*Carlina* cf. *gummifera*), page 119 on blue paper 171
'Corniolo,' Cornelian Cherry (*Cornus mas*) and Coronopo Domestica,' Buck's-horn Plantain (*Plantago coronopus*), page 149 on white paper 173
Detail: Grapes (*Vitis vinifera*) drying on a table, page 483 on blue paper 175
Six medicinal plants: 'Pavate,' not identified (*Morinda citrifolia?*); 'Persicaria' (*Persicaria maculosa*), Red Shank; 'Pedicularia' (*Pedicularis* sp.), Lousewort; 'Phillitide' (*Phyllitis scolopendrium*), a variety of Hart's-tongue fern; 'Pellosela' (*Hieracium pilosella*), Mouse-ear Hawkweed; 'Piramidale' (*Ajuga pyramidalis*), Pyramidal Bugle; Vv4 176

PAUL DE RENEAULME, *Specimen Historiæ Plantarum*, 1611. Common mullein (*Verbascum thapsus*), page 102 179

JACQUES AND PAUL CONTANT, *Les Oeuvres*, 1628.
Title-page, *7 181
90 plants described in the text with their reference numbers, *8 183
Title-page to the second part, containing 24 plants along the outer borders with reference numbers that are described in the text 185

FRENCH SCHOOL, Eight medicinal plants.
Mandrake (*Mandragora officinarum*) 187
Silver thistle (*Carlina acaulis*), Carline thistle (subsp. *caulescens*) 189

FRENCH SCHOOL, Eight medicinal plants (*cont.*)
Aloe, Common aloe or Medicinal aloe 190
False Helleborine (*Veratrum album*) 191
Laurel, Butcher's Broom (*Ruscus hypoglossum*) 194
Lavender (*Lavandula* cf. *stoechas*) 195
Datura metel 196
Fritillaria pallidiflora 197

ANTONIO DONATI, *Trattato de Semplici,* 1631.
Sea Bindweed (*Calystegia soldanella*), page 82 201

WILLIAM COLES, *Adam in Eden,* 1657. Title-page 203

JOHN HILL, *The British Herball,* 1756.
A page of *Umbelliferae:* 'Great Hercules Allheal'; 'Black Libanotis'; 'Common Dill'; 'Broad leav'd Thapsia,' Penny Cress; 'Narrow leav'd fennell Giant'; 'Broad leav'd fennell Giant'; 'Laserwort'; 'Common Cummin'; 'Masterwort'; 'Lovage'; 'Sermountain'; 'Common Skirret'; 'Common Anise'; 'Common Parsley'; 'Common Bishopweed'; 'Spanish Toothpick'; 'Candy Daucus'; 'Umbelliferous Pellitory'; 'Black Masterwort.' Plate 60 207
A page of ferns and allied plants: 'Common Harts-tongue' (*Phyllitis scolopendrium*); 'Polypody' (*Polypodium vulgare*); 'Rough Spleen-wort' (*Blechnum spicant*); 'Smooth Spleen-wort' (*Ceterach officinarum*); 'Dwarf Sea Fern'; 'English Maiden-hair'; 'Forked Maiden-hair' (*Asplenium viride*); 'Common Male Fern' (*Dryopteris filix-mas*); 'White Maiden-hair' (*Asplenium ruta-muraria*); 'The True Maiden-hair' (*Adiantum pedatum*); 'Common Female Fern' (*Athyrium filix-femina*); 'Black Maiden-hair' (*Asplenium adiantumnigrum*); 'Winged Maiden-hair' (*Polystichum* sp.); 'Osmund Royal' (*Osmunda regalis*); 'Adders Tongue' (*Ophioglossum vulgare*); 'Moonwort' (*Botrychium lunaria*); 'Common Duckweed' (*Lemna minor*); 'Large Duckweed' (*Spirodela polyrhiza*); 'Great Water Horsetail' (*Equisetum fluviatile*); 'Wood Horsetail' (*Equisetum sylvaticum*). Plate 74 209

JOHANN WILHELM WEINMANN, *Phytanthoza-iconographia,* 1736.
Frontispiece portrait of Johann Weinmann, volume I 213
Five varieties of Lavender (*Lavandula*), volume III, plate 632 214
Two varieties of *Solanum: S. pensile* and *S. melongena*, volume IV, plate 933 215

FRANÇOIS ALEXANDRE PIERRE DE GARSAULT, *Description, Vertus et Usages de … Plantes,* 1767.
'Vanilla' (*Vanilla planifolia*) and 'Cardamomum elettari,' True cardamom or Green cardamom (*Elettaria cardamomum*), vol. I, plate 55 219
'Adiantum ruta-muraria,' Wall-rue (*Asplenium ruta-muraria*), and 'Adiantum Polytrichum,' Maidenhair Spleenwort (*Asplenium trichomanes*), vol. II, plate 128 221
'Lactuca sativa,' Garden lettuce (*Lactuca sativa*), vol. III, plate 314 222

JOHN EDWARDS, *The British Herbal,* 1770.
Gas plant or Burning bush (*Dictamnus albus*), plate 19 225

ANDREAS FRIEDRICH HAPPE, *Herbarium pictum, c.* 1780.
'Le Dragonier,' Dragon's blood (*Calamus draco*), volume II, folio 149 227
'Papaver hortense,' Opium poppy (*Papaver somniferum*), volume III, folio 272 228

JOSHUA WEBSTER, *Webster's Distribution of English Medicinal Plants, c.* 1780.
Stinging Nettle (*Urtica dioica*), with two mosquitoes (*Culicidae*), fol. 70 231
'Sweet Fennel' (*Foeniculum vulgare*), folio 23 233
'Asparagus vul. Common Asparagus' (*Asparagus officinalis*), folio 164 234

NICHOLAS CULPEPER, *Culpeper's Complete Herbal,* 1828.
Frontispiece portrait and image of his home, part 1 238
Printed wrapper, part 2 239
'Common Alder,' Alder (*Alnus glutinosa*); 'Garden Arrach,' Orache (*Atriplex hortensis*); 'Alehoof,' Ground Ivy (*Glecoma hederacea*); 'Agrimony' (*Agrimonia eupatoria*); 'Water Agrimony' (*Bidens tripartita*); 'Alexander' (*Angelica* sp.); 'Amaranthus Flower Gentle' (*Amaranthus hypochondriacus*); 'Anemone' (*Anemone* sp.); 'Asparagus' (*Asparagus officinalis*); 'Alacost' (*Lysimachia* sp.?); 'Angelica,' Wild Angelica (*Angelica sylvestris*) or (*A. archangelica*); 'Allheal,' an unidentified member of *Lamiaceae* family. Plate 1 240

JOACHIM CAMERARIUS, *Hortus medicus et philosophicus,* 1588.
'Alcea arborescens' (*Althaea* sp.), Aa3 247
'Nasturtium Indicum' (*Tropaeolum majus*), Ee 249

GUY DE LA BROSSE, *Description du Jardin Royal des Plantes,* 1636.
Folding plate of the 'IARDIN DU ROY POUR LA CULTURE DES PLANTES MEDECINALES, A PARIS, 1636' 252–53
One leaf from index, G1, listing some plants cultivated in the garden 255

DENIS DODART, *Mémoires pour servir à l'Histoire des Plantes*, 1676.
'Millefolium odoratum,' Yarrow (*Achillea millefolium*), EEe verso 259

JAN AND CASPAR COMMELIN, *Horti Medici Amstelodamensis Rariorum ... Plantarum*, 1697–1701.
Frontispiece, volume I 262
Jan Commelin's coat-of-arms, volume I 263
'Aster Africanus,' possibly a member of *Polyarrhena* or *Felicia*, South African Compositae, volume II, plate 28 265

CAROLUS LINNAEUS, *Hortus Cliffortianus*, 1737.
Title-page 267
Kaempferia sp., plate 3 269

GIOVANNI BATTISTA MORANDI, *Collectio Plantarum*, 1737.
Title-page with specimens of *Gagea, Viola, Vinca, Knautia, Scilla*, fol. 1 271
Index leaf, fol. 16, with Morandi's signature 273
Blood lily (*Haemanthus* sp.), folio 30 275

GIORGIO BONELLI, *Hortus Romanus*, 1772–1793.
Title-page, volume I 279
'Prospectus Horti Romani,' volume I 280–81
Rhubarb (*Rheum palmatum*), volume I, plate 34 283
Water-prince (*Nymphoides peltata*), volume II, plate 67 284

GAETANO SAVI, *Materia Medica Vegetabile Toscana*, 1805.
Ramping fumitory (*Fumaria capreolata*), Common fumitory or Earth Smoke (*Fumaria officinalis*), plate 1 289

JOHN LINDLEY, *Flora Medica*, 1838. Title-page 291

GIOVANBATTISTA DELLA PORTA, *Phytognomonica*, 1588.
Title-page 296
Toothwort (*Lathraea squamaria*); Pomegranate (*Punica granatum*); *Pinus pinea*; all showing resemblance to teeth, page 138 301

JOACHIM CAMERARIUS, *Symbolorum et Emblematum*, 1590.
Man harvesting balsam resin (*Commiphora opobalsamum*), vol. I, page 38 305
Cluster of clover (*Trifolium* sp.) with two serpents, vol. I, page 76 306
Crocus sp., vol. I, page 77 309

CLAUDE DURET, *Histoire Admirable des Plantes et Herbes*, 1605.
Barnacle Goose (*Branta leucopsis*) hatching from a seashell and growing on a tree (*Lepas anatifera*, Pelagic gooseneck barnacle, a Crustacean), page 303 313

'Boramet' (*Agnus sciticus*) growing on a stem from the mother plant, page 330 314

PIERRE POMET, *A Compleat History of Druggs*, 1712.

'Peruvian Bark' (*Cinchona* sp.); 'The Male Mandrake' (*Mandragora officinalis*); 'The Female Mandrake' (*Mandragora autumnalis*); 'The Cork Tree' (*Quercus suber*), from 'Book the Fourth. Of Barks,' plate 30 317

Natives 'stringing & rolling tobacco,' from 'Book the Fifth. Of Leaves,' plate 37 318

JEAN PIERRE RAMBOSSON, *Histoire et Légendes des Plantes*, 1869.

Man drinking water from the cut stem of the 'Arbre du voyageur' (*Ravenala madagascariensis*), plate 26 321

Cityse commun ou false ébénier (*Laburnum anagyroides*), plate 42 322

Harvesting of tea, plate 102 323

JOHANNES HARDER, *Historia stirpium, c.* 1595.

Three species of violets (*Viola canina, V. alba, V. odorata*) and a species of felwort (*Gentianella* sp.), folio 20 326

Arum maculatum, (Cuckoo-pint), folio 21 328

ENGLISH SCHOOL, Nature prints of plants and trees from England, early 1700s.

Double-page spread of nature prints: 1. 'Asrabacca,' or European wild ginger (*Asarum europaeum*); 2. 'Silver Weed' (*Potenilla anserina*); 3. 'Indian Cresses,' or Indian Cress (*Tropaeolum majus*); 4. 'Sweet Brier' (*Rosa rubiginosa*); 5. 'Fool's Parsley' (*Aethusa cynapium*), volume 1 332–33

CARLO SEMBERTINI, *Pharmaco, c.* 1720.

Title-page 335

Leaves of two not identified monocots and *Ipomoea quamoclit* at the center, folio 29 337

Two specimens of *Leontopodium alpinum, Geranium argenteum* at the center, two specimens on each side of *Antennaria dioica* s.l. and *Micropus erectus* (*Bombycilaena erecta*) at the center, folio 53 338

CHRISTIAN GOTTLIEB LUDWIG, *Ectypa Vegetabilium,* 1760–1764.

Black salsify or Spanish salsify (*Scorzonera hispanica*), plate 96 341

Common thorn apple (*Datura stramonium*), plate 108 342

SAMUEL HENRY, *A New and Complete American Medical Family Herbal,* 1814.

Title-page 346

SAMUEL HENRY, *American Medical Family Herbal*, 1814 (*cont.*)

Common Sweet Flag (*Acorus calamus*), page 9 351

Elderberry (*Sambucus nigra*), page 102 353

JACOB BIGELOW, *American Medical Botany*, 1817–1821.

Thorn apple (*Datura stramonium*), volume I, part I, plate I 359

Skunk cabbage (*Ictodes foetidus*), volume II, part I, plate 4 361

Elliotts' gentian, or American gentian, or Bottle gentian (*Gentiana catesbaei*), volume II, part II, plate 34 363

CONSTANTINE SAMUEL RAFINESQUE, *Medical Flora*, 1828–1830.

Blue Cohosh (*Caulophyllum thalictroides*), volume I, plate 19 369

Twinleaf (*Jeffersonia diphylla*), volume II, plate 55 371

American ginseng (*Panax quinquefolius*), volume II, plate 71 373

PETER PEYTO GOOD, *The Family Flora and Materia Medica Botanica*, 1845–1854.

Common dandelion (*Taraxacum officinale*), volume I, plate 29 377

Chart of the vegetable kingdom, according to Linnaeus, volume II 379

'Helonias dioica,' Blazing-star or Rattlesnake root (*Chamaelirium luteum*), volume II, plate 76 381

FOREWORD

AN OAK SPRING HERBARIA, the fourth catalogue in the Oak Spring Garden Library series, describes the Library's remarkable books and artwork on herbs and herbals. From the beginning of early civilization to the present, herbs have played an extraordinary role. Whether gathered for curing the sick from all walks of life, for cooking or seasoning of foods, or for aromatic and aesthetic purposes, herbs continue to be grown and studied to benefit our basic needs and wants.

An herb, as defined by Liberty Hyde Bailey in *The Standard Cyclopedia of Horticulture*, 'is a plant that dies to the ground each year, or at least that does not become woody. It may be annual, as bean, pigweed; biennial, as mullein, parsnip; perennial, as dictamnus, rhubarb.' As one can clearly see, this type of classification covers a wide spectrum of plants!

During the warmer months, many herbs grow wild around our library, which is located in the Virginia Piedmont. The most common ones are the Common Mullein, Dandelion, Plantain, Poke Weed, and Yarrow:

– COMMON MULLEIN (*Verbascum thapsus*), a very majestic plant, has been used over time to treat earaches and help with asthma. It is also called the Candlestick Plant; when the stalk is dipped in suet it will burn like a candle.

– DANDELION (*Taraxacum officinale*), which is considered a nuisance to most gardeners, is a very powerful diuretic. In the early spring, the fields here are covered with its bold yellow flowers. An infusion of the root can cleanse the body of impurities and the flowers are used to make wine.

– PLANTAIN (*Plantago major*), a very pervasive plant, has been used to treat cuts and open wounds. The leaves act as an antibacterial agent that prevents infection.

– POKE WEED (*Phytolacca americana*) is a plant that has been used by Native Americans for centuries. This plant is edible in the early spring, but if not used or cooked correctly it can be very toxic. The beautiful berries are also poisonous and can be used to make a purple dye. Today Poke Weed is being studied as a treatment in curing cancer.

– YARROW (*Achillea millefolium*) has been used for centuries to stop bleeding. It is presently used in the treatment of common colds, kidney disorders, and toothaches. *Achillea* comes from the mythical Greek character, Achilles, who carried this plant with his army as a treatment for battle wounds.

Watching these charming plants grow truly helps to bring our books, drawings, and manuscripts to life by blending the past with the present and future. It is also very interesting to see how many of the old-fashioned remedies that are mentioned in these early books and manuscripts are still in use today.

This diverse library of books and drawings is forever giving us hope and new ideas. Lucia Tongiorgi Tomasi, a loyal and dear friend who wrote *An Oak Spring Flora*, has once again come from Italy to America to work with the Library staff on this book. Lucia's knowledge of herbs and herbals is extraordinary. It has been an incredible and fun journey. I am very grateful for her help.

My dear friend and librarian of twenty-eight years, Tony Willis, has worked with Lucia and he has kindly helped me with this foreword. Tony started working here in 1981 and he has helped to build this Library's collection with great enthusiasm and spirit.

Mark Argetsinger, who designed our other catalogues in the series, *An Oak Spring Sylva*, *Pomona*, and *Flora*, has helped to produce another beautiful volume as the editor and book designer. My talented grandson, Stacy Lloyd, and the staff—Susan Leopold, Kimberley Fisher, Gloria Woodson, and Ricky Willis—help so much to keep this Library moving towards the future. I am grateful to all for their artistic talents, knowledge, and friendship.

Just like the early herbalists and historians who were constantly seeking knowledge, hope, and treatment from herbs, we at the Library are always learning and making new discoveries along the way. This, in itself, is the real and genuine inspiration for *An Oak Spring Herbaria*.

RACHEL LAMBERT MELLON
Oak Spring
1 April 2009

ACKNOWLEDGEMENTS

DURING the preparation of this catalogue, many people helped make its publication possible. A very special and heart-felt thanks must first be given to Rachel Lambert Mellon. As with the previous three catalogues, *An Oak Spring Sylva*, *Pomona*, and *Flora*, she was actively involved and always interested in its progress. Without her passion for learning and collecting, there would be neither a Library nor catalogue series. Mrs. Mellon began collecting printed and manuscript herbals at a very young age. She has always been interested in and fascinated with plants and their mystical qualities. Even today, she continues to learn with great enthusiasm about plant-based and holistic medicines and their remedies. We shall always be grateful to her.

Lisa Chien from Lucca, Italy, fluidly translated the text. As she did for *Flora*, Lisa helped to bring life and vision to each entry written by Lucia. Also, a great deal of thanks and recognition need to be given to Mark Argetsinger, book designer and editor. Mark oversaw every aspect of the project from its inception with efficiency and eagerness. His research and writing on the American section have brought great interest and clarity to the book. Sophie Argetsinger, his daughter, entered the book's text corrections with great accuracy. David Lorczak of Finlay Printing helped with the checking of the color proofs. His advice, along with Jim Morris from Digital Color Productions, on photography and lighting was very helpful.

The botanists, Professor Garbari from Italy and Professor Marion Lobstein of Virginia, helped with the botanical nomenclature. Professor Garbari was able to identify several of the plants with their modern or current botanical names.

Colleagues here at the Oak Spring Garden Library also worked diligently. Susan Leopold, with her vast knowledge of plants and botany, helped with the editing of the manuscript. She also learned how to work with the Library's new digital camera and equipment and taught and trained her co-workers, Kimberley Fisher and Stacy Lloyd. Gloria Woodson helped with proof-reading. We are also grateful to other friends and colleagues, both in the United States and in Italy for their work and input: Luca Ciancio, Melissa Fritts, Jay Keys, Mark Oliver, Giuseppe Olmi, Therese O'Malley, Milu Rodrigues, George Shaffer, Charles Thomas, Alessandro Tosi, and Ricky Willis. Tony Willis and Susan Leopold began the researches on the American entries and Christopher Hoolihan, History of Medicine Librarian at the Edward G. Miner Library, University of Rochester Medical Center, in Rochester, New York, was instrumental in setting them on a sound footing.

Finally, we should recognize all the botanists, scholars, doctors, and artists who are the subject of this publication—from Konrad von Megenberg to Peter Peyto Good. Their pioneering spirit, knowledge, and desire to help humanity have greatly influenced medicine and technology in the twenty-first century.

Natura pinxit remedia,
in floribus, in seminibus, visuque ipso
animos invitavit

NATURE HAS PAINTED HER
REMEDIES IN FLOWER AND SEED, BY THE
VERY SIGHT OF WHICH SHE HAS
ATTRACTED
THE MINDS OF MEN

★

GIOVANNI DOMENICO CIVININI
Della Storia e Natura del Caffè
1731

JOSHUA WEBSTER, *Webster's Distribution of English Medicinal Plants*, *c.* 1780. Frontispiece depicting a shepherd with an injured leg sitting beneath a tree; a radiant angel by his side gestures toward the pastoral landscape with one hand, while in the other is held a banner that reads *Hic san*[*itas*] ('Here is healing')

INTRODUCTION

TOWARDS the end of the sixteenth century in Rome, when Federico Zuccari—*principe* of the Accademia di San Luca, which counted the most renowned artists of the period among its members—set out the inspired decorative program for his private residence in Via Gregoriana, it was his wish to dedicate one of the rooms to a celebration of the art of *disegno*. A convinced proponent of the 'primacy of drawing' among the visual arts, one of the arguments that he used to demonstrate the divine origin and intellectual nature of the art of drawing was the pivotal contribution made by this skill to a variety of other fields, including medicine and the sciences. For the allegorical painting that was to decorate the ceiling, therefore, he conceived a *Wunderkammer* where, amidst alembics and the texts of the classical authorities, various personages are shown studying models and conducting anatomical dissections. Presiding over the scene is the Greek physician Galen who holds a caduceus, the symbol of medicine, in one hand whilst with the other he points to an illustrated herbal, the essential tool for those engaged in the healing arts.

In this fresco the herbal takes the form, in accordance with codified tradition, of a manuscript filled with drawings of various plants endowed with healing 'virtues,' but, in fact, over the centuries the word assumed a variety of formats and meanings—manuscript herbal, painted herbals, printed herbal, herbarium, dried herbal, and so on—and would take concrete form in specific typologies that presented scholars with intricate problems of a historical, textual, philological, and iconographic nature.

The first botanical-medical treatises date to the classical period, and were based on an even more ancient knowledge of the healing virtues and poisonous properties of different plants. Mastery over these powerful substances—believed to be subject to astrological influences and often connected with magical practices—was considered to be the special prerogative of women. It is no coincidence that a knowledge of plants and the ability to brew lethal poisons, miraculous cures, and palliative nostrums were traditionally attributed to witches and *signore delle erbe*, or women versed in plant lore.

From the civilization of ancient Egypt to those of the Middle East and the Orient, from the classical age to the first decades of the sixteenth century, the history of the botanical sciences was closely linked to that of the herbal. Initially in manuscript form and then—with the invention of the printing press—in published volumes, herbals provided morphological descriptions of medicinal plants, the names by which each was known in different languages, and (sometimes) information on their habitat. Preparations made from a single vegetable (or animal or mineral) component were known as 'simples' and the same term was used to refer to the plants themselves. Frequently, however, identification was difficult on the basis of a written description alone, and to help those gathering herbs in the wild or engaged in the study of unfamiliar species, authors began to supplement their texts with more or less accurate representations of the entities being described.

Given the essentially practical ends for which these herbals were destined, their authors and owners were generally physicians, herbalists, spice vendors, rhizotomists ('root cutters'), and druggists. During the Middle Ages, from their activities would spring the first chemists' shops and pharmacy-laboratories (from *pharmakon*, the Greek word for medicine); initially connected with hospitals and convents, in these busy ateliers medicines in the form of syrups, unguents, and pills were prepared and sold. In *The Name of the Rose*, Umberto Eco offers a fascinating and historically accurate description of the pharmacy in the Benedictine monastery at the foot of the Alps where the novel's mystery unfolds, and of the diligent activities of the friar and herbalist Severinus. Herbals, therefore, constitute historical documents of extraordinary significance because of the light they can shed on the development of the botanical, medical, and pharmacological sciences, as well as on the history of the scientific illustration. But who were the *veteres auctores* (ancient authors) to whom we owe the birth of the herbal?

The author of the first catalogue of plants mentioned in the ancient sources appears to have been Diocles of Carystus, a physician who lived and worked in Athens around 350 B.C., and whose writings on natural history would influence his contemporary, the philosopher Aristotle (384–322 B.C.). A large corpus of botanical texts that has survived can be ascribed to the physician Theophrastus of Eresos (*c.* 370–288 B.C.), who studied under Aristotle and succeeded him as director of the Lyceum. Compiled following his period of service in the army of Alexander the Great (356–323 B.C.), of particular interest are several sections in the ninth book of Theophrastus' Περὶ φυτῶν ἱστορίας (*Peri phutōn historias*, or *Historia plantarum*), which are in fact configured as an herbal. It can therefore be affirmed that during the course of the fourth and third centuries B.C. a specific format for the text of the herbal—and perhaps also for the illustrations, although no concrete evidence of this has come down to us—was established, a typology that acquired the authoritative weight of tradition and would not undergo substantial modification for centuries.

In fact, the codification over time of beliefs in the therapeutic properties of specific plants led to a reluctance to modify their descriptions or illustrations in any way, and knowledge was transmitted through the passive copying of texts and illustrations from one manuscript to the next. Gradually the images underwent a process not only of simplification, but also of symbolic stylization, and as a result the connection between the drawing and the plant as it actually appeared in nature sometimes was lost. All the same, it is important to recall that the herbal was nearly always viewed by its owner as a 'living text,' a work-in-progress that could be carried along on herborizing expeditions or consulted in the workroom, to which he would add annotations and glosses based on his own observations with a frequency and regularity that is not encountered in the manuscripts or printed books in any other field.

Other outstanding personages from Antiquity would leave their marks on the science of botany and the production of herbals. These included the Roman author Pliny the Elder (A.D. 23–79), several chapters of whose *Naturalis historia* were devoted to the subject of plants; Mithridates VI Eupator, the king of Pontus (120–63 B.C.), famed for his knowledge of poisons and their antidotes; and Mithridates' personal physician and rhizomatist Crateuas (111–64 B.C.), the author of a work on the nature and therapeutic uses of plants, Ῥιζοτομικόν (*Rhizotomikon*) and, most importantly, an herbal accompanied by colored drawings (as recorded by Pliny, *Nat. hist.* XXV.8).

The classical author, however, who had the greatest influence on the science of botany was certainly Dioscorides Pedanius of Anazarbus in Anatolia (present-day Turkey), who lived between the years A.D. 40 and 90. Dioscorides, like Theophrastus, began his career as a military physician; attached to the Roman army of Nero, he had the opportunity to visit many European and Asian countries in Rome's vast empire. In A.D. 65 Dioscorides produced his *opus magnum*—Περὶ ὕλης ἰατρικῆς (*Peri hulēs iatrikēs*), or *De materia medica* (On medicinal substances), as it is generally known in its Latin title—in which he describes over six hundred plants and more than one thousand medicinal preparations. *De materia medica* was destined to become the most authoritative text on the subject; translated first into Latin and then into many other languages, integrated with other texts, and published in innumerable editions, it served as the primary source of botanical knowledge and terminology throughout the Middle Ages and well into the Renaissance.

While Crateuas' illustrated herbal has, unfortunately, been lost, some notion of it can be reconstructed, on the textual side, from Dioscorides' quotations and, on the iconographical side, from the *Codex Vindobonensis*, the most important and earliest illustrated herbal to have survived from Antiquity. Also known as Codex Juliana Anicia, it was commissioned in Constantinople around A.D. 512 and is conserved today in the Austrian National Library (*Österreichische Nationalbibliothek, Vienna, Med. Gr. 1*). Many of its images may derive from Crateuas' herbal. This imposing manuscript was much prized by its successive owners, as is attested to by the numerous annotations penned next to the text and illustrations in Greek, Arabic, Hebrew, Turkish, and Latin. Eventually, in 1568 the codex was acquired by the Hapsburg emperor Maximilian II, at the urging of Ogier Ghiselin de Busbecq, ambassador to the Sublime Porte, that is, to the court of Süleyman I the Magnificent. The *Codex Vindobonensis* of Dioscorides, which has been placed by UNESCO on its Heritage of Humanity list, is composed of a series of texts on medicine and natural history based in large part on Dioscorides' *De materia medica*. Arranged in alphabetical order are a series of plants, splendidly depicted in color on large parchment pages, and in this case the unknown artist has not limited himself to copying earlier prototypes, but has based his portrayals on the study of the live plants themselves.

Another significant contribution to botany was made by Galen, or Claudius Galenus (the central figure in the fresco commissioned by Federico Zuccari), a physician born in Pergamum around A.D. 130 whose medical theories would dominate Western medicine for over a millennium. Among his prolific output were several books, written in Greek and generally known as *De simplicibus*, on the subject of drugs and their uses. Galen's work was translated into Latin at the end of the fifth century A.D. and, like many of the herbals of Antiquity, was subjected to re-elaboration and interpolation with other texts over time. Galen's work on *materia medica* is of a lower order than that of Dioscorides, since he describes medicaments not as 'simples' but as a complex polypharmacy of herbs.

The most celebrated herbal in the Latin-speaking world was compiled in the fifth century A.D. and has been attributed to a certain Apuleius Platonicus, a figure whose exact historical outlines are still vague. Apuleius drew amply on the traditions of Pliny, Dioscorides, and Galen, but at the same time introduced a novel component that had little counterpart in the Greek herbals of the period—the marked presence of magical elements (although Galen himself endorsed the use of magical amulets). This can be seen, for example, in his description of the mandrake, the fabulous plant *par excellence*, and the legends associated with it. Apuleius' work would exert an enormous influence over the herbals of the Middle Ages and in addition would be instrumental in the propagation throughout Europe of many medical-alchemistic notions, such as the Galenic theory of the four humors of man over which, it was believed, many plants could exercise an influence. The spread of Christianity as well contributed in no small measure to enrich the iconography of the plant world through its repertoire of sacred symbols.

With the collapse of the Roman Empire in the sixth century A.D., it fell to Muslim scholars to continue the tradition of botanical study and the dissemination of knowledge through their own herbals. Beginning in the ninth century, the work of Dioscorides was elaborated upon and expanded by Arab physicians, and then disseminated during the twelfth and thirteenth centuries in herbals whose text and illustrations reflected a completely new typology. The most important contribution in this sense was an herbal of about eight hundred plants compiled by the illustrious Muslim scientist Abdallah Ibn al-Baitar, who was born in the Spanish city of Málaga, and died in Damascus in 1248. While he based his work on the texts of Theophrastus, Dioscorides, and Galen, Ibn al-Baitar corrected many of their errors and added numerous Persian and Indian species that were up to then completely unknown in the West.

Another family of illustrated herbals originated in southern Italy, in the milieu of the celebrated medical school of Salerno, which reached its apex in the twelfth century and succeeded in assimilating and unifying the Greek, Latin, and Arab schools of medicine. The herbal of Salerno, known by its opening phrase *Circa instans*, was translated beginning in the thirteenth century into all the languages of Europe, as well as into Hebrew, and contributed decisively to the botanical knowledge connected with medicine; authors of the most important encyclopaedias of the Middle Ages, from Albertus Magnus and Thomas de Cantimpré to Vincent of Beauvais and Konrad von Megenberg, borrowed many notions from it. In 1422 the illustrious faculty of medicine at the University of Paris decreed that *Circa instans* should serve as the codex for all herbalists. The *Scuola Salernitana* introduced many innovations in its herbals; not only were mineral and animal components added to the pharmacopeia, but artists were encouraged to express themselves with greater liberty in their botanical illustrations and in the scenes and vignettes that enliven many of the pages.

Nevertheless, next to the *veteres auctores*, whose names and teachings were handed down by scholarly tradition, there seems to have existed a host of more modest 'herbalists' who by dint of their own efforts managed to acquire an empirical knowledge of medicinal plants. Indeed, the fourteenth and fifteenth centuries witnessed the proliferation of manuscript herbals of the most varied provenance, sometimes written in Latin but more often in the vernacular. Once again we find ourselves considering works that were essentially empirical in nature, compiled by or for persons who looked upon them as instruments necessary to their work, which could then be handed on, at each stage augmented with new annotations and comments. At the same time during this crucial period we find artists beginning to pay closer attention to natural reality, progressively shedding the figurative schemata imposed by tradition.

With the spread of the printing press, the successful format of the manuscript herbal was transferred more or less unchanged to the first incunabula and these works played an important role in the dissemination of botanical knowledge. The first herbal to be produced using the new technology was the compendium of Apuleius, which was printed in Rome between 1481 and 1483. In this volume the botanical entities are presented in alphabetical order in chapters, each one furnished with a description and an illustration printed from a woodblock or metal-plate engraving, with results that—at least in the earliest works—appear somewhat primitive, especially when compared with the remarkable naturalism of the paintings in many of the finest manuscript herbals. These incunabula were quickly followed all over Europe by printed works that, although at first accompanied by somewhat inexact illustrations, would contribute to the spread of this typology of the herbal and open the way for experiments with new ways of describing and portraying plants. The herbal, therefore, represents a chapter of immense significance in the history of printing, engraving, and techniques for coloring illustrations.

With the cultural revolution brought about by the Renaissance—which encompassed not only the *humanae litterae* of history, literature, and the arts, but also the scientific study of nature—the botanical sciences, and, as a consequence, the herbal, underwent profound transformations. To the practical ends of the herbal were added a pronounced theoretical and intellectual component and, at the same time, the obligation to adhere more closely to natural reality. The texts of the classical authorities were subjected to critical analysis on the basis of the plants as they actually existed in nature, and priority was given to the systematic description of species unknown to Antiquity. At the same time, traditional stylized schemata were set aside in favor of the minute and faithful representation of plants *dal vivo*. A particular interest in the study of local flora emerged (indeed, the botanical term 'flora' seems to have been coined in this period) and, as the insufficiency of presenting plants in simple alphabetical order became clear, more rational methods for their classification were elaborated. While such topical issues seemed to have the most urgent claim on their attention, botanists never lost their interest in the medical properties of plants. A case in point is provided by the physician and naturalist Pietro Andrea Mattioli; his *Commentarii* on Dioscorides' *De materia medica* became the most important botanical text of the sixteenth century, and was still regarded as an authoritative source during the entire course of the following century. In his work Mattioli added the descriptions of many 'modern' species, including a careful review of their medical properties.

Alongside these *historiae plantarum*, whose texts were characterized by a more rigorous, theoretical approach to the science of botany, herbals intended for everyday use by physicians and healers continued to proliferate, genuine *vademecums* produced in smaller, more manageable formats. This period, furthermore, witnessed the spread of the dried herbal (*hortus siccus, hortus mortus*), which allowed botanists to study plants even during those months of the year when they could not be found in nature. At the same time, the painted herbal (*hortus pictus*), with illustrations rigorously drawn from life, reached new heights of artistry and scientific accuracy.

Beginning in the second half of the sixteenth century, another development would exert a profound influence on the typology of the herbal. This was the study and dissemination by the botanical community of exotic plants brought by traders and explorers from distant lands, in particular from the New World and also from the Far East and the continent of Africa. Many of these species were completely new to the floral panorama of Europe and as physicians began to use them in the treatment of patients, a vigorous debate was sparked off regarding the efficacy of medicines prepared from native, as opposed to foreign, plants. The increasingly rigorous scientific approach to the study of plants led to the establishment of the first *giardini dei semplici* or botanical gardens in Europe where, under the supervision of university professors or learned *studiosi rei herbariae*, native and exotic species were cultivated and their characteristics, including their medicinal properties, were systematically studied. For this reason, the format of the first catalogues produced by many of these gardens closely resembled that of the traditional herbal.

Furthermore, in the Botanical Garden of Pisa, and similarly in other European cities, a *fonderia* was set up—a chemical laboratory, before they were know as such—in which plants were subjected to chemical procedures to extract and distil their active components. Here the first serious attempts were made to supersede the antiquated, pseudo-scientific practices of alchemy, which combined notions of metallurgy and physics with elements of mysticism and magic in the quest not only for a process that would transform base metals into gold, but also for a universal panacea that could cure all the illnesses of mankind. Once the pastime of a select few, as the success of these laboratories spread many were established and run privately by herbalists who prepared expensive remedies from rare plants, the precursors to the modern, state-licensed pharmacy.

Among the exotic species whose discovery provoked the greatest interest in Europe and whose impact extended well beyond their supposed curative properties, were the coffee, tea, cacao, and tobacco plants. Fortunes were built upon the products made from these plants, which also provided the occasion for new forms of socializing (one need only consider the *cafeteries* and *chocolateries* that enjoyed an enormous vogue in France and England of the eighteenth century).

Towards the end of the seventeenth century, as the new field of chemical processes emerged upon the scene, significant strides were made in the field of botanical pharmacology as well. In this period the new science of pharmaceutical botany, which sought to investigate the active principles of medicinal plants and their underlying chemical composition, was created. Studies on the transformation of matter, beginning with the extraction of 'volatile salts' from plants, would lead to the discovery of opium and other key substances. This period saw the birth of the science of chemistry, and it was Louis XIV's personal pharmacist, Nicolas Lémery (1645–1715), who first made the crucial distinction between organic chemistry (based on plants) and inorganic chemistry (based on minerals).

Also not to be forgotten is the increasing use, beginning in the second half of the seventeenth century, of the microscope, the extraordinary instrument that finally allowed scientists to study the complex structure of plants in minute detail. Nevertheless, revolutionary developments in methodology did not entirely sweep away the magical component that had accompanied the study of plants from its very origins. For example, the 'theory of signatures' was still being expounded in botanical treatises and herbals well into the eighteenth century, just as lively discussion continued regarding the influence of the stars and planets on certain medicinal plants.

Herbals would play an important role in the young country of America, whose luxuriant flora included many plants that had been used for centuries by the indigenous peoples. These provided a precious new resource for settlers whose homesteads were situated in the middle of vast, empty territories, far from any centers of habitation with their pharmacies and hospitals.

The complicated historic context that we have attempted to sketch out can be retraced through the extraordinary collection of manuscripts and printed herbals in the Oak Spring Garden Library, almost every one of them beautifully illustrated. A selection of the most outstanding works are presented here, divided into eight sections, although we have been constrained—with immense regret—to leave out many noteworthy examples.

⋆

The first section is devoted to *Late Medieval Herbals*, works that marked the passage from the medieval to the early modern age. We begin with *Buch der Natur*, written between 1349 and 1351 by the German theologian and teacher Konrad von Megenberg, which constitutes one of the most interesting and influential 'encyclopaedias' of the late Middle Ages. Von Megenberg dedicates many pages to the world of nature, writing a precise and detailed excursus on plants, especially those known for their medicinal properties. His text circulated widely in manuscript form and was eventually published in 1475 in Augsburg. Various copies of *Buch der Natur* have survived, among which the splendid manuscript dating from the middle of the fourteenth century and conserved in the Oak Spring Garden Library constitutes one of the earliest.

Another text of supreme importance is the *Hortus Sanitatis*, which may be considered the first true herbal compiled in Europe. Published in Mainz in 1491, it was quickly followed by other editions in every country of Europe. Two particularly fine examples are presented here, both of which are embellished with a large number of pleasing and realistic woodcut illustrations—the circa 1500 Paris edition and the 1511 Venice edition, the latter of which was once owned by the bibliophile Odorico Pillone.

The second section, *The Great Age of Renaissance Botany*, reviews the splendid and variegated production of herbals during the High Renaissance, beginning with *Herbarum vivæ eicones* (1530). Compiled by the Carthusian monk Otto Brunfels, this was the first work to describe not only plants well known from the classical tradition, but also many species indigenous to the author's own country of Germany. Furthermore, with its beautifully

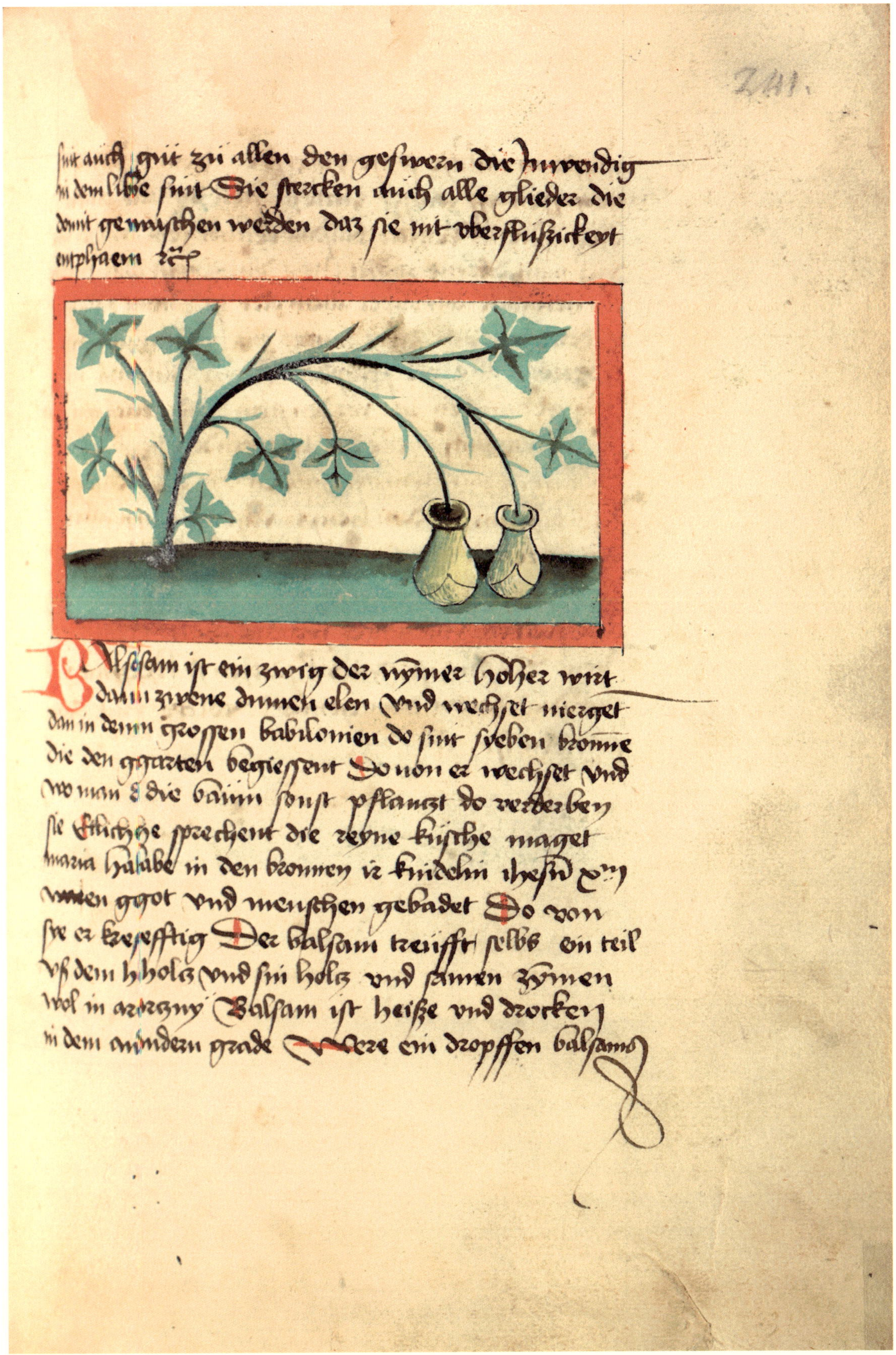

sint auch gut zu allen den geswern die inwendig
in dem libe sint Die sterken auch alle glieder die
damit gemusthen werden daz sie mit uberflussikeyt
entphaem etc

Balsam ist ein zweig der nymer hoher wirt
dann zwene dumen elen vnd wechset niergent
dan in deme grossen babilonien do sint syben bronne
die den garten begiessent Do von er wechset vnd
wo man die baum sunst pflanzt do verderben
sie Etliche sprechent die reyne kusche maget
maria habe in den bronnen ir kindelin ihesu crist
unsern got und menschen gebadet Do von
sie er krefftig Der balsam treifft selbs ein teil
vß dem holcz vnd sin holcz vnd samen zymen
wol in arczny Balsam ist heisse vnd drocken
in dem andern grade Were ein dropffen balsams

KONRAD VON MEGENBERG, *Buch der Natur, c.* 1350. Plant secretions: *balsam*, extracted into two small bottles, folio 241 r

[*H*]*ortus Sanitatis*, Venice, 1511. Herbalists discussing a plant, folio a1^{v}

LEONHART FUCHS,
De historia stirpium, 1542.
Oak tree (*Quercus robur*),
page 229

executed woodcut engravings, Brunfels' herbal inaugurated the new practice of depicting plants from life.

The herbals discussed in this section made a fundamental contribution to the development of the botanical sciences in Europe, providing the vehicle for men of science now engaged in the systematic study of plants to record and share the results of their work. The visual component of these texts is of extraordinary importance; the clarity of their layouts and the beauty of their illustrations, which were often the fruit of a close collaboration with well-known artists, marked a crucial step forward in the art of book production. A magnificent series of illustrated herbals by some of the most distinguished botanists of the period, including Leonhart Fuchs, William Turner, Carolus Clusius, and John Gerard, are presented here. One of the most remarkable is a copy of the 1565 Venice edition of Pietro Andrea Mattioli's *Commentarii* on blue paper; produced for an eminent personage at the Hapsburg court, many of its woodcuts have been embellished with silver and gold highlights.

The third section, *Herbals and Plants from Distant Lands*, is dedicated to a phenomenon that contributed greatly to the influence and popularity of the herbal from the Renaissance to the nineteenth century—the discovery and study of plants from faraway lands. One of the earliest examples is Prospero Alpini's *De plantis Aegypti liber* (1592), which describes species from the furthermost corners of the Mediterranean basin. Soon botanists were immersing themselves in the study of plants from the Americas, and the new science of botany advanced rapidly with the publication of texts by Nicolas Monardes, Garcia da Orta, Carolus Clusius, Jean Robin, and others, in which illustrations of such hitherto unknown species as the pepper plant, the Peruvian fig, and the Virginia narcissus appeared for the first time.

The study of exotic plants continued during the following centuries, with botanists dedicating ample space in their herbals, treatises, and monographs to this fascinating subject. One need only consider the enthusiastic interest prompted by the discovery of tobacco, coffee, and cacao, which at first were used only for medicinal purposes, as is demonstrated by such works as Nicolas de Blégny's *Le bon usage du Thé du Caffé et du Chocolat* (1687) and Niccolò Gavelli's *Storia Distinta, e Curiosa del Tabacco* (1758).

Also discussed in this section are two collections of botanical illustrations. The first is a series of paintings in gouache of exotic flowering plants by Baldassare Cattrani, a recently rediscovered, late-eighteenth-century artist (see *An Oak Spring Flora*), who also produced drawings for the Botanical Garden of Padua and for Josephine Beauharnais, the owner of a magnificent garden at Malmaison. The second is a unique collection of paintings of Chinese plants on paper executed by artists from Canton and Macao between 1744 and 1773. Bound in four volumes, these vivid sketches in watercolor and gouache are accompanied

DIALOGVS. 78

BALSAMVM.

X 2 Cur

PROSPERO ALPINI, *De Balsamo dialogus*, 1592. 'Balsamum,' Balm of gilead (*Commiphora opobalsamum*), page 78

by a manuscript glossary in English and Chinese that provides the names of more than three hundred different items, including many plants.

The fourth section, *Herbals by Herbalists, Pharmacists, and Physicians*, examines the works produced by or for those with a professional interest in medicinal plants over the course of three hundred years, beginning with the sixteenth century herbals of Antoine Du Pinet and Castore Durante. The latter was a celebrated Italian physician whose *Herbario Nuovo* (1585) was widely consulted and reprinted many times up until the end of the seventeenth century.

These volumes, particularly the earliest examples, were generally quite modest in size—Du Pinet's *Historia Plantarum*, for instance, was produced in a 'pocket-sized' duodecimo format—so that they could be carried by botanists on plant-gathering expeditions. They are nevertheless of considerable interest because they were furnished with sets of illustrations that were indispensable for the correct identification of plants. Furthermore, in some of these works we can retrace the threads of debates over various controversial issues within the medical community, such as the merits of astrological botany, which found a supporter in Nicholas Culpeper but was dismissed out of hand by William Coles.

The edition of the four volumes of botanical illustrations prepared for Johann Wilhelm Weinmann's *Phytanthoza-iconographia* (1737) deserves particular attention because it constitutes an early experiment with color printing, the publisher using a novel technique to produce the mezzotint etchings in color. This work was greeted with considerable interest by naturalists, who were seeking mechanized alternatives to the costly and time-consuming process of coloring illustrations by hand.

An interesting variant on the herbal is provided by a series of eight oil paintings of medicinal plants dating from the seventeenth century that, before the French Revolution, hung on the walls of a pharmacy in the town of Saumur. In these works the anonymous artist has portrayed a series of medicinal plants (including the aloe, the centaury, the date palm, and the hellebore) with astonishing realism against landscape backgrounds that evoke their original habitats.

The works in the fifth section, *Herbals of the Botanical Gardens and Private Gardens of Europe*, were compiled by the directors and owners of the most important gardens in Europe, which were celebrated for their collections of rare plants. We begin with the catalogue of the private garden of Nuremburg by Joachim Camerarius, *Hortus medicus & philosophicus*. Camerarius was a physician and botanist who commissioned such prestigious artists as Jost Amman and Joachim Jungermann to illustrate this important work. Guy de la Brosse, the *intendant* of the Jardin des Plantes of Paris under Louis XIII, in 1636 published *Description du Jardin Royal des Plantes Médecinales,* an elegant volume describing the garden's design, layout and collection of 2,000 medicinal plants. Another important project that was

XLIX.

NON INFERIORA SECUTUS.

Solis ut hunc florem radiantia lumina versant;
Dirige sic mentem Christe benigne meam.

JOACHIM CAMERARIUS, *Hortus medicus et philosophicus*, 1588. Sunflower (*Helianthus annuus*), volume 1, page 51

DENIS DODART, *Mémoires pour servir à l'Histoire des Plantes*, 1676. Vignette on page 1 showing a group of academics watching a chemistry experiment

sponsored by the royal court was *Mémoires pour servir à l'Histoire des Plantes* (1676) by Denis Dodart, a physician at the Faculty of Medicine in Paris and a member of the Académie Royale. Nicolas Robert, who painted a series of magnificent *vélins du Roy* for the Jardins des Plantes in the capacity of 'peintre ordinaire de Sa Majesté pour la miniature' (see *An Oak Spring Flora*), collaborated in the illustration of this work.

Other catalogues produced during the course of the eighteenth century document the collections of the botanical gardens of Amsterdam (J. Commelin), Turin (G. B. Morandi), Rome (G. Bonelli), and Nuremburg (C. J. Trew). Among these Carolus Linnaeus' *Hortus Cliffortianus* is particularly impressive, with its splendid illustrations by the botanical artist Georg Dionysius Ehret.

In the vast panorama of herbals produced over the centuries, not a few helped to propagate curious episodes, legends, and beliefs linked to the plant world, as can be seen from the *Curious and Strange Herbals* in the sixth section. For example, in the late sixteenth century the physician-magician Giovanbattista della Porta, an advocate of Paracelsus' doctrine of signatures, explained in *Phytognomonica* how one could determine from the form of a plant the specific medicinal use for which it had been destined by Nature. Camerarius, the author of the *Hortus medicus* discussed in the previous section, also compiled a fascinating volume of emblems inspired by the plant world—*Symbolorum & Emblematum* (1590)—in which different plants were assigned specific symbolic meanings. The seventeenth-century botanists Claude Duret, author of an illustrated work on the flora of the New World, and

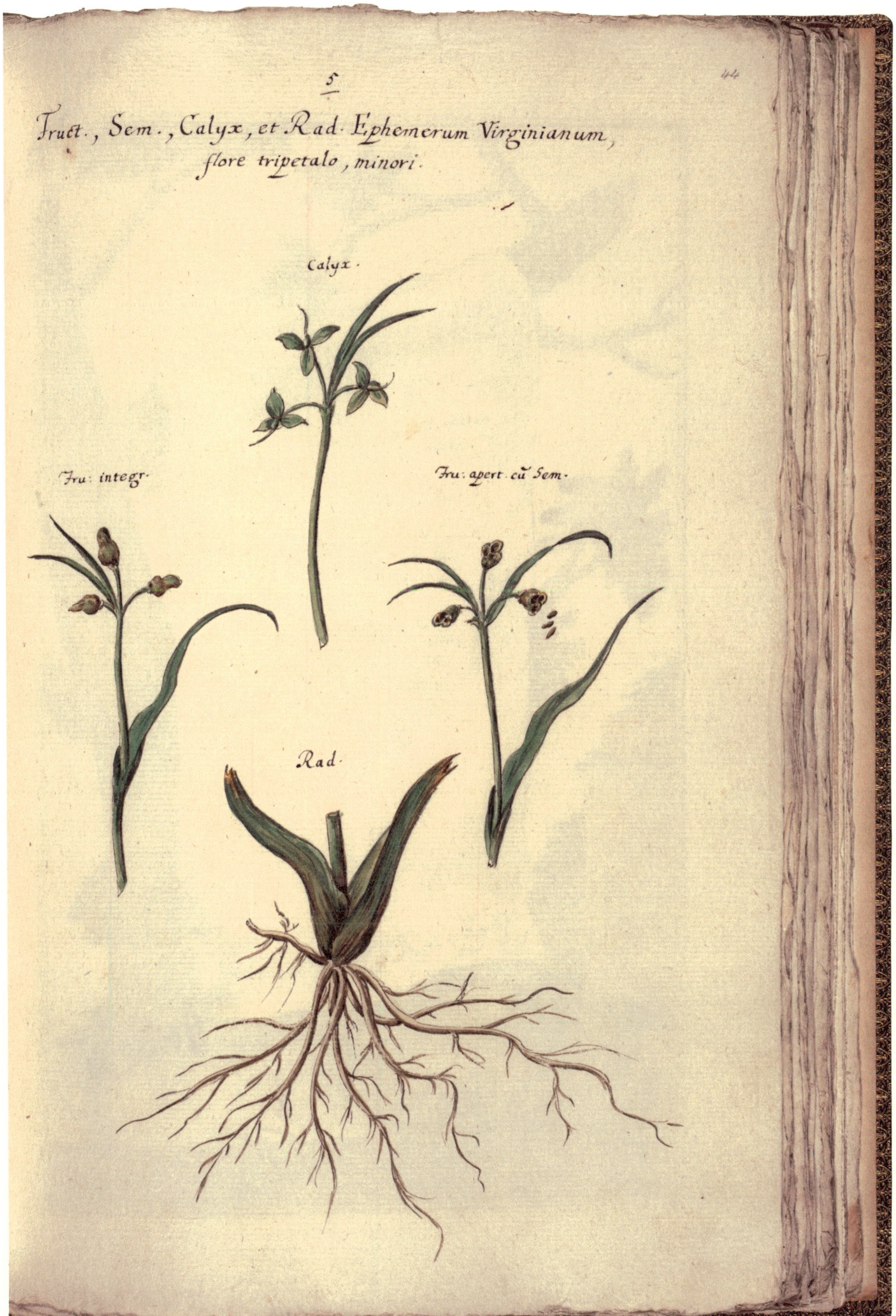

GIOVANNI BATTISTA MORANDI, *Collectio plantarum*, 1737. Various parts of *Tradescantia virginiana*, 'Ephemerum virginiana,' folio 5

GIOVANNI BATTISTA MORANDI, *Collectio plantarum*, 1737. Bulb and leaves, possibly of a Squill (*Scilla* sp.), with inflorescence of a Yucca (*Yucca* sp.), folios 92 and 93

Pierre Pomet, who compiled an herbal on exotic medicinal plants, did not hesitate to relay the most extravagant legends and hearsay, thus keeping alive traditions that would continue well into the nineteenth century with such works as Jean Pierre Rambosson's *Histoire et Légends des Plantes*.

The seventh section, *Dried Specimens and Nature Printing*, discusses two particular typologies of the herbal—the *hortus siccus* and the 'nature print.' The practice of gathering plants and carefully drying them between sheets of paper emerged around the middle of the sixteenth century, contemporaneously with the art of botanical painting and for the same purposes of study and teaching. The earliest dried herbal in the Oak Spring Garden Library was assembled towards the end of the sixteenth century by Johannes Harder, son of the pharmacist Hieronymus and the first botanist in Germany to adopt the practice of systematically drying plants for the aims of research. Between the pages of this valuable album a total of 578 dried plants are conserved, most of them supplemented with details added by hand in watercolor. The volumes of nature prints in the Oak Spring Garden Library date from the eighteenth century. They reflect the new spirit of scientific inquiry that motivated botanists to experiment with new methods for the documentation and conservation of plants, meant to offer practical alternatives to the traditional painted and dried herbals.

The last section is dedicated to *American Herbals*, all dating from the nineteenth century. Outstanding examples include Samuel Henry's *A New and Complete American Medical Family Herbal* and Jacob Bigelow's *American Medical Botany*. Bigelow was a professor of *materia medica* at Harvard University and his herbal not only describes the properties of indigenous medicinal plants, but also their uses 'in diet and the arts.' Another interesting work is Constantine Samuel Rafinesque's *Medical Flora*, published in Philadelphia, which is illustrated with 100 woodcuts printed in green ink. Rafinesque was a true eclectic who during the course of an adventurous life found the time to write poetry as well as to engage in the study of botany, natural history, and economics.

★

It is appropriate to recall that—alongside the exceptional examples presented here of manuscript and printed herbals of historic importance from both a scientific and an artistic perspective—a constant stream of more modest volumes were produced in Europe and the United States that served to disseminate the still popular tradition of herbal medicine. A typical exemplar is *A Brief Treatise on Various Ailments and their Treatment by Nature's Remedies*, published in London by an anonymous author around 1890. The recipes it provides for numerous remedies, with the appropriate dosages, are all made from plants in accordance with knowledge transmitted from generation to generation, in large part by

BRADFORD MEDICAL INSTITUTE, *A Brief Treatise on Various Ailments and Their Treatment by Nature's Remedies*, London, *c.* 1890. Elder or Elderberry (*Sambucus nigra*), page 12

ELDER.

Elder plant and berries; medicinal properties of the flower, Emollient, Diaphoretic; colour of the flower, pale lemon tint. Any grandmother will tell you how to make elderberry wine; and the best judges will tell you not to forget to add a few raisins and a little brandy.

women herbalists. In this little volume the fearsome witch has been replaced by the kindly grandmother. We read, for example, that elder is an 'emollient [and] diaphoretic' plant from which 'any grandmother will tell you how to make elderberry wine: and the best judges will tell you not to forget to add a few raisins and a little brandy.' The author even offers to send 'the herbs (and pills to serve one week) complete, postage paid, on the receipt of Postal Order for one shilling, or 12 penny stamps.'

LUCIA TONGIORGI TOMASI

DESCRIPTIVE METHOD

EACH ENTRY begins with a description of the item (or items) concerned. Manuscripts and works of art are given a brief physical description. Simple transcriptions of the title-pages of printed works (and of manuscripts, when they are formally laid out) are supplied in full, but without indication of typographical variations or of line endings. Words that in the original are set in full capitals or small capitals are rendered with capitalization that approximates the style of the work itself or that of the age and country in which it was printed. The use of *sic* following oddities in spelling is severely restricted. Square brackets are employed when the title or date of a work has a source other than the work itself; to indicate inserted material and variants and supplied letters when special marks of contraction have been employed by the printer; and to describe vignettes, ornaments, printer's rules, and devices. Earlier forms of letters and diacritical marks are converted to modern form. The colophon is given only when it contains information not provided on the title-page.

The collation section states the format, measurements in centimeters, gatherings by signatures (for books of the hand-press period), pages or foliation, and the number of leaves of plates. Italics indicate inferred pagination and signatures; errors in numbering or signings are noted. 'Binding' briefly describes material and decoration. The 'Plates' paragraph provides the illustration processes, whether printed separately from the text or not; artists, where known; and the presence of hand coloring or hand finishing. 'Provenance' gives information, where available, of previous owners and on their bookplates, inscriptions, and marginalia. Also noted are enclosures, as for example, personal correspondence. Finally, 'References' supplies details of relevant published materials, full details of which will be found in the Bibliography.

BIBLIOGRAPHY

This bibliography includes all abbreviated titles referred to in the 'References' and in descriptions of items.

ACKERKNECHT:
Erwin H. Ackerknecht, *Therapeutics from the Primatives to the 20th Century*. New York: Hafner Press, 1973 [translated from *Therapie von den Primitiven bis zum 20. Jahrhundert*. Ferdinand Enke Verlag: Stuttgart, 1970].

ADANSON:
Michel Adanson, *Familles des plantes*. Lehrte and New York: Stechert-Hafner and J. Cramer, 1966; first published in Paris in 1763 by Vincent, Imprimeur-Libraire de Mgr le Compte de Provence.

Age of the Marvelous:
The Age of the Marvelous. Exhibition catalogue, edited by Joy Kenseth. Hood Museum of Art, Dartmouth College, Hanover, N.H., 1991.

ALLIONI:
Carlo Allioni, *Flora pedemontana*. 2 vols. Florence: Olschki, 2003; reprint of 1785 edition.

American Register:
American Medical & Philosophical Register, Vol. IV (1814), Art. X, pp. 597 ff.

Analectic Magazine:
[Notice of Samuel Henry's 'An American Family Medical Herbal'] *The Analectic Magazine*, Vol. IV, November (1814), pp. 435–436; W.B. [William Baldwin], 'Review of Henry's American Herbal,' *The Analectic Magazine and Naval Chronicle*, Vol. VII, No. 39, March (1816), pp. 248–264.

ANDERSON:
Frank J. Anderson, *An Illustrated History of the Herbals*. New York: Columbia University Press, 1977.

ARBER:
Agnes Arber, *Herbals, Their Origin and Evolution: A Chapter in the History of Botany 1470–1670*. Cambridge, 1990 [3rd ed. 1986].

BARBERI:
Francesco Barberi, *Per una storia del libro: Profili, note, ricerche*. Rome: Bulzoni, 1981.

BARBON:
Caterina Barbon, editor, *L'Erbario di Udine, Biblioteca Civica Vincenzo Joppi, Ms. 1161*. Udine: Roberto Vattori Editore, 2007.

BARKER 1994:
Nicolas Barker, *Hortus Eystettensis: The Bishop's Garden and Besler's Magnificent Book*. London: The British Library, 1994.

BARKER 1999:
Nicolas Barker, *Treasures from the Libraries of National Trust Country Houses*. New York: The Royal Oak Foundation / Grolier Club, 1999.

BERÈS:
Pierre Berès, *Un groupe de livres Pillone*. Catalogue 67. Paris [*c.* 1970].

BERMAN:
Alex Berman and Michael A. Flannery, *America's Botanico-Medical Movements: Vox Populi*. New York: Pharmaceutical Products Press, 2001.

BIGELOW:
Jacob Bigelow, *Nature in Disease; Illustrated in Various Discourses and Essays. To Which are Added Miscellaneous Writings, Chiefly on Medical Subjects*. 2nd ed. Boston: Phillips, Sampson and Company, 1859.

BLUNT and RAPHAEL:
Wilfrid Blunt and Sandra Raphael, *The Illustrated Herbal*. London: Thames and Hudson, 1994 [1979, rev. ed. 1994].

BLUNT and STEARN:
Wilfrid Blunt, with the assistance of William T. Stearn, *The Art of Botanical Illustration*. London, 1950, revised by W. T. Stearn, Woodbridge, Suffolk: Antique Collectors' Club / Royal Botanic Gardens, 1994.

BM CAT. (*Nat. Hist.*)
Catalogue of the Books, Manuscripts, Maps and Drawings in the British Museum (*Natural History*). 8 vols. London: British Museum, 1903–1940 [Reprint: Maurizio Martino, New York].

BIBLIOGRAPHY

Botany in the Low Countries:
Botany in the Low Countries: End of the 15th Century – ca. 1650. Exhibition catalogue. Antwerp: The Plantin-Moretus Museum and the Stedelijk Prentenkabinet, 1993.

Brignoli:
Giovanni Brignoli, 'Biografia compendiosa del Cavaliere Gaetano Savi,' in *Istituzioni Botaniche.* G. Savi, Parma: Fiaccadori, 1848, pp. v–xv.

Brunet:
J. C. Brunet, *Manuel du libraire et de l'amateur de livres.* 6 vols. Paris: Didot, 1860–65.

Bruni and Nicoletti:
Alessandro Bruni and Marcello Nicoletti, *Dizionario ragionato di erboristeria e di fitoterapia.* Padua: Piccin, 2003.

Bush-Brown:
Louise Bush-Brown, *Men With Green Pens: Lives of the Great Writers on Plants in Early Times.* Philadelphia, Pa.: Dorrance, 1964.

Call:
Richard Ellsworth Call, *The Life and Writings of Rafinesque.* Louisville, Kentucky: Filson Club Publications, 1895.

Calmann:
Gerta Calmann, *Ehret: Flower Painter Extraordinary.* Oxford: Phaidon Press, 1977.

Cambridge:
Catalogue of Books Printed on the Continent of Europe, 1501–1600 in Cambridge Libraries, compiled by Herbert Mayow Adams. 2 vols. Cambridge: Cambridge University, 2000 [first published in 1967].

Camerarius:
Joachim Camerarius the younger, *Symbola et emblemata: (Nürnberg, 1590 bis 1604),* facs. edition edited by Wolfgang Harms and Ulla-Britta Kuechen. 2 vols. Graz, Austria: Akademische Druck- u. Verlagsanstalt, 1986–88.

Capparoni:
Pietro Capparoni, *Profili bio-bibliografici di medici e naturalisti celebri Italiani dal sec. XV al sec. XVII.* 2 vols. Rome: Nazionale Medico Farmacologico, 1925–1928.

Cave and Wakeman:
Roderick Cave and Geoffrey Wakeman, *Typographia naturalis.* Wymondham, Leicestershire: Brewhouse Press, 1967.

Chapman and Tweddle:
George T. L. Chapman and Marilyn N. Tweddle, editors, *A New Herball by William Turner.* 2 vols. Cambridge: Cambridge University Press, 1995.

Chastel:
André Chastel, 'Petrarque et son illustrateur devant la peinture' in *Études d'art médiéval offerts à Louis Grodecki.* S. McK. Crosby et al. Paris: Ophrys, 1981, pp. 343–352.

Cleveland:
Stanley H. Johnston, Jr., *The Cleveland Herbal, Botanical, and Horticultural Collections: A Descriptive Bibliography of the Pre-1830 Works from the Libraries of the Holden Arboretum, the Cleveland Medical Library Association, and the Garden Center of Greater Cleveland.* Kent, Ohio: Kent State University Press, 1992.

Collins:
Minta Collins, *Medieval Herbals: The Illustrative Traditions.* London: British Library, 2000.

Collins and Raphael:
Minta Collins and Sandra Raphael, *A Medieval Herbal: A Facsimile of British Library Egerton ms.747.* London: The British Library, 2003.

Conihout:
Isabelle de Conihout, editor, *Botanica in originali: Livres de botanique réalisés en impression naturelle du XVIe Au XIXe siècles.* Exhibition catalogue. Paris: Bibliothèque Nationale, 5 May – 12 June 1993.

Contant:
Paul Contant, *Le jardin et cabinet poétique,* edited by M. Marrache-Gourard and P. Martine. Rennes: Presses Universitaires de Rennes, 2004.

Conti and Livi:
Paola Barbara Conti and Paola Livi, '*L'illustrazione botanica in Italia nella prima metà del XVIII secolo: Note introduttive allo studio e alla figura e dell'opera di Giovanni Battista Morandi*' in *Il fondo antico della Biblioteca del Museo Civico di Storia Naturale di Milano. Catalogo delle Collezioni,* Andrea di Pasquale and Paola Livi, editors, published in *Natura,* Vol. 92, No. 2, (2003).

Corbett:
Sir Julian Stafford Corbett, *Sir Francis Drake.* London and New York: Macmillian, 1922.

Cox:
E.H.M. Cox, *Plant-Hunting in China.* Hong Kong: Oxford University Press, 1986.

CUISENIER:
Robert Cuisenier, *Jehan Bauhin: Médecin et Botaniste Montbéliardais 1541–1612*. Montbéliard, 1991.

CULPEPER 1947:
Nicholas Culpeper, *Culpeper's English Physician & Complete Herbal: Arranged for Use as a First Aid Herbal by Mrs. C. F. Leyel*. London: H. Joseph, 1947.

CULPEPER 1995:
Nicholas Culpeper, *Culpeper's Complete Herbal: A Book of Remedies for Ancient Ills*. Ware: The Wordsworth Collection Reference Library, 1995.

DELATTE:
Armand Delatte, *Herbarius: Recherches sur le cérémonial usité chez les anciens pour la cueillette des simples et des plantes magiques*. Paris: Société d'édition 'Les Belles lettres,' 1938.

DE TONI:
Giovanni Battista De Toni, 'Pierandrea Mattioli' in *Gli Scienziati Italiani dall'inzio del Medio Evo ai Nostri Giorni*, Aldo Mieli, editor. Rome, 1923, Vol. 1, ii, pp. 382–387.

DESMOND:
Ray Desmond, *Wonders of Creation: Natural History Drawings in the British Library*. London: British Library, 1986.

DIBNER:
Heralds of Science as Represented by Two Hundred Epochal Books and Pamphlets in the Dibner Library, Smithsonian Institution, preface and notes by Bern Dibner. New York: Walson Academic Publications, 1980.

Dictionary of Scientific Biography:
Dictionary of Scientific Biography. New York: Charles Scribner's Sons, 1970–1990, Vol. 14, pp. 212–213.

DI NOTO and WINTER:
Andrea Di Noto and David Winter, *The Pressed Plant: The Art of Botanical Specimens, Nature Prints, and Sun Pictures*. New York: Stewart, Tabori & Chang, 1999.

DBI:
Dizionario biografico degli Italiano. 61 vols. Rome, 1982.

DOBAT:
Klaus Dobat, *Leonhart Fuchs (1501–1566). Mitbegründer der modernen Botanik*. Stuttgart: Schwäbische Forscher und Gelehrte, 1992.

DODGSON:
Campbell Dodgson, 'An Alphabet by Hans Weiditz,' *The Burlington Magazine*, (1907–1908), pp. 289–293.

DROUIN:
Jean-Marc Drouin, *L'herbier des philosophes*. Paris: Seuil, 2008.

DUNTHORNE:
Gordon Dunthorne, *Flower & Fruit Prints of the 18th and Early 19th Centuries*. New York, 1970 [London, 1938].

EAMON:
William Eamon, *Science and the Secrets of Nature: Books of Secrets in Medieval and Early Modern Culture*. Princeton, N.J.: Princeton University Press, 1994.

EGMOND:
Florike Egmond, 'Clusius, Cluyt, Saint Omer: The Origins of the Sixteenth Century Botanical and Zoological Watercolors' in 'Libri picturati a 1630' in *Nuncius-Journal of History of Science*, Vol. XX, 1 (2005), pp. 11–67.

EGMOND, HOFTIJZER and VISSER:
Florike Egmond, Paul Hoftijzer, and Robert Visser, editors, *Carolus Clusius: Towards a Cultural History of a Renaissance Naturalist* (History of Science and Scholarship in the Netherlands, vol. 8). Amsterdam: Koninklijke Nederlandse Akademie van Wetenschappen, 2007.

ELLIS:
George E. Ellis, 'Memoir of Jacob Bigelow,' *Proceedings of the Massachusetts Historical Society*, Vol. XVII (1879–1880), pp. 383–467.

EVELYN:
John Evelyn, *Elysium Britannicum, or the Royal Gardens*, edited by John E. Ingram. Philadelphia, Pa.: University of Philadelphia Press, 2001.

FAT and JONG:
Leslie Tjon Sie Fat and Erik de Jong, *The Authentic Garden: A Symposium on Gardens*. Leiden: Clusius Foundation, 1991.

FAVARO:
Giuseppe Favaro, 'Contributo all biografia di A. Spiegeli (Adriaan van den Spiegel) nel terzo centenario della sua morte (1625–1925),' *Atti dell'Istituto veneto di scienze, lettere ed arti*, 85, part 2, (1925–1926), pp. 213–252.

FERRI:
Sara Ferri, editor, *Pietro Andrea Mattioli: Siena, 1501 – Trento, 1578: la vita e le opere*. Perugia: Quattroemme, 1997.

FORNERIS and PISTARINO:
Giuliana Forneris and Annalaura Pistarino, editors, *Herbario Nuovo de Castore Durante, Venetia, MDCC-XVII, conservato presso la Biblioteca del Museo Regionale di Scienze Naturali di Torino*. Torino, 2000.

FOSTER:
J. Foster, 'The Art of Simpling,' *Alumni Oxonienses*, I, early series, 302 (1891), No. 266.

GARBARI and TONGIORGI TOMASI:
Fabio Garbari and Lucia Tongiorgi Tomasi, *The Paper Museum of Cassiano dal Pozzo, Flora: The Erbario Miniato and Other Drawings*. 2 vols. Turnhout: The Royal Collection, Harvey Miller Pub., 2007.

GARBARI, TONGIORGI TOMASI, and TOSI:
Fabio Garbari, Lucia Tongiorgi Tomasi, and Alessandro Tosi, *Giardino dei semplici: L'Orto botanico di Pisa dal XVI al XX secolo*. Pisa: Pacini editore, 1991.

GABRIELI:
Giuseppe Gabrieli, *Contributi all storia dell'Accademia dei Lincei*. Rome: Accademia Nazionale dei Lincei, 1989.

GEISBERG 1932:
Max Geisberg, 'Der deutsche Einblatt-Holzeschnitt in der ersten Hälfte des XVI' in *Speculum*, Vol. 7, No. 3, (July 1932), pp. 431–432.

GEISBERG 1974:
Max Geisberg, *The German Single-Leaf Woodcut, 1500–1550*, revised and edited by Walter L. Strauss. 4 vols. New York: Hacker Art Books, 1974.

GEUS:
Aramin Geus, 'Nature Self-prints as a Methodical Instrument in the History of Botany' in *Natura-cultura: L'interpretazione del mondo fisico nei testi e nelle immagini: Atti del convegno internazionale di studi, Mantova, 1996*, Giuseppe Olmi, Lucia Tongiorgi Tomasi, and Attilio Zanca, editors. Florence: Olschki, 2000, pp. 245–254.

GIVENS, REEDS and TOUWAIDE:
Jean A. Givens, Karen M. Reeds, and Alain Touwaide, editors, *Visualizing Medieval Medicine and Natural History, 1200–1550*. Aldershot, England: Ashgate, 2006.

GOOD:
Peter Peyto Good, *A Materia Medica Animalia. Containing the Scientific Analysis, Natural History, and Chemical and Medical Properties and Uses of the Substances that are the Product of Beasts, Birds, Fishes, and Insects*. Cambridge, Mass.: Published by the Author, 1853.

GOODMAN:
Nathan G. Goodman, *Benjamin Rush, Physician and Citizen, 1746–1813*. Philadelphia: University of Pennsylvania Press, 1934.

GOTTSCHALL:
Dagmar Gottschall, *Konrad von Megenbergs Buch von der natürlichen Dingen*. Leiden: Brill, 2004.

GRAY:
Asa Gray, 'Notice of the Botanical Writings of the late C. S. Rafinesque,' *American Journal of Science*, Vol. XL (1841), pp. 221–241.

GREENE:
Edward Lee Greene, *Landmarks of Botanical History*, edited by Frank N. Egerton. 2 vols. Palo Alto, Calif.: Stanford University Press, 1983.

GRIECO:
Allan Grieco, Odile Redon, Lucia Tongiorgi Tomasi, editors, *Le monde végétal, XIIe–XVIIe siècles: Savoirs et usages sociaux*. Saint-Denis: Presses universitaires de Vencennes, 1993.

GROLIER CLUB:
One Hundred Books Famous in Medicine. Exhibition catalogue, organized and with an introduction by Haskell F. Norman. New York: Grolier Club, 1995.

GUARNERINUS DE PADUA:
Antonius Guarnerinus de Padua, *Herbe Pincte, Codice MA 592 della Biblioteca Civica di Bergamo*, edited by Giovanni Silini. Palazzago, 2000.

GUERRA:
Francisco Guerra, *Nicolás Bautista Monardes: Su vida y su obra, ca. 1493–1588*. Mexico City: D. F. Compañia Fundidora de Fierro y Acero de Monterrey, 1961.

GRÜNWALD:
Michael Grünwald, 'Die Beziehungen des Jungen Hans Weiditz zu Hans Frank' in *Jahrbuch der Preussischen Kunstsammlungen*, 44 (1923), pp. 26–36.

HAMY:
T. Hamy, 'La famille de Guy de la Brosse' in *Bulletin du Musée d'Histoire Naturelle*, 6 (1900), pp. 13–16.

HENREY:
Blanche Henrey, *British Botanical and Horticultural Literature Before 1800*. 3 vols. Oxford and New York: Oxford University Press, 1975.

HOLMES:
Oliver Wendell Holmes, 'Current and Counter-Currents in Medical Science,' in *Medical Essays, 1842–1882*, Boston: Houghton Mifflin, 1861, pp. 173–208.

HOPPER:
F. Hopper, 'Clusius' World: The Meeting of Science and Art' in *The Authentic Garden: A Symposium on Gardens*, Leiden: Clusius Foundation, 1991.

HUNGER:
Frederich W. T. Hunger, *Charles de L'Escluse (Carolus Clusius): Nederlandsch kruidkundige 1526–1609*. 2 vols. The Hague: Martinus Nijhoff, 1927–1943.

HUNT:
Catalogue of Botanical Books in the Collection of Rachel McMasters Miller Hunt, compiled by Jane Quinby and Allan Stevenson. 3 vols. Pittsburgh, Pa.: Hunt Botanical Library, 1958–1961.

IVINS:
William M. Ivins, Jr., *Prints and Visual Communication*. Cambridge, Mass.: M.I.T. Press, 1953.

JONES:
Whitney Richard David Jones, *William Turner: Tudor Naturalist, Physician, and Divine*. London and New York: Routledge, 1988.

KLEBS:
Arnold C. Klebs, *A Catalogue of Early Herbals*. L'Art Ancien, Bulletin XII, Lugano, 1925.

KREMERS and URDANG:
Edward Kremers and George Urdang, *History of Pharmacy*. Philadelphia, Pa., 1946.

LACK:
H. Walter Lack, *Garden Eden: Masterpieces of Botanical Illustration*. Exhibition catalogue [in German, English, and French]. Köln: Taschen, 2001.

LOWNES:
A. E. Lownes, 'The Strange Case of Coles versus Culpeper,' *Journal of the New York Botanical Garden*, 41 (1940), pp. 158–166.

LEUNG and FOSTER:
Albert Y. Leung and Steven Foster, *Encyclopedia of Common Natural Ingredients Used in Food, Drugs, and Cosmetics*. New York and Chichester: Wiley, 1995.

LINNAEUS:
Carl von Linnaeus, *Genera plantarum*. New York, 5th edition, facsimile of 1754 edition, 1960, with an introduction by William T. Stearn.

LUPO:
Michelangelo Lupo, *L'erbario di Trento: il manoscritto n. 1591 del Museo provinciale d'arte*. Calliano: Manfrini, 1982.

MANCHESTER:
Catalogue of Medical Books in Manchester University Library, compiled by Ethel Midgley Parkinson. Manchester: Manchester University Press, 1972.

Mattioli Woodblocks:
William Patrick Watson, Sandra Raphael, and Iain Bain, *The Mattioli Woodblocks*. London: Bernard Quaritch / Hazlitt, Gordon & Fox / Antiquariaat Junk, [1989].

MEERBEECK:
Phillips Jacob van Meerbeeck, *Recherches historiques et critiques sur la vie et les ouvrages de Rembert Dodoens (Dodonaeus)*. Utrecht: Hes & De Graaf, [1841] 1980.

MENTEN:
Theodore Menten, *Plant and Floral Woodcuts for Designers & Craftsmen: 419 Illustrations from the Renaissance Herbal of Carolus Clusius*. New York: Dover, 1974.

MEYER FREDERICK:
Frederick G. Meyer, *The Great Herbal of Leonhart Fuchs: De historia stirpium Commentarii Insignes, 1542 (Notable Commentaries on the History of Plants)*. Stanford, Calif.: Stanford University Press, 1999.

MEYER ERNEST:
Ernest H. F. Meyer, *Geschichte der Botanik*. 4 vols. Königsberg: Bornträger, 1854–1857.

MINELLI:
Alessandro Minelli, editor, *L'Orto botanico di Padova, 1545–1995*. Venice: Marsilio, 1995.

MOGGI:
Guido Moggi, 'L'erbario: Origine, evoluzione storica, significato' in *Erbari e iconografia botanica*, Franco Montacchini, editor. Turin: Allemandi, 1986, pp. 24–28.

MONTACCHINI:
Franco Montacchini, editor, *Erbari e iconografia botanica*. Turin: Allemandi, 1986.

MORAT, AYMONIN, and JOLINON:
Philippe Morat, Gérard Aymonin, and Jean-Claude Jolinon, *L'herbier du monde: Cinq siècles d'aventures et de passions botaniques au Muséum National d'Histoire Naturelle*. Paris: Le Grand livre du mois, 2004.

Morren:
Charles Morren, 'Adrien Spiegel: Extrait d'une Histoire Inédite de la botanique belge depuis les temps les plus reculés jusqu' à nos jours,' *Revue de Bruxelles* I (1938), pp. 51–79.

Mortimer:
Ruth Mortimer, *Harvard College Library Department of Printing and Graphic Arts, Catalogue of Books and Manuscripts,* Part I: *French 16th-Century Books,* 2 vols.; Part II: *Italian 16th Century Books,* 2 vols. Cambridge, Mass.: Belknap Press, 1964–74.

Morton:
A. G. Morton, *History of Botanical Science: An Account of the Development of Botany from Ancient Times to the Present Day.* London: Academic Press, 1981.

Morton and Garrison:
Leslie Thomas Morton and Fielding H. Garrison, *A Medical Bibliography: An Annotated Check-List of Texts Illustrating the History of Medicine*—incorporates entries from *A Revised Students' Check-List of Texts Illustrating the History of Medicine.* Aldershot, Hampshire, and Lexington, Mass.: Scolar Press, 1983.

Murray–*French*:
Hugh Wm. Davies, editor, *Catalogue of a Collection of Early French Books in the Library of C. Fairfax Murray.* 2 vols. London, 1910.

Murray–*German*:
Hugh Wm. Davies, editor, *Catalogue of a Collection of Early German Books in the Library of C. Fairfax Murray.* 2 vols. London, 1913.

Musper:
Theodor Musper, 'The Petrarch Master (alleged to be Hans Weiditz): An Angel in a Mandorla with the Arms of Mattheus Lang von Wellenburg' in *Old Master Drawings,* 8 (1933–1934), pp. 30–31.

Nelson and McKinley:
E. Charles Nelson, *Aphrodite's Mousetrap: A Biography of Venus's Flytrap with facsimiles of an original pamphlet and the manuscripts of John Ellis, F.R.S.,* with a 'Tipitiwichet' postscript by Daniel L. McKinley. Aberystwyth, Wales: Boethius Press / Linnean Society, 1990.

Nissen 1951:
Claus Nissen, *Die botanische Buchillustration, ihre Geschichte und Bibliographie.* Stuttgart: Hiersemann, 1951; Supplement, 1966.

Nissen 1958:
Claus Nissen, *Herbals of Five Centuries: 50 Original Leaves from German, French, Dutch, English, Italian, and Swiss Herbals,* with an introduction and bibliography. Zurich: L'Art ancien, 1958.

An Oak Spring Flora:
Lucia Tongiori Tomasi, *An Oak Spring Flora: Flower Illustration from the Fifteenth Century to the Present Time: A Selection of the Rare Books, Manuscripts, and Works of Art in the Library of Rachel Lambert Mellon.* Upperville, Va.: Oak Spring Garden Library, 1997.

An Oak Spring Garland:
Sandra Raphael, *An Oak Spring Garland: Illustrated Books, Prints, and Drawings for the Oak Spring Garden Library.* Exhibition catalogue, Princeton University Library. Upperville, Va.: Oak Spring Garden Library, 1989.

An Oak Spring Pomona:
Sandra Raphael, *An Oak Spring Pomona: A Selection of the Rare Books on Fruit in the Oak Spring Garden Library.* Upperville, Va.: Oak Spring Garden Library, 1990.

An Oak Spring Sylva:
Sandra Raphael, *An Oak Spring Sylva: A Selection of the Rare Books on Trees in the Oak Spring Garden Library.* Upperville, Va.: Oak Spring Garden Library, 1989.

Olmi:
G. Olmi, 'Molti amici in vari luoghi: Studio della natura e rapporti epistolari nel secolo XVI,' *Nuncius,* VI (1991), pp. 3–31.

Olmi, Tongiorgi Tomasi, and Zanca:
Giuseppe Olmi, Lucia Tongiorgi Tomasi, and Attilio Zanca, eds., 'Natura-Cultura. L'interpretazione del mondo fisico nei testi e nelle immagini,' *Atti del Convegno Internazionale di Studi.* Mantova, 1996; Florence: Olschki, 2000.

Ongaro and Mariani:
Giuseppe Ongaro and Paola Mariani, 'Prospero Alpini' in *L'Orto botanico di Padova, 1545–1995,* Alessandro Minelli, editor, Venice: Marsilio, 1995, pp. 64–69.

Osler:
Sir William Osler, *Bibliotheca Osleriana: A Catalogue of Books Illustrating the History of Medicine and Science.* Oxford: Clarendon Press, 2000.

Parshall:
Peter Parshall, 'Imago Contrafacta: Images and Facts

in Northern Renaissance' in *Art History*, (December 1993), pp. 554–579.

PEDRAZZINI:
Carlo Pedrazzini, *La farmacia italiana nella storia e nell'arte*. Milan: IGAP, [1964].

PINAULT:
Madeleine Pinault, *The Painter as Naturalist from Dürer to Redouté*. Paris: Flammarion, 1991.

PIROTTA and CHIOVENDA:
Romualdo Pirotta and Emilio Chiovenda, 'Flora Romana,' in *Annuario del R. Istituto Botanico di Roma*, X, 2 (1901), Rome, pp. 244–247.

PLESCH:
Arpad Plesch, *Mille et un livres botaniques: Répertoire bibliographique de la Bibliothèque Arpad Plesch*, by Henri Pierre Gourry. Brussels: Arcade, 1973.

PORTER:
Charlotte M. Porter, *The Eagle's Nest: Natural History and American Ideas, 1812–1842*. Tuscaloosa, Alabama: University of Alabama Press, 1986.

POTTERTON:
David Potterton, editor, *Culpeper's Color Herbal*. New York: Sterling, 1983.

PRAZ:
Mario Praz, *Studies in Seventeenth-Century Imagery*. Rome: Edizione di storia e letteratura, 1975.

PRITZEL:
Thesaurus literaturae botanicae, omnium gentium [1851], edited by G. A. Pritzel. 2nd edition. Leipzig, 1872–7 (reprinted Milan, 1950).

RAFINESQUE:
Constantine Samuel Rafinesque, '"A Life of Travels," being a verbatim and literatim reprint of the original and only edition (Philadelphia, 1836),' Foreword by Elmer D. Merrill, *Chronica Botanica*, Vol. VIII, No. 2 (1944).

RAPHAEL and WATSON:
Sandra Raphael and William Watson, 'The Camerarius Florilegium.' Chistie's, London, Catalogue, May 2002.

REEDS:
Karen Meier Reeds, *Botany in Medieval and Renaissance Universities*. New York & London: Garland, [1975] 1991.

RIDOLFI:
Cosimo Ridolfi, 'Elogio del Prof. Gaetano Savi morto in Pisa il 28 aprile 1844 scritto dal socio attuale Signor Marchese Cosimo Ridolfi inserito nella parte matematica del tomo XXIII delle Memorie della Società Italiana delle Scienze residente in Modena,' *Memorie della Società Italiana delle Scienze*, XXIII (1845), I–XXIV.

ROHDE:
Eleanour Sinclair Rohde, *The Old English Herbals*. London: Longmans, Green and Co., 1922 [reprint 1972].

ROSENBERG:
Charles E. Rosenberg, *'Every Man His Own Doctor': Popular Medicine in Early America*. Exhibition catalogue. Philadelphia, Pa.: The Library Company of Philadelphia, 1998.

ROWLANDS:
John Rowlands, *German Drawings from a Private Collection*. London: British Museum, 1984, pp. 72–73.

RYDÉN:
Mats Rydén, *The English Plant Names in the Grete Herball, 1526: A Contribution to the Historical Study of English Plant-Name Usage*. Stockholm: Almqvist & Wiksell, 1984.

SABIN:
Joseph Sabin, *A Dictionary of Books Relating to America, From its Discovery to the Present Time*. 2 vols. Mansfield Center, Conn., [1998].

SACCARDO 1895:
Pierandrea Saccardo, *La botanica in Italia: Materialia per la storia di questa scienza*. Venice: Carlo Ferrari, 1895.

SACCARDO 1902:
Pierandrea Saccardo, 'La Iconografia botanica dell'Ab. Angelo Franciosi, Veneto. Notizie, storiche e revisione botanica' *Atti e Memorie della R. Accademia di scienze, lettere ed arti di Padova*, XVIII (1902), pp. 249–294.

SAINT-LAGER 1885:
Jean-Baptiste Saint-Lager, *Histoire des herbiers*. Paris: Baillière, 1885.

SAINT-LAGER 1886:
Jean Baptiste Saint-Lager, *Recherches sur les anciens 'herbaria.'* Paris: Baillière, 1886.

SAUNDERS:
Gill Saunders, *Picturing Plants: An Analytical History of Botanical Illustration*. London, Berkeley, and Los Angeles, Calif.; Victoria & Albert Museum and University of California Press, 1995.

SCHIAVONE:
Mario Schiavone, 'Il frontespizio dell'Erbario di Otto Brunfels' in *L'Esopo*, 6 (1984), pp. 47–56.

SCHNAPPER:
Antoine Schnapper, *Le Géant, la licorne et la tulipe: Collections et collectionneurs dans la France du XVIIe siècle*. Paris: Flammarion, 1988.

SCHREYL and FREITAG-STADLER:
Karl-Heinz Schreyl and Renate Freitag-Stadler, editors, *Die Welt des Hans Sachs: 400 Holzschnitte des 16. Jahrhunderts*: [*Ausstellung im Kemenatenbau der Kaiserburg vom 30.7.–30.10.1976: Katalog*]. Stadtgeschichtlichen Museen, Nürnberg: H. Carl, 1976.

SCRASE:
David Scrase, *Flower Drawings*, (Fitzwilliam Museum Handbooks). New York: Cambridge University Press, 1997.

SÉGUIER:
Jean-François Séguier, *Bibliotheca botanica*. Hagæ-Comitum [The Hague]: Jean Néaulme, 1740.

SHAFER:
Henry Burnell Shafer, *The American Medical Profession, 1783 to 1885*. New York: Columbia University Press, 1936.

SHAFFER:
Ellen Shaffer, *The Garden of Health: An Account of Two Herbals, the Gart der Gesundheit and the Hortus Sanitatis*. [San Francisco, Calif.]: Book Club of California, 1957.

SHRYOCK:
Richard Harrison Shryock, *Medicine and Society in America, 1660–1860*. New York: New York University Press, 1960.

SINGER 1923:
Charles Singer, 'Herbals' in *Edinburgh Review*, 237 (January 1923), pp. 95–112.

SINGER 1927:
Charles Singer, 'The Herbal in Antiquity and its Transmission to Later Ages' in *Journal of Hellenic Studies*, XLVII (1927), pp. 1–52.

SITRAN REA:
Luciana Sitran Rea, editor, *L'Orto rappresentato: Scienza, didattica e immagine a Padova tra sette e ottocento*. Cittadella (Padua): Biblos, 2002.

SITWELL and BLUNT:
Sacheverell Sitwell and Wilfrid Blunt, *Great Flower Books 1700–1900: A Bibliographical Record of Two Centuries of Finely-Illustrated Flower Books*. New York: Atlantic Monthly Press, 1990 [London: Collins, 1956].

SOLINAS:
Francesco Solinas, editor, *Fiori: Cinque secoli di pittura floreale*. Rome: Campisano Editore, 2004.

SPYRA:
Ulrike Spyra, *Das Buch der Natur, Konrad von Megenberg: Die illustrierten Handschriften und Inkunabeln*. Köln: Böhlau, 2005.

STANNARD 1969:
Jerry Stannard, 'The Herbal as a Medical Document' in *Bulletin of the History of Medicine*, 43 (1969), pp. 8–79.

STANNARD 1999:
Jerry Stannard, *Herbs and Herbalism in the Middle Ages and Renaissance*. Aldershot: Ashgate, 1999.

STAFLEU and COWAN:
Frans A. Stafleu and Richard S. Cowan, *Taxonomic Literature*. 7 vols., 2nd edition. Utrecht: Bohn, Scheltema & Holkema, 1976–88.

STAMMEN and WEBER:
Theo Stammen and Wolfgang E. J. Weber, editors, *Wissenssicherung, Wissensordnung und Wissensverarbeitung: Das europäische Modell der Enzyklopädien*, Vol. 18 in 'Colloquia Augustana' series, Berlin: Akademie Verlag, 2004.

STEARN:
William T. Stearn, *Botanical Latin: History, Grammar, Syntax, Terminology and Vocabulary*. Newton Abbot: David and Charles, 1992.

STERN:
William Stern, editor, *John Lindley, 1799–1865: Gardener, Botanist, and Pioneer Orchidologist*. Woodbridge: Antique Collectors' Club / Royal Horticultural Society, 1998.

Stuffing Birds, Pressing Plants, Shaping Knowledge:
Stuffing Birds, Pressing Plants, Shaping Knowledge: Natural History in North America, 1730–1860, edited by Sue Ann Prince. Exhibition catalogue. Philadelphia, Pa.: American Philosophical Society, 2003.

SUDHOFF:
Karl Sudhoff, *Deutsche medizinische Inkunabeln – bibliographisch-literarische untersuchungen*. Mansfield Center,

Conn.: Martino Fine Books; first published in Leipzig: J. A. Barth, 1908.

Sur la terre comme au ciel:
Sur la terre comme au ciel: Jardins d'Occident à la fin du Moyen Âge. Exhibition catalogue. Paris: Musée de Cluny, 2002, p. 228.

SWAN 1998:
Claudia Swan, *The Clutius Botanical Watercolors: Plants and Flowers of the Renaissance.* New York: Harry N. Abrams, 1998.

SWAN 2005:
Claudia Swan, *Art, Science, and Witchcraft in Early Modern Holland, Jacques de Gheyn II (1565–1629).* Cambridge: Cambridge University Press, 2005.

TAVONI, TONGIORGI TOMASI, and TONGIORGI:
Maria Gioia Tavoni, Lucia Tongiorgi Tomasi, Paolo Tongiorgi, *Immagine e natura: l'imagine naturalistica nei codici e libri a stampa delle Biblioteche Estense e Universitaria, secoli XV–XVII.* Exhibition catalogue, 21 March–15 May 1984. Modena: Panni, 1984.

TOGNONI:
Federico Tognoni, 'Nature Described: Fabio Colonna and Natural History Illustration' in *Nuncius,* XX, 2, (2005), pp. 347–370.

TONGIORGI TOMASI 1990 a:
Lucia Tongiorgi Tomasi, 'Dall'essenza vegetale agglutinata all'impressione a stampa: il percorso dell'illustrazione botanica nei secoli XVI–XVII,' in *Luca Ghini: Cinquecento anni di scienze botaniche (1490–1990: Convegno Internazionale, Imola 1990,* [Imola]: Comune del Imola / Cassa di Risparmio di Imola, pp. 271–295.

TONGIORGI TOMASI 1990 b:
Lucia Tongiorgi Tomasi, *The Visual Arts and the Science of Horticulture in Tuscany from the 16th to the 18th Century.* The 23rd I.H.C Plenary Lectures. Florence: Parretti, 1990.

TONGIORGI TOMASI 1997:
Lucia Tongiorgi Tomasi, 'Il problema delle immagini nei "Commentarii,"' in *Pietro Andrea Mattioli: Siena, 1501 – Trento, 1578: la vita e le opere,* Sara Ferri, editor, Perugia: Quattroemme, 1997, pp. 369–376.

TONGIORGI TOMASI 1993:
Lucia Tongiorgi Tomasi, editor, *I ritratti di piante di Iacopo Ligozzi.* Pisa: Pacini, [1993].

TONGIORGI TOMASI 2002 a:
Lucia Tongiorgi Tomasi, 'L'illustrazione botanica nell'orto patavino e le tavole di Baldassarre Cattrani: Un caso esemplare di fine Settecento' in *L'Orto rappresentato: Scienza, didattica e immagine a Padova tra sette e ottocento,* Luciana Sitran Rea, editor, Cittadella (Padua): Biblos, 2002, pp. 43–70.

TONGIORGI TOMASI 2002 b:
Lucia Tongiorgi Tomasi, 'Tra arte e scienza: Note su Badassarre Cattrani illustratore dell'orto botanico di Padova' in *Lezioni di metodo: Studi in onore di Lionello Puppi,* Loredana Olivato and Giuseppe Barbieri, editors, Venice: Terra ferme, 2002, pp. 405–412.

TONGIORGI TOMASI 2003 a:
Lucia Tongiorgi Tomasi, 'L'illustrazione naturalistica: Tecnica e invenzione' in 'Natura-Cultura. L'interpretazione del mondo fisico nei testi e nelle immagini,' Giuseppe Olmi, Lucia Tongiorgi Tomasi, and Attilio Zanca, editors, *Atti del Convegno Internazionale di Studi,* Mantova, 1996; Florence: Olschki, 2000, pp. 133–152.

TONGIORGI TOMASI 2003 b:
Lucia Tongiorgi Tomasi, 'Immagine e scienze della natura nell'età dei lumi: l'orto botanico di Torino, Carlo Allioni e la sua "Flora Pedemontana"' in Carlo Allioni, *Flora pedemontana,* 2 vols., Florence: Olschki, 2003, I: IX–XXXI.

TONGIORGI TOMASI and HIRSCHAUER:
Lucia Tongiorgi Tomasi and Gretchen A. Hirschauer, *The Flowering of Florence: Botanical Art for the Medici.* Exhibition catalogue. Washington, D.C.: National Gallery of Art, 2002.

TONGIORGI TOMASI and TOSI:
Lucia Tongiorgi Tomasi and Alessandro Tosi, *Flora e Pomona: L'orticoltura nei disegni e nelle incisioni dei secoli XVI–XIX.* Florence: Olschki / Gabinetto disegni e stampe degli Uffizi, 1990.

TOSI 1995:
Allesandro Tosi, 'L'illustrazione naturalistica tra neoclassicismo e romanticismo,' *Atti e memorie della Accademia nazionale di scienze, lettere e arti di Modena,* Serie VII; XI (1993–1994), pp. 59–78.

TOSI 1997:
Tosi Alessandro, 'In Matthioli effigem,' in *Pietro Andrea Mattioli: Siena, 1501 – Trento, 1578: la vita e le opere,*

Tosi 1997 (*cont.*):
Perugia: Quattroemme, 1997, Sara Ferri, editor, pp. 377–390.

Turrentine:
Herbert C. Turrentine, 'Hans Weiditz's "Emperor Maximilian at Mass": An Intriguing Liturgical Scene in the Chapel of Annakirche in Augsburg' in *Explorations in Renaissance Culture*, 27, No. 1 (Summer 2001), pp. 21–30.

Ullmann:
Ernst Ullmann, editor, 'Hans Weiditz, Illustrateur de la Réforme à Strasbourg,' in *Von der Macht der Bilder. Beiträge des C.I.H.A.-Kolloquium Das Macht der Bilder*, Leipzig: Seemann, 1983, pp. 299–318.

Uphof:
Johannes Cornelis Theodorus Uphof, *Dictionary of Economic Plants*. New York: J. Cramer, 1968, 2nd edition.

Veendorp and Becking:
H. Veendorp and L.G.M. Baas Becking, *Hortus academicus Lugduno-Batavus 1587–1937*. Leiden: Rijksherbarium / Hortus Botanicus, 1990.

Virville:
Ad. Davy de Virville, *Histoire de la botanique en France*. Paris: Société d'édition d'enseignement supérieur, 1954.

Voigt:
Johannes Voigt, *Briefwechsel der berühmtesten Gelehrten des Zeitalters der Reformation mit Herzog Albrecht von Preussen*. Königsberg, 1841.

Warren:
Leonard Warren, *Constantine Samuel Rafinesque: A Voice in the American Wilderness*. Lexington: University Press of Kentucky, 2003.

Warner:
John Harley Warner, *The Therapeutic Perspective: Medical Practice, Knowledge, and Identity in America, 1820–1885*. Cambridge, Mass.: Harvard University Press, 1986.

Weimerskirch:
Philip J. Weimerskirch, 'The Beginning of Color Printing in America,' *Printing History: The Journal of the American Printing History Association* 48, Vol. xxiv, No. 2 (2005), pp. 383–467.

Wellcome:
A Catalogue of Printed Books in the Wellcome Historical Medical Library. 5 vols. London: Wellcome Historical Medical Library, 1962–2006.

Wheelwright:
Edith Grey Wheelwright, *Medicinal Plants and Their History*. New York: Dover, [1935] 1974.

Wickes:
Stephen Wickes, *History of Medicine in New Jersey, and of Its Medical Men, from the Settlement of the Province to A.D. 1800*. Newark, N.J.: Martin R. Dennis & Co., 1879.

Wijnands:
D. O. Wijnands, *The Botany of the Commelins*. Rotterdam: A. A. Balkema, 1983.

Wolfe:
Richard Wolfe, *Jacob Bigelow's American Medical Botany, 1817–1821*. North Hills, Pa.: Bird & Bull Press, 1979.

Young:
James Harvey Young, *The Toadstool Millionaires: A Social History of Patent Medicines in America before Federal Regulation*. Princeton, N.J.: Princeton University Press, 1961.

Yates:
Frances Yates, *Giordano Bruno and the Hermetic Tradition*. London: Routledge & K. Paul, 1964.

Zucchi:
Luca Zucchi, 'Brunfels e Fuchs: l'illustrazione botanica quale ritratto della singola pianata o immagine della specie' in *Nuncius*, xviii (2003), Florence, pp. 411–465.

I

LATE MEDIEVAL HERBALS

KONRAD VON MEGENBERG, *Buch der Natur,* 14th century. Binding

1. Konrad von Megenberg (*c.* 1309–1374)

Buch der Natur, 14th-century German manuscript, perhaps compiled in Regensburg

28.5 × 20.5 cm. 332 leaves, numbered subsequently in pencil, with gouache drawings and text in German *lettre bâtarde*.

PLATES: 304 total: 163 of plants, 137 of animals, two of insects, one of minerals, one scene with mountains, all gouache drawings on paper.

BINDING: 15th-century deerskin over wooden boards with remaining brass studs and bosses on covers. Various small sheets of paper with manuscript text pasted on inside of boards.

PROVENANCE: Library of Count Erbach-Fürstenau [possibly Philipp Karl (1677–1736)].

REFERENCES: Anderson, pp. 73–81; Arber, pp. 13–14; Gottschall.

This is a rare manuscript copy of one of the first scientific encyclopaedias to appear in Europe, *Buch der Natur,* a work that enjoyed an immediate success that is reflected in its diverse publishing history. The first printed edition was produced by Johann (Hans) Bämler in Augsburg in 1475 and represented a significant event, for it was among the earliest illustrated incunabula to appear in Germany. Other editions came out within the space of a few years: Bämler reprinted the work in 1478 and 1481; a fourth edition was produced in 1481, also in Augsburg, by Johann Schönsperger; and another Augsburg edition was published by Anton Sorg in 1482. While *Buch der Natur* cannot be considered an herbal in the strict sense of the term, both the manuscript copies and printed editions were embellished with a rich sequence of botanical images, resulting in a work that bears many affinities to the typology of the herbal. In addition, the *editio princeps* contains the first instance of a botanical woodcut.

The author was a German who was born around the year 1309 in the town of Megenberg close to Schweinfurth. After studying at Erfurt, and then at the university in Paris—where he earned the title of *Magister* and taught philosophy and theology from 1329 to 1337—he moved to Vienna to serve as director of the celebrated St. Stephen's School. His career culminated in his appointment as bishop of the city of Regensburg (Ratisbon), where he died in 1374.

Some of Konrad's other writings that should be acknowledged are a small German compilation dealing with physics and astronomy, *Sphære,* from the Latin work of Johannes Sacrobosco; *Planctus ecclesiæ in Germania,* a poem written in 1337; *Speculum felicitatis humanæ,* a work from 1348 on morals; *De erroribus Begehardorum et Beguinarum*; *De translatione imperii* (On the Transfer of the Empire, 1355); *Œconomica,* a large work written between 1353 and 1363; *Tractatus contra mendicantes ad Papam Urbanum V,* a work on managing beggars; a number of biographies of saints; several historical treatises on the local history of Regensburg; and a hymn in praise of the Virgin. Konrad's writings strongly reflect his support of the pope, as against William of Occam, who questioned the Church's temporal power, as Konrad also opposed Occam's reforms of Scholasticism.

During his sojourn in Paris, Konrad translated a text on the natural sciences—*De natura rerum* by Thomas de Cantimpré—into a German dialect spoken in a region of what is today Austria and Bavaria. Thomas de Cantimpré, a student of Albertus Magnus and a Dominican friar from an area of the Low Countries that now corresponds to Belgium, lived and preached during the first half of the thirteenth century. Konrad's aim with this vernacular translations was to present the whole of human knowledge concerning the nature of things in a single volume that would appeal to a general public made up of both men and women who had no acquaintance with Latin, the language of the cultivated elite. Thomas' philosophy was based on the conviction that God manifested Himself both in the tangible aspects of the world (*res*) and in the more profound signs that could be attributed to these things (*significationes*).

Guided by this conception and using *De natura rerum* as his model, Konrad set out to compile his own encyclopaedia, *Buch der Natur*, dedicating it to 'ein gutes Freund,' perhaps a colleague at the St. Stephen's School in Vienna. Like Thomas de Cantimpré, Konrad's purpose was to conduct the reader through a study that could be apprehended by the senses toward a more exalted contemplation of the invisible and spiritual that could only be apprehended by the mind. In obedience to the duty of the Christian scientist in the Middle Ages, he sought to unveil the divine plan that lay hidden beneath the myriad aspects of the material world and to demonstrate how the physical world reflected this divine order.

Buch der Natur is made up of a large number of short chapters. There is an extensive description of the human body, the heavens, and the seven planets, explained in accordance with the concept of the four humors that were thought to be responsible for a person's nature and physical characteristics. The humors were directly linked to the four seasons of the year, and to the four elements or 'roots of all things'—earth, fire, air, and water—as described by the Greek philosopher Empedocles. And, as expounded by the Greek physician Galen (A.D. 129–201), earth was associated with autumn and a melancholy character; fire with summer and a choleric temperament; air with spring and a sanguine humor; and finally water was linked to winter and a phlegmatic temperament. Illnesses were caused by an imbalance between these four humors, or by an excess of one of them. This theory formed the cornerstone of medieval medicine and would continue to influence the thinking of many physicians until the end of the eighteenth century.

Ample space in Konrad's encyclopaedia is dedicated to animals (birds, fishes, serpents, worms, and insects) and plants and simples, but the author also discusses spices from the Orient, precious stones, minerals, and metals. The work presents an overview of the secular, popular knowledge of the period, and in many chapters begins to show a direct rather

KONRAD VON MEGENBERG, *Buch der Natur,* 14th century. Five small trees: apple, pear, hazel, cherry, and one in center unidentified, folio 194^{v}

KONRAD VON MEGENBERG, *Buch der Natur*, 14th century. *Crocus* sp., folio 246[v]

Crocus heisset safferan vnd ist gar ein woil/
smackende krut vnd heisset sin blome auch
zu latin safferan Crocus ist heisse vnd drocken
mittelmessig Der safferan hut die krafft zu kreff/
tigen vnd zu strecken Vnd darvmb ist er gut
wieder des magen kranckheit vnd wedder des
menschen amachte die zu latin sinpis heisset
vnd wiedder die augen zoit die von blut kompt
Oder von der ~~ko~~ colera Man sol den safferan
hitzigen in eyner schierben vnd sol yne puluern
Vnd daz puluer misthen mit feistem wasser
daz enslusset vnd weichet den liep Vnd ist auch
gut zu den sachen die vorgenant sin Aber man
sol Jnne den luten nit geben die heisse vnd
drocken sint vnd die zu latin Colera heissent
Wan die machet es vnlustig vnd bringet

Cucurbita heysset kirbs daz krüte ist heisse vnd
fiuchte mittelmessig als plateariús spricht Der
same der in des kirbes fiuchte ist güt in erczenyen
wan man yne sudet vnd mit zoret Er ist güt vor
der lebern verstopffünge vnd vor die geistlichen
glieder Der kirbs ist güt in süchten wan man
yne sudet oder bret one alles gemeche vnd yne de
siechen gibt zü essen wan man yne brackt yne
teigke vnd in daz zuclot in wasser Vnd zücker dut
zü dem wasser Daz ist gar ein güt strophel den
die in siechtüm ligent Wan man den kirbes
sudet vnd daz wasser zü trincken gibt dem die
leber erhitzet ist das hilffet yne gar wole zü

KONRAD VON MEGENBERG, *Buch der Natur*, 14th century. Cucubits (*Cucurbita pepo*) growing on poles, folio 247^{v}

than an *a priori* approach to the study of physical phenomena. This attitude emerged as a response to the demand for concrete knowledge on the part of a new middle class. *Buch der Natur* therefore was conceived as a veritable compendium of knowledge in a period of transition between the Middle Ages and the early modern age.

As with the other encyclopaedias of the period, the material presented by Konrad was not original, but rather was collated from many other sources, beginning with Thomas de Cantimpré's *De natura rerum*, entire parts of which he appears to have translated and transposed to his own work. Further sources that he drew upon include the works of the Greek philosopher Aristotle, the medical texts of Galen, a heritage that had been preserved by the Arabs and subsequently transmitted to medieval Europe, the 'Canon' of the Muslim physician and philosopher Avicenna, who died in 1037, and the writings of the medical school at Salerno.

Konrad's vast erudition, however, allowed him to ransack other works from late Antiquity and the early Middle Ages, such as Isidore of Seville's *Etymologiæ* (A.D. 570–636), and, for his section on animals, the celebrated anonymous work *Physiologus*. The primary sources for the chapters dedicated to plants were *De vegetabilibus* by the Dominican theologian and philosopher Albertus Magnus, who also served as archbishop of Regensburg one century prior to Konrad, and the work of Matthaeus Platerius, a physician associated with the medical school of Salerno, which flourished between the eleventh and thirteenth centuries, who is credited with a compilation of a series of pharmacological manuscripts widely read all over Europe, known by its opening phrase *Circa instans*.

Konrad was not merely a passive conduit, content to disseminate the ideas and knowledge of others. He carried out many studies of his own and was the author of various original observations on, for example, the properties of the magnet and the processes of evaporation and circulation of water, which anticipated by many years the speculations of Leonardo da Vinci and the experiments of Bernard Palissy, a French ceramist, natural historian, and designer of gardens who was active during the second half of the sixteenth century. Konrad's knowledge of botany was equally profound and much of it was clearly the fruit of direct study, for he amplified the section on plants in his encyclopaedia from the 114 described by Thomas in *De natura rerum* to a total of 173, by adding his own descriptions of fifty-nine plants.

No less than fifty-six manuscript copies of *Buch der Natur* are known to exist, and of these forty-six are illustrated, a fact that demonstrates the importance ascribed to the scientific illustration in the encyclopaedic works being written at the dawn of the modern age. Among the manuscripts that have survived, the work in the Oak Spring Garden Library is without a doubt one of the most outstanding, not only because of the pristine state of its pages and the beautifully legible calligraphy in German *lettre bâtarde*, but also because of the sheer number of the drawings (304) and their vivid coloring and refined execution. Indeed, this is one of the most lavishly illustrated copies of Konrad's encyclopaedia extant; the exemplar at the Library of Würzburg has a mere twenty-nine illustrations and the one in Frankfurt forty. The library at the University of Heidelberg owns two manuscript copies of *Buch der Natur*, one of which is embellished with sixty-one illustrations and the other 304. The latter is therefore the only exemplar comparable to the work described here in terms of the magnificence of its visual component.

In the Oak Spring Garden Library manuscript we find interspersed throughout the text 163 vignettes of plants, 139 of animals, and one of a mineral; sometimes two images appear on a single page. The drawings, set off by frames in alternating colors, are quite lively and convincing. The plants are almost always portrayed in a realistic setting—growing out of the ground or, in the case of aquatic species, floating in the water. Sometimes their roots are also shown, perhaps in the interests of scientific documentation and to help physicians and naturalists identify the plant.

The unknown artist has applied his dense colors with great skill and appears to have based his drawings on the direct study of nature rather than merely relying on the codified and repetitive imagery of the contemporaneous herbal.

2. Herbal Manuscript, *Italian School* (*c.* 1425)

28 × 21.3 cm. 121 leaves of plants with manuscript text in early Italian and Latin, including a manuscript index of the drawings; five blank leaves.

PLATES: 116 watercolors of various plants, some with pencil and ink, all on paper.

BINDING: Contemporary vellum, which was subsequently re-backed in old vellum.

PROVENANCE: Leaf 117 states that the manuscript belonged to a member of the Paradiser family of Trisoz, in the Austrian Duchy of Carniola. Christoph Ruaber (1466–1536), son of Helena Paradiser who served in the courts of Maximilian I and Ferdinand I, is also mentioned.

THE HERBAL represented an essential tool for physicians, pharmacists, and herbalists who gathered or cultivated their own plants for the treatment of patients. This manuscript, which was compiled sometime during the first quarter of the fifteenth century, constitutes a typical example of an herbal that might have belonged to such a practitioner. Works of this kind were in use all over Europe and given their purpose, which was one of paramount utility, they would continue to be in great demand even after the invention of the printing press.

ITALIAN SCHOOL,
An herbal, circa 1455.
Viper's bugloss
(*Echium vulgare*),
folio 65[r]

The writing in this manuscript seems to date to the first quarter of the fifteenth century. A later annotation in Latin (fol. 117) suggests that the herbal belonged at one time to a member of the aristocratic Paradiser family of Trisoz, in what was formerly the Austrian Duchy of Carniola. The text also mentions Christoph Ruaber (1466–1536), a son of Helena Paradiser who held important positions at the courts of Maximilian I (1459–1519) and Ferdinand I (1452–1516). These manuscript herbals served during the late Middle Ages as veritable manuals of practical medicine. Their text and images were usually mechanically copied from earlier works of the most varied provenance, drawing upon the traditions of both the classical and Arab civilizations. The contents, however, were often integrated with glosses and personal observations by the successive owners, who introduced additions and corrections that were the fruit of their own concrete experiences.

This herbal, with its somewhat miscellaneous organization, evidently served as a working tool for a series of diligent medieval scientists, for in addition to the extensive notes of the original author it contains annotations in Latin and Italian, added at different periods by subsequent owners of the book. The alphabetical index that closes the work was drawn up at some later date. The vocabulary employed does not provide any clear clues as to the provenance of the author, although it is possible that he originally came from the central-northern part of Italy.

Like many other herbals painted in Italy during this period (analogous works include MS. 106 in the Botanical Library of the University of Florence, MS. 1591 in the Museo Provinciale d'Arte di Trento, and MS. 1161 of Joppi Library in Udine), this manuscript consists of a series of botanical drawings executed in a markedly geometricized style, the artist sometimes incorporating anthropomorphic elements from medieval herbal lore. The plants presented are species that were common to the Italian peninsula in the fifteenth century. Each drawing portrays a specimen in a frontal view, often with its roots, in a simple gouache drawing with no shadowing. Herbals of this type generally had few artistic pretensions, but the archaic, ingenuous, and spontaneous style of the drawings lend these works great charm to modern eyes. Despite the abbreviated style of the artist, the plants can be easily identified. To aid the herbalist, each is also labelled with the name or names (usually in Italian) by which it was known, and sometimes information on its habitat—for example, 'l'herba santi pauli nascit in pratis arenosis' (the herb of Saint Paul grows in sandy soil) (fol. 13)—or recipes for its use in the preparation of simple remedies.

The manuscript opens with the illustration of a yarrow plant (*Achillea millefolium*), portrayed in the artist's characteristically schematic style. The herbalist has filled the empty spaces on the page with notes describing the plant and its properties; his scribbling sometimes strays onto parts of the drawing itself, confirming the essentially practical use to which this handbook was devoted. One of the notes informs the reader in antique

ITALIAN SCHOOL,
An herbal, circa 1455.
Flowers of *Viola* sp.,
folio 79[v]

vernacular Italian that 'E' bona a persona che avesse febre terzana' (This is good for the person who has the third fever). Beneath the drawing in another, more educated hand, which recurs on many of these pages, is a notation in Latin identifying the specimen as an *achillea alias mellefolium, sive millephilon.* In some of the drawings, such as that of the narcissus (fol. 40[r]), the artist has colored in the background around the flower with a series of rapid brushstrokes. In others, such as the mallow (fol. 40[v]), the artist has added details of the flower, albeit somewhat loosely and inaccurately drawn. In accordance with established iconographic tradition, the mandrake plant is shown with its root in human form in fol. 44[v].

ITALIAN SCHOOL,
An herbal, circa 1455.
Details of various plants: top left, Lady's mantle (*Alchemilla vulgaris*); top right, Grape vine (*Vitis vinifera*); center, Belbine (*Calystegia sepium*); below, Monkshood (*Aconitum* sp.), folio 118^{v}

3. [H]ORTUS SANITATIS, PARIS [*c.*1500]

[*H*]*ortus Sanitatis*, Paris, *c.* 1500. Colophon, volume II, B8[v]

[*Xylographic titles with calligraphic initial L for Part Two and initial L with grotesque head for Part Three. Part One, Volume I,* a1]: Ortus sanitatis translate de latin en francois. [*Part Two, Volume II,* A1]: Le traictie des bestes, oyseaux, poissons, pierres precieuses et orines, du jardin de sa[n]te. [*Part Three, Volume II,* 3a1]: Le traictie des urines.

[*Imprint from Antoine Vérard's device, last leaf in second volume*]: Anthoine Verard humbleme[n]t te recorde ce qu['] il a il tient de toi par don por provocquer ta gra[n]t misericorde de tous pecheurs faire grace et pardon.

2° 31.7 × 22.3 cm. Three parts in two volumes. Volume I: a^8 b–z^6 &^{6} 2a–$2x^6$ $2y^4$ (-2y4) a–b^6 c^4 [582 pp.]. Volume II: A–X^6 2A–$2F^6$ $2G^8$ $3a^6$ $3b^8$ (-3b8) A^6 B^8 [394 pp.]. Numerous errors in foliation in both volumes.

PLATES: Volume I: 430 woodcuts within text; one full-page: presentation of book to prelate (on verso of title-page). Volume II: 541 woodcuts within text; four full-page: (1) human skeleton (on verso of title-page), (2) presentation of book to king on horseback with falconer scene above (preceding section on birds), (3) presentation of book to prelate—repeated from Vol. I—(preceding section on fishes), (4) apothecary shop (preceding section on urines).

BINDING: Modern red morocco; gilt tooled borders on covers; gilt tooling on spines. Manuscript annotations throughout volumes.

PROVENANCE: Bookplate of Alfred Petit in both volumes.

REFERENCES: Anderson, pp. 106–112; Arber, p. 13; Blunt & Raphael, pp. 118–119; Hunt, 12; Klebs, 52; Murray, *French* 227; Nissen 2373; Reeds, pp. 30, 185; Rohde, pp. 66–67; Shaffer.

THE [*H*]*ortus Sanitatis* was one of the most celebrated early printed herbals to appear in Europe. This incunabulum was first published in Latin in 1491 by Jacob Meydenbach of Mainz and it is assumed that Meydenbach was the compiler as well as the printer; translations and new editions with various modifications—particularly of the full-page illustrations—would continue to be produced during the entire course of the sixteenth century. The Oak Spring Garden Library possesses two paragons of this important text: the fifth edition, circa 1500, printed by Antoine Vérard, which we shall examine here, and the first Italian edition of the Latin original, printed in 1511 in Venice by Bernardino Benalio and

Giovanni Tacuino. The fifth edition is exceedingly rare and is the last incunabulum of the work to be printed in France, as well as being the first complete vernacular translation.

Two other important incunabula herbals, however—both issuing from the Mainz press of Peter Schöffer, successor to Gutenberg—preceded the [*H*]*ortus Sanitatis*: the small quarto *Latin Herbarius* of 1484, with an adapted medieval text and with stylized but decorative woodcuts, and the folio *German Herbarius* (*Der Gart der Gesundheit*) of 1485, with a new vernacular text and new illustrations, a portion of which, as a seminal advance, were drawn from life. The *German Herbarius* served as the author's initial model in compiling the [*H*]*ortus Sanitatis*, but the work was conceived as a considerably more extended and comprehensive medical encyclopaedia, one that described not only plants but also other natural substances that might be used to prevent or cure illnesses. Thus, for instance, the second volume of Vérard's French edition is entitled *Le traictie des bestes, oyseaux, poissons, pierres precieuses et urines, du jardin de santé,* augmenting the botanical material with a medieval bestiary of sorts, as well as with treatises on mineralogy and urinalysis. The lengthy section on medicinal plants in particular appears to have been profoundly modified with respect to the *German Herbarius*, with more than ninety plants added. Each chapter begins with a description of the plant, noting its various names (and sometimes the region where a given name was current), and concluding with a list of its *operationes,* or pharmacological properties, based on the medieval concept of the four humors or temperaments that were believed to determine a person's character and physical constitution. Indices were added as well, including therapeutical indices of diseases.

The Meydenbach edition of the [*H*]*ortus Sanitatis* distinguishes itself from the other texts of the period because it is accompanied by an extraordinarily fine set of illustrations that include seven full-page woodcuts, which were imitated in all subsequent printed editions, and nearly one thousand smaller vignettes scattered through the text. A large number were copied from the *German Herbarius*—some of them in greatly simplified form—but fully a third of the 430 botanical illustrations are original. The images appear on the page, generally without a border, beneath the chapter heading. Most of the specimens are shown cut off at the stem, although in some cases the artist includes the root, or portrays the plant growing out of the ground. One rare exception is the illustration of basil, which is depicted growing in a pot. As Agnes Arber has perceptively observed, many of the botanical images demonstrate 'a liveliness of imagination which one misses in modern botanical books.' The artist's fancy is expressed in various details, such as the tiny human figures that can be seen creeping out of the corolla of the *narcissus*, or the serpent twining through the branches of the *Arbor vel lignum scientiae* (the tree of knowledge). Other woodcuts present charming genre scenes: in the chapter on the olive, for instance, we find a laborer surrounded by large vessels filled with oil (the same image appears in the chapter on butter in

[H]ortus Sanitatis, Paris, *c.* 1500. Title-page, volume II

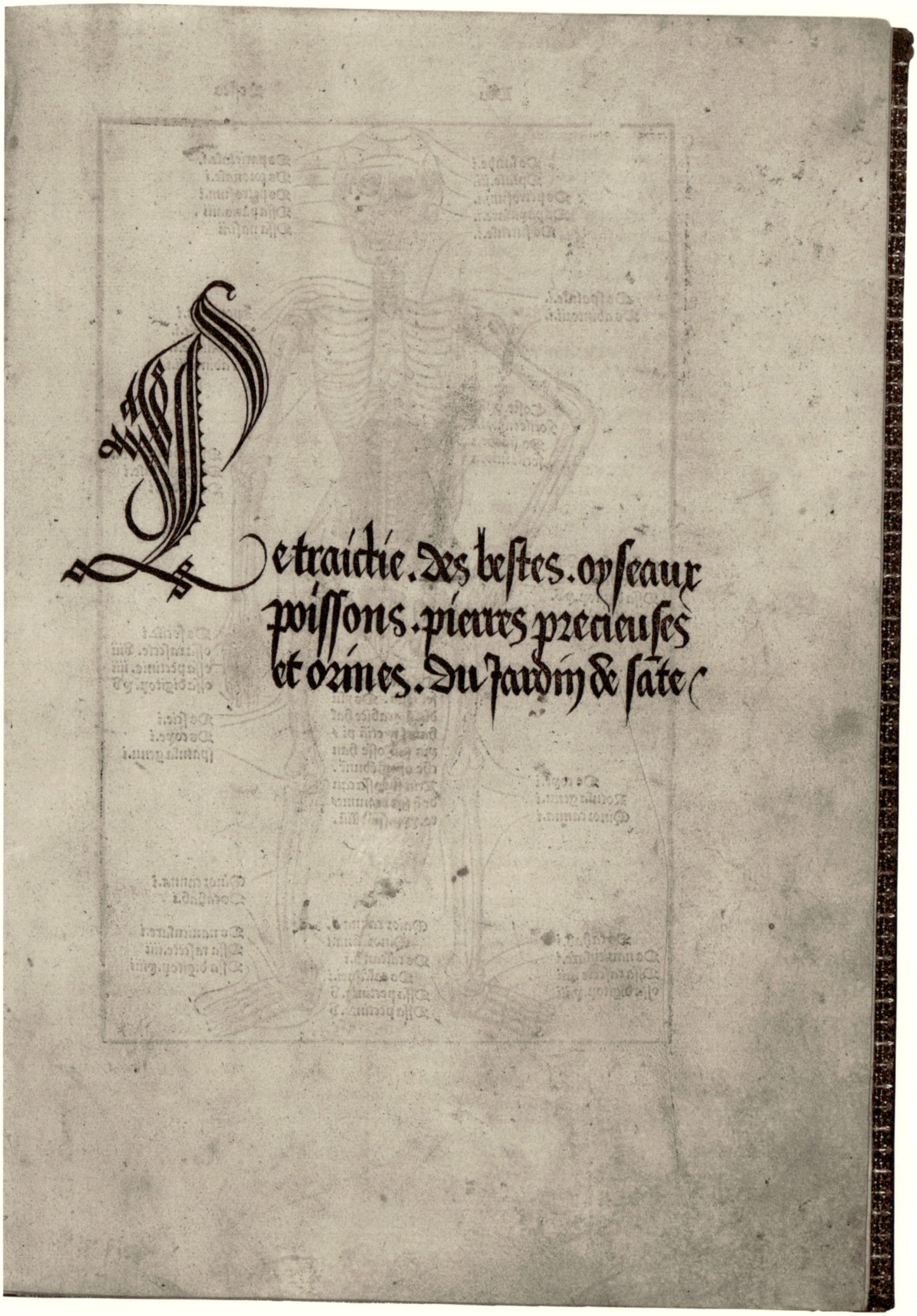

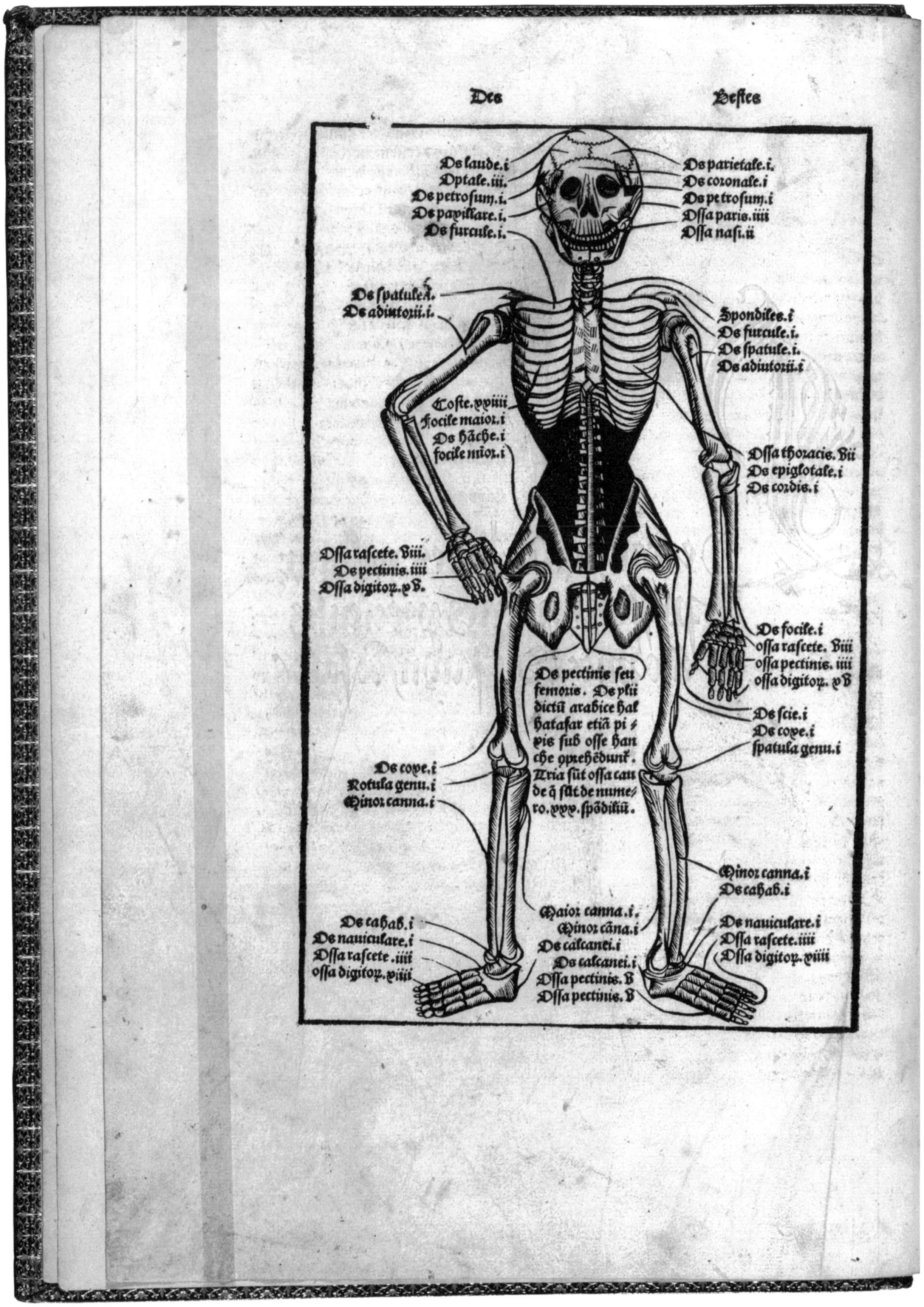

[H]ortus Sanitatis, Paris, *c.* 1500. Verso of title-page, skeleton with the bone names in Latin, volume II

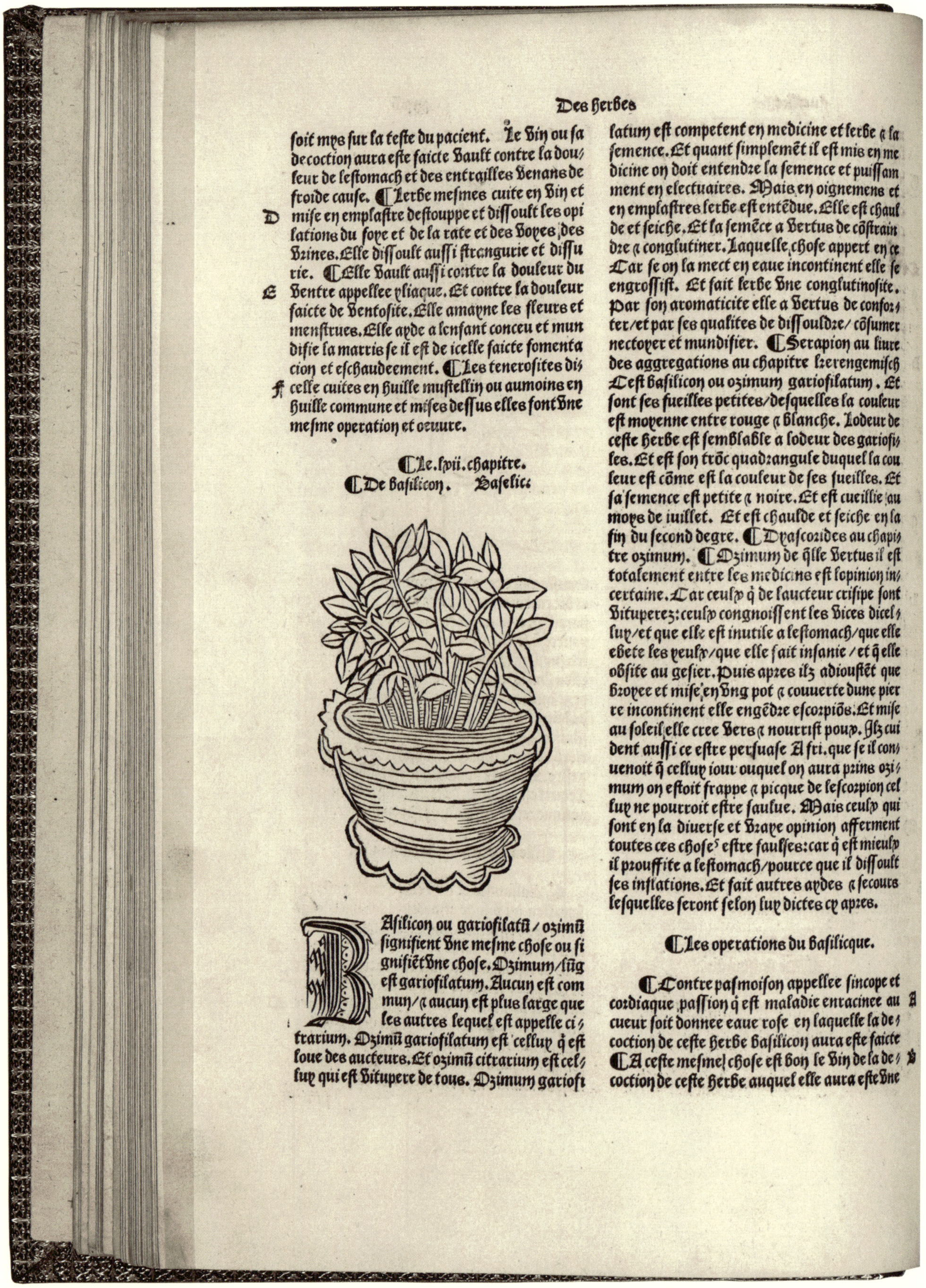

Des herbes

soit mys sur la teste du pacient. Le vin ou sa decoction aura este faicte vault contre la douleur de lestomach et des entrailles venans de froide cause. Lerbe mesmes cuite en vin et

D mise en emplastre destouppe et dissoult les opilations du foye et de la rate et des voyes des urines. Elle dissoult aussi strangurie et dissurie. Elle vault aussi contre la douleur du

E ventre appellee yliaque. Et contre la douleur faicte de ventosite. Elle amayne les fleurs et menstrues. Elle ayde a lenfant conceu et mundifie la matris se il est de icelle faicte fomentacion et eschaudeement. Les tenerosites di-

F celle cuites en huille mustellin ou aumoins en huille commune et mises dessus elles font une mesme operation et oeuure.

Le .lxvii. chapitre.

De basilicon. Baselic.

Basilicon ou gariofilatũ / ozimũ signifient une mesme chose ou signifiẽt une chose. Ozimum / sũg est gariofilatum. Aucun est commun / et aucun est plus large que les autres lequel est appelle citrarium. Ozimũ gariofilatum est celluy q̃ est loue des aucteurs. Et ozimũ citrarium est celluy qui est vitupere de tous. Ozimum gariofilatum est competent en medicine et lerbe et la semence. Et quant simplemẽt il est mis en medicine on doit entendre la semence et puissamment en electuaires. Mais en oignemens et en emplastres lerbe est entẽdue. Elle est chaulde et seiche. Et la semẽce a vertus de cõstraindre et conglutiner. Laquelle chose appert en ce. Car se on la mect en eaue incontinent elle se engrossist. Et fait lerbe une conglutinosite. Par son aromaticite elle a vertus de conforter / et par ses qualites de dissouldre / cõsumer nectoyer et mundifier. Serapion au livre des aggregations au chapitre hzerengemisch. Cest basilicon ou ozimum gariofilatum. Et sont ses fueilles petites / desquelles la couleur est moyenne entre rouge et blanche. Lodeur de ceste herbe est semblable a lodeur des gariofiles. Et est son trõc quadrangule duquel la couleur est cõme est la couleur de ses fueilles. Et sa semence est petite et noire. Et est cueillie au moys de iuillet. Et est chaulde et seiche en la fin du second degre. Dyascorides au chapitre ozimum. Ozimum de q̃lle vertus il est totalement entre les medicins est lopinion incertaine. Car ceulx q̃ de lauteur crisipe sont vituperez: ceulx congnoissent les vices dicelluy / et que elle est inutile a lestomach / que elle ebete les yeulx / que elle fait insanie / et q̃ elle obsite au gesier. Puis apres ilz adioustẽt que broyee et mise en ung pot et couverte dune pierre incontinent elle engẽdre escorpiõs. Et mise au soleil elle cree vers et nourrist poux. Ilz cuident aussi ce estre persuase A fri. que se il convenoit q̃ celluy iour ouquel on aura prins ozimum on estoit frappe et picque de lescorpion celluy ne pourroit estre saulve. Mais ceulx qui sont en la diverse et vraye opinion afferment toutes ces choses estre faulses: car q̃ est mieulx il prouffite a lestomach / pource que il dissoult ses inflations. Et fait autres aydes et secours lesquelles seront selon luy dictes cy apres.

Les operations du basilicque.

Contre pasmoison appellee sincope et cordiaque passion q̃ est maladie entracinee au A
cueur soit donnee eaue rose en laquelle la decoction de ceste herbe basilicon aura este faicte.
A ceste mesme chose est bon le vin de la de- B
coction de ceste herbe auquel elle aura este une

[*H*]*ortus Sanitatis*, Paris, *c.* 1500. Basil (*Ocimum basilicum*) growing in a pot, volume 1, f3 v

the second volume). The discussion of incense is illustrated with a figure holding a vessel from which the fragrant fumes of the burning resin can be seen rising into the air. The second part of *[H]ortus Sanitatis* is dedicated to animals and contains an even larger number of woodcuts of the same type, including many animated genre pictures.

This French edition of the *[H]ortus Sanitatis* is of particular interest, for it was produced by Antoine Vérard (*fl.* 1485–1512), a printer who counted King Charles VIII of France and King Henry VII of England among his patrons. Vérard printed the first typographical Books of Hours and specialized in *éditions de luxe*, often printed on vellum. His edition of the *[H]ortus Sanitatis* is embellished with his calligraphic woodcut titles and elaborate calligraphic initials, while the text is set in his fine *lettre bâtarde*. Its illustrations present many divergences from the original German edition and are based instead on the Strasbourg editions of Johann Prüss. The first full-page woodcut, which depicts a prelate seated in state at his desk being presented a volume by a group of learned colleagues, is unchanged. The human skeleton, however, which embellishes the verso of the title-page to the second volume, appears with legends in Latin identifying the most important bones. Toward the end of the work is another full-page woodcut depicting a scene in an apothecary's shop, where a teacher holding a pointer is lecturing on various drugs to another figure, perhaps a student. These three illustrations are of immense significance for they carry a new 'scientific' message, reflecting the commitment of the emerging science of plants and simples to the objective study of the natural world, with the aim of gaining a greater knowledge of the healing arts.

3. Hortus Sanitatis, Venice (1511)

[*Within woodcut border*]: Ortus Sanitatis. De herbis & plantis. De Animalibus & reptilibus. De Auibus & volatilibus. De Piscibus & natatilibus. De Lapidibus & in terre venis nascentibus. De Urinis & ea[rum] speciebus. De Facile acquisibilibus. Tabula medicinalis cum directorio generali per omnes tractatus.

[*Colophon, V8, verso*]: Impressum Venetijs per Bernardinum Benalium.

[*Second Colophon, 2f5, verso*]: Impressum Venetijs per Bernardinum Benalium: Et Joannem de Cereto de Tridino alias Tacuinum. Anno Domini. M.ccccxi. Die. xi. Augusti. Regnante Inclyto Duce Leonardo Lauredano Venetiarum Principe. Laus Deo.

2° 30.7 × 21.3 cm. a^8 b–k^6 l^8 m–r^6 s^8 t–z^6 $2a^6$ $2b^8$ $2c$–$2e^6$ $2f^8$ $2g$–$2i^6$ A^8 B–C^6 D^8 E–H^6 I^8 K–Q^6 R^8 S–T^6 V^8 $2a$–$2f^6$ [736 pp.].

PLATES: Three full-page woodcuts: (1) congress of doctors (verso of title-page), (2) human skeleton (section on animals), (3) doctors with urine flasks (section on urines). 1,055 woodcuts within text.

BINDING: Contemporary wooden boards half covered in blind-stamped calf, decorated with fleurons and geometric patterns, and furnished with four clasps (two on the fore-edge and one each on the head and tail), one preserved. Back board exhibits vertical break, which is repaired with contemporary string stitching. Manuscript annotations on endpapers. Fore-edge painting of a Numidian crane, roses with stem and foliage, and a lion by Titian's nephew, Cesare Vecellio (1521–1601).

PROVENANCE: Pillone coat-of-arms drawn on inside of front cover, with label: 'Livre No. '69' de la Bibliothèque Pillone.' Bookplate: Thomas Brooke, F.S.A., Armitage Bridge.

REFERENCES: Anderson, pp. 106–112; Blunt & Raphael, pp. 118–119; Hunt 12; Klebs 49; Mortimer, *Italian* 1.238; Nissen 2,368; Shaffer.

[*H*]*ortus Sanitatis*, Venice, 1511. Fore-edge painting by Titian's nephew, Cesare Vecellio, of a Numidian crane, roses with stem and foliage, and a lion

THE Oak Spring Garden Library also possesses a copy of the first edition of [*H*]*ortus Sanitatis* to be printed in Italy, which was published in Latin in 1511 by the celebrated Venetian printers Bernardino Benalio and Giovanni Tacuino. This edition was—like Vérard's—modelled upon previous ones, in particular the Strasbourg editions of Johann Prüss; the 1,055 woodcuts were certainly copied from Prüss' work, although with considerable variations. 'De facile acquisibilibus,' the second book of Galen's pharmacological work *De remediis facile parabilibus* has been added in a translation by Nicolaus de Regio.

While the human skeleton makes its appearance unmodified on the verso of the title-page to the volume on animals (even retaining some German labels of bones), the opening woodcut in the first volume and that for the *tractatus de Urinis* have been completely transformed. Rather than a professor lecturing, we find a group of colleagues engaged in a discussion regarding the plant that one of them is holding, and the group of physicians examining urinals includes a young boy holding a flask covered by a basket for safe transport. Even their attire has been modernized to fit the period of this later edition. This edition's dolphin title-page border was—in a tribute to its excellence—widely copied by other printers (see Mortimer, *Italian*, II), and two of the full-page woodcuts are surrounded by handsome white-on-black ornamented borders.

This copy is unique because it originally belonged to the Venetian aristocrat and bibliophile Odorico Pillone and his son Giorgio, and has been decorated with naturalistic motifs painted along its fore-edge. Furthermore, the Pillone coat of arms can be found drawn in ink on the inside front cover; the only other work from their collection known to carry the family's heraldic device is conserved in the Bibliothèque

Capitulum.ccliij.

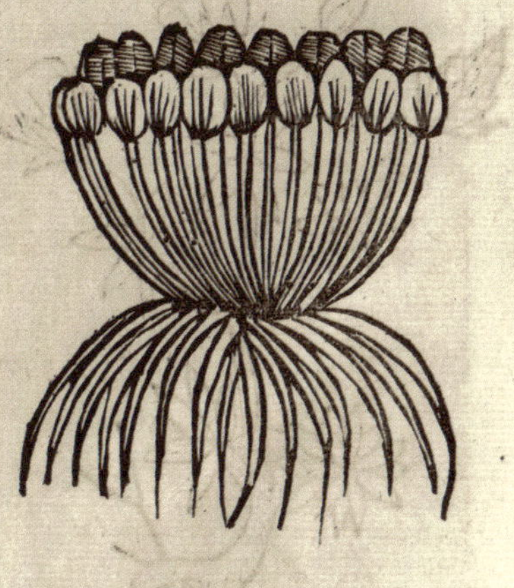

l Enticula aque vel Lētigo aque. Serap.li.ag.c.gabaleb.i.lēticula aque: ex ea est marina: ⁊ ex ea est aque dulcis: sed illa que est aque dulcis est lēticula que reperit super aquas erecta hñs corticem viridem: ⁊ est rotunda sicut lēticula: silis est ei in figura sua. Et repitur in lapidibus sup aquā erecta: natās super faciē aque. Gal. viij. sim. far. c. facus. i. lēticula aquatica: humide ⁊ frigide complexionis in.ij.gradu: existens ſm vtrunqz. ¶ Paulus cap. Stratiotis. Et est herba ⁊ nascitur in superficie aque sine radice cum folijs similibus semperuiue. Est autē virtutis redarguentis ac remordētis. Dyas. cap. de lenticula aque. Lenticula colorem viridem habens: virtus est ei frigida.

Operationes.

A ¶ Gal. Lenticula aque habet quid stipticitatis propter quod vulnera potest consolidare: ⁊ vlceribus est vtilis: quidam autem eo ⁊ ad emorrosagias vtuntur ⁊ ad fistulas.

B ¶ Paulus. Potui data valet ad fluorem sanguinis ex renibus: ⁊ cum aceto contrita: atqz vt cathaplasma adhibita.

C ¶ Dyas. Oēs tumores spargit. Ignē sacrū extinguit: podagricis necesse miscet: medicaminibus sola: ⁊ cū polēta apponit īfantibus.

Capitulum.ccliiij.

l Ens. Ser. li. ag. c. zedēs. auc. Dyas. meliores ex lentibus sunt ille que velocius maturant: ⁊ qñ infundunt in aqua ad mollificanduz non nigrescunt. Gal. vij. sim. far. c. de lentibus ſm transla. gre. inquit. Lentes stipticant non fortiter: in medio vero ca. ⁊ fri. consistunt: siccāt autē in.ij. gra.

Operationes.

A ¶ Gal. eodē li. ⁊ ca. ſm transla. ara. Sapor earū est stipticus stipticitate pauca: virtus lētis pma est media inter ca. ⁊ fri. ⁊ exiccat in.ij. gra. Et ipsa stringit vētrē qñ coqtur bis ⁊ piicitur aqua pma: ⁊ aqua in qua coquuntur laxat vētrē.

B ¶ Et dixit Ypo. vj. epidi. ꝙ est tēpata in ca. ⁊ fri. ⁊ declinat ad calorē parūper. Et signū in hoc est ꝙ nō apparet per eā qñ comeditur aut qñ moratur in stō: aut qñ descēdit a ventre infrigidatio corpori: ⁊ est cōposita ex substantia que stringit ventrē qñ coqtur bis: ⁊ proijcit aqua pma: ⁊ aqua in qua coquū tur laxat vētrē.

C ¶ Idē auc. Dyas. Qñ comeditur ex ea assidue facit obscuritatē visus: ⁊ ē dure digestionis: mala stomacho ⁊ accidunt ex esu eius somnia mala: ⁊ est mala neruis ⁊ pulmoni.

D ¶ Et idem Serap. auct. Mesarugie. Excorticate ex his sunt frigide: grosse stringunt ventrē ⁊ vrinā: ⁊ ingrossant sanguinem in venis

[H]ortus Sanitatis, Venice, 1511. Two woodcuts: 'Capitulum. ccliij,' a member of the Bladderwort family (*Lentibulariaceae*), and 'Capitulum.ccliiij,' possibly a Lentil (*Ervum lens*), P6[v]

Operationes.

ℭ Myas. Eius folia in sale cōponuntur. ventrem stringunt. ℭ Radix eius venenis occurrit tussionibus ⁊ ortonoicis vrinam ꝓuocat. ℭ Folia eius cathaplasmatibus adhibita tumores ⁊ duricies spargunt. ℭ Radix eius similiter facere potest. semen eius vrinam ꝓuocat. ℭ Nefreticis ⁊ spleneticis ⁊ dolori vesice maximum presidium est: menstruis operat. ℭ Secundinas expellit.

Capitulum cccccxxix.

Ma siue muza grece. Ex sinonomis Muza est fructus paradisi: vt dicunt quidam: in quo comedendo peccauit adam. Sed alii verius dicunt ꝙ peccauit in ficu. Fertur autem muza crescere in babylonia alio quoqꝫ nomine muza ē opiata. Ysaac. Muza est calida in medio primi gradus ⁊ humida in fine ipsius.

[*H*]*ortus Sanitatis*, Venice, 1511. Detail, woodcut of Adam and Eve with serpent around an apple tree (*Malus domestica*), T6v

Nationale de Paris. The Pillones amassed a valuable library and had the edges of many of their books painted in order to enhance their beauty and value. Thus we find on this volume sketches in gouache of a Numidian crane, a trailing stem of roses, and a prancing lion. The great painter Titian Vecellio was a close friend of the family and the task of decorating the collection was entrusted to his nephew Cesare Vecellio (1521–1601). Whilst most of the volumes are embellished with classical subjects, vistas of Venice, or scenes portraying intellectuals at work in their *studioli* or conversing with colleagues, the decoration of the *[H]ortus Sanitatis* reflects the naturalistic themes treated in the text.

The Pillone Library, which comprised 160 works, remained in Venice until the nineteenth century, when it was acquired by the English nobleman and bibliophile Thomas Brooke. The collection was subsequently dispersed; three works can be found in the Bibliothèque Nationale de Paris, while the others are scattered among the most important libraries in the world. The volume in the Oak Spring Garden Library does not appear to be cited in Pierre Berès' study of the collection, which was published in 1957.

4. The Grete Herball (1526)

[*Title-page in red and black; first three words xylographic*]: The grete herball whiche geueth parfyt knowlege and under standyng of all maner of herbes & there gracyous vertues whiche god hath ordeyned for our prosperous welfare and helth, for they hele & cure all maner of dyseases and sekenesses that fall or mysfortune to all maner of creatoures of god created, practysed by many expert and wyse maysters, as Auicenna & other. &c. Also it geueth full parfyte understandynge of the booke lately prentyd by me (Peter treueris) named the noble experiens of the vertuous hand warke of surgery. [*Large woodcut*].

[*Colophon 2E6 verso, woodcut borders surrounding large printer's device*]: Imprentyd at London in Southwarke by me Peter Treueris, dwellynge in the sygne of the wodows. In the yere of our lorde god. M.D. xxvi. the xxvii. day of July.

2° 25 × 17.2 cm. ✠^{6}–2E^{6} (-✠6 C4 N1 T5 2D6 2E3 2E4) [334 pp.].

PLATES: Woodcut on title-page, 466 woodcuts within text.

BINDING: Modern tan polished calf. Last leaf inscribed: 'Peter Treveris, a foreigner first erected a printing press in Southwark 1514 and continued till 1532. He lived at the sign of the Widows [*sic*] and printed several books for Wm Rastele, John Reyner, R. Coplar[?] and others in the city of London.'

REFERENCES: Anderson, pp. 98–105; Cleveland 36, Henrey I.15–19 and No. 168; Hunt 25; Klebs 59; Nissen 2,296; Pritzel 10,762 (*Le grant herbier*); Rohde, pp. 65–74, 207–208; Rydén.

THE *Grete Herball* was the first illustrated herbal published in England. In keeping with the tradition of incunabula and works printed in the earliest part of the sixteenth century, the colophon supplies the publication detail that the book was 'Imprentyd at London ... by me Peter Treueris ... in the yere of our lorde god' 27 July 1526. The title, *The great herbal which giveth perfect knowledge and understanding of all manner of herbs and their gracious virtues which God hath ordained for our prosperous welfare and health ...*, is quite significant, as is the fact that Treveris chose to publish his book in English. In this period all over Europe most scientific texts were written in Latin, and the appearance of *The Grete Herball* in English testifies to the practical ends for which it was intended and to the widespread interest—even among those who did not have the benefit of an academic education—in works of this type. Peter Treveris, who was perhaps German by birth, was a prolific printer between 1514 and 1532. He owned a printing shop 'in the sygne of the Wodows,' as inscribed on the colophon of this herbal, over the bridge of Southwark in London. The work was evidently very well received, for he published a completely revised version of the work in 1529; editions by other London printers appeared in 1539 and 1561. *The Grete Herball* had been preceded a year earlier by a rather more modest *Herball*, perhaps based on a medieval manuscript, that was published without illustrations by Richard Banckes. This work was quite successful and went through numerous reprintings.

The herbal published by Treveris was not original, however, since, as is stated in a note at the foot of the index, a large part of it actually consisted of a translation from the French. The bulk of the text derives from *Arbolayre* which was first printed in Besançon by Pierre Metlinger around 1487 and was re-published in Paris around 1498 as *Le grant herbier en françoys*: hence the English title. *Arbolayre* is a French version of Matthaeus Platerius' antidotarium *Circa instans* (see No. 1). The treatise on urines at the end of the text and the preface, which explains the medical aims of the work and underlines the importance of plants in the treatment of illnesses, appear to have been borrowed from the *German Herbarius* (Mainz, 1485) and Meydenbach's *Hortus Sanitatis* (see No. 3). The publisher also pays tribute to the great authorities of the past who made possible the compilation of 'this noble worke,' among them Avicenna, Matthaeus Platerius, Joannes Mesuë the Younger (an Arab physician who was active during the eighth century A.D.), Albertus Magnus, and the Franciscan Friar Minor Bartolomaeus Anglicus, the author of another celebrated encyclopaedia, *De proprietatibus rerum*, which was written sometime before 1283.

The Grete Herball opens with a splendid page bearing the title and a charming vignette on the subject of gardening. To the left a man leaning on a spade gathers grapes from a vine, while in the center another figure pours flowers from an apron into a large basket. On the right are three trees growing in a basket. A row of plants occupies the foreground: a rosebush and another flowering species, with a male and a female mandrake plant on either side. In accordance with medieval tradition the mandrakes are shown in humanized form; the female plant can be seen modestly covering her pudendum with her hands.

The text is preceded by an index or table that lists the plants in alphabetical order by their Latin and English names beginning with 'Aloe' and ending with 'Zuccaru[m].Sugre.' This is followed by a shorter index of the other substances and animals described in the

The grete herball
which geueth parfyt knowlege and under
standyng of all maner of herbes & there gracyous vertues whiche god hath
ordeyned for our prosperous welfare and helth/for they hele & cure all maner
of dyseases and sekenesses that fall or mysfortune to all maner of creatoures
of god created/practysed by many expert and wyse maysters/as Auicenna &
other. &c. Also it geueth full parfyte understandynge of the booke lately pren
tyd by me(Peter treueris)named the noble experiens of the vertuous hand
warke of surgery.

The Grete Herball, London, 1526. Title-page

PETRVS TREVERIS

¶ Imprentyd at London in South-
warke by me Peter Treueris/dwel-
lynge in the sygne of the wodows.
In the yere of our lorde god. M. D.
xxvi. the xxvii. day of July.

The Grete Herball, London, 1526. Colophon, 2E6v

text. The text ends with a glossary of terms (the exposicyo[n] of wordes obscure and euyll knowen) and an index of remedies.

The format of the *The Grete Herball* reflects that of the contemporaneous manuscript herbal and confirms the direct ties linking the first printed herbals with their illuminated predecessors. Each of the 472 chapters is preceded by a small woodcut vignette of the plant in a quadrangular frame. Like the text, these images were copied from other works, in particular the *German Herbarius,* or its first Dutch edition of 1514, and the first edition of the *[H]ortus sanitatis,* which appeared in Mainz in 1491. In the Oak Spring Garden Library's copy, the full-page woodcut of the human skeleton (borrowed from *[H]ortus sanitatis,* see No. 3) is missing. There are also occasional genre cuts of animals, landscapes, objects of work and commerce, and human activities. The woodcuts, with their simple, deeply incised outlines and sober shadowing, are striking, and the plants portrayed are quite recognizable.

The full-page woodcut on the colophon deserves particular attention. It presents the coat of arms of the publisher flanked by a 'woodwose' or 'Homo sylvestris' and his mate. The iconographic image of 'the wild man of the forest,' covered with hair and carrying a bow and arrow, was quite diffuse in France, Italy, and Germany in this period, as is demonstrated by the heraldic image which appears in the *Portrait of Oswold Krell,* painted by Dürer in 1499 and today conserved in the Alte Pinakothek of Münich. Treveris's device seems to be derived indirectly from that of the Parisian printer Philippe Pigouchet (*fl.* 1483–1515), which also pictures a wildman and woman.

II

THE GREAT AGE OF RENAISSANCE BOTANY

OTTO BRUNFELS,
Herbarum vivæ eicones,
1530. Title-page

HERBARUM
VIVAE EICONES
ad naturę imitationem, suma cum diligentia et artificio effigiatę, unà cum EFFECTIBUS earundem, in gratiam ueteris illius, & iamiam renascentis Herbariæ Medicinæ, PER OTH. BRUNF. recens editæ. M. D. XXX.

¶ Quibus adiecta ad calcem, APPENDIX isagogica de usu & administratione SIMPLICIUM. Item Index Contentorũ singulorum.

Argentorati apud Ioannem Schottũ, cum Cæs. Maiest. Priuilegio ad Sexennium.

Inuentũ Medicina meũ est, opiferq; p orbem Dicor & Herbarũ subiecta potentia nobis.

5. OTTO BRUNFELS (1488–1534)

[*Title-page in red and black with a woodcut surround (within which are typographic labels)*]: Herbarum vivæ eicones ad naturæ imitationem, sum[m]a cum diligentia et artificio effigiatæ, unà cum effectibus earundem, in gratiam ueteris illius, & iamiam renascentis Herbariæ Medicinæ, per Oth. Brunf. recens editæ. M.D.XXX. ¶ Quibus adiecta ad calcem, Appendix isagogica de usu & administratione simplicium. Item Index Contentoru[m] singulorum. Argentorati apud Joannem Schottu[m], cum Caes. Maiest. Privilegio ad Sexennium.

[*Colophon, G5 verso*]: Argent. per Io. Schottum, Carol. Imp. V. Anno.10. Christi uero Seruatoris, 1530.

2° 31.5 × 21 cm. A^4 a^6 b^4 c–z^6 A–F^4 G^6 *i–viii* 1–266 [*66*] (97 as 96, 100 as 99, 103 as 89, 135 as 137, 136 as 138, 183 as 181, 196 as 236, 245 as 249, 257 as 251, 258 as 254) [340 pp.].

PLATES: Coat-of-arms of Strasbourg printed in red and black on A4 verso; 86 illustrations of plants, all woodcuts by Hans Weiditz.

BINDING: Original vellum flap binding with two paper labels on spine.

PROVENANCE: A contemporary signature dated 1559 [?]; bookplate of Dr. Juan Carlos Ahumada (b. 1890), Buenos Aires, plus a manuscript note by him laid in discussing the volume. Blue stamp on title-page: 'Juan Carlos Ahumada Biblioteca.'

REFERENCES: Anderson, pp. 121–129; *Botany in the Low Countries*, pp. 11–12; Blunt & Raphael, pp. 120–123; Cleveland 42; Hunt 30; Klebs 69; Lack 2; Murray, *German* 462; Nissen 257; Pritzel 1,283; Reeds, 104–105; Stafleu & Cowan 852.

OTTO BRUNFELS' *Herbarum vivæ eicones* constitutes a genuine milestone in the history of the botanical sciences and the date of its publication—1530—is generally acknowledged as the birth of modern botany. The title itself makes explicit the profound cultural innovation contained in the work: *Herbarum vivæ eicones ad naturae imitationem, summa cum diligentia et artificio effigiatæ* (living images of plants, portrayed with great care and artistic skill, faithfully following nature).

The author, Otto Brunfels, was born in the town of Braunfels near Mainz, the city where he began his studies. He entered a Carthusian monastery in Strasbourg, but he left the order in 1521, renounced his Catholic faith and embraced Protestantism. He was influenced more by the theology of Thomas Muntzer than Martin Luther and devoted himself for a period to preaching in southern Germany. In 1524 he opened a school in Strasbourg, and during the period that followed published several religious, botanical, and medical texts. In 1530 he received his degree in medicine from the university in Basel and shortly before his death in 1534 was nominated chief physician of the city of Berne.

Many and varied were the sources that contributed to Brunfels' formation as a naturalist, beginning with the classical texts and continuing with the works of various Italian authors, such as the humanists Ermolao Barbaro (1513–1593) and Marcello Virgilio (b. 1464), both of whom edited and wrote commentaries on Dioscorides' *Materia medica.* In his work *Castigationes*, Barbaro expunges various errors contained in Pliny's *Naturalis historia.* Brunfels also drew upon the experience of contemporaries, such as his friend the herbalist Hieronymus Bock (known as Tragus, 1498–1554), and did not scorn the practical knowledge accumulated by 'herb-women' and that contained in country lore. *Herbarum vivæ eicones*, however, was a conscious expression of the period's humanistic movement and for this reason is described in the opening *Encomium medicinæ* as an entirely 'new' work whose aim was to overcome the limits of the contemporaneous manuscript herbals, which more often than not were prepared by almost unlettered herbalists.

The title-page of the work is connotative of this renewed cultural climate because—probably in keeping with the suggestions of the author himself—it is filled with learned references to classical mythology and inscriptions in Greek and Latin that were intended to aid the reader in deciphering the scenes depicted. Surrounding the title printed in red and black is a magnificent, highly articulate engraving printed from several woodblocks. Above the space containing the title appears the Garden of Adonis, with Venus, his consort, to the left and Silenus to the right, accompanied—in accordance with classical tradition—by the two putti Mnasilus and Chromis. A scene representing the luxuriant Garden of the Hesperides fills the lower part of the page, with the Hesperides on the left and on the right Hercules battling the dragon that had been placed there by the gods to defend the tree of golden apples. Flanking the title are the two patrons of the medical arts standing on pedestals—on the right Apollo, the inventor of the art, who is shown with an aureole, holding a musical instrument and a handful of arrows, and on the left Dioscorides, clad in a doctoral gown and holding a plant in one hand while he peruses a volume held in the other.

The originality of *Herbarum vivæ eicones* is immediately evinced in the title and in the complex construction of its title-page, for its iconographic apparatus bespeaks the new outlook that dominates the work. Brunfels, with great perspicacity, took the unprecedented step of commissioning his botanical illustrations from an artist who had the capacity to study the living plant directly, grasp its specific characteristics, and depict it objectively and realistically, leaving behind the symbolic and allegorical tradition that had weighed so heavily upon the illustrated herbal since the Middle Ages, a tradition aggravated by the widespread practice of mechanically copying the same images from one herbal to the next.

The work, therefore, reflects a new conception of the relationship between the artist and the world of nature, one that had been inaugurated just a few years earlier by Leonardo da Vinci in Italy and France, and by Albrecht Dürer in Germany. Brunsfels wrote that his illustrations were 'durch den hochberühmten meyster Hans Weiditz von Strasbourg gerissen und contrafeyt' (by the gifted master from Strasbourg Hans Weiditz drawn and reproduced). In the Latin poem addressed *Sapido lectori* (to the learned reader) he reiterates his praises, going so far as to compare Weiditz with the legendary Greek painter Apelles:

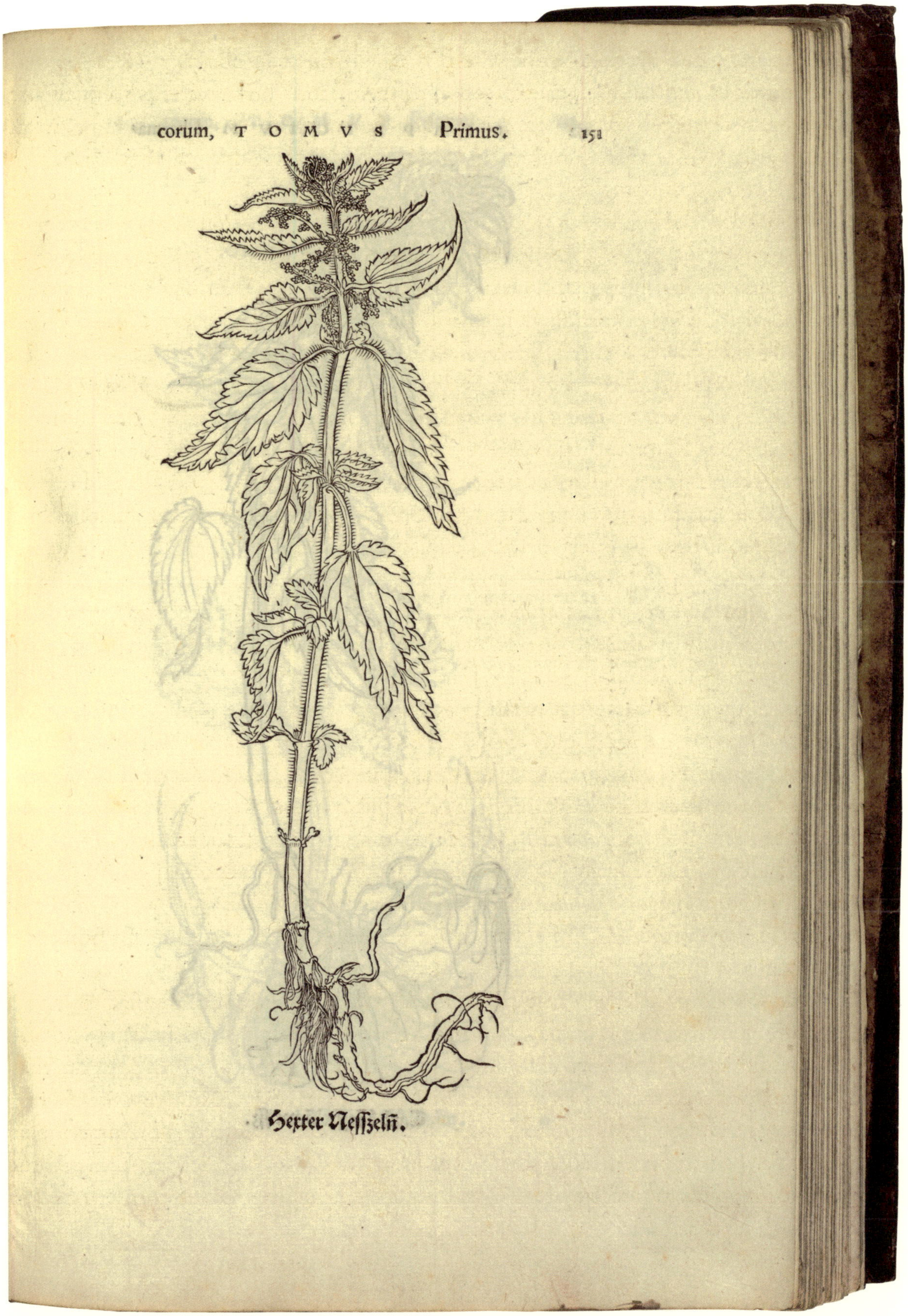

OTTO BRUNFELS,
Herbarum vivæ eicones,
1530. Stinging nettle
(*Urtica dioica*), page 151

'Nunc & Iöannes pictor Guidictius ille | Clarus Apelleo non minus ingenio, | Reddidit adfabras acri sic arte figuras, | Ut non nemo Herbas dixerit esse meras.' (Also that painter Hans Weiditz, whose skill was no less that that of Apelles, and who succeeded by means of a difficult artistic technique to render images of plants that no one could vow were not true to life).

The author's flattering opinion is borne out by the discovery of a large number of drawings in watercolor by Weiditz's hand conserved at the Felix Platter Herbarium Systematisch-Geobotanisches Institut of the University of Berne. Although the drawings have been cut and glued to new sheets, they still allow us to analyze the style of the artist, which was strongly influenced by Dürer, whose naturalistic studies enjoyed a great vogue and were widely copied and reproduced beginning in the first decades of the seventeenth century. Weiditz was born around 1495 in Freiburg or Augsburg and died in Augsburg in 1537. He remained in his native town until 1522, and then moved to Strasbourg. His talents as an illustrator were in great demand among the printers of that city, as is demonstrated by his fine woodcut illustrations in various editions of Pliny and Petrarch; he produced drawings that were then transferred to woodblocks, and often included his monogram 'H.W.' at the foot of the engraving. It may be hypothesized that both the original botanical drawings and the woodcuts in *Herbarum vivæ eicones*—product of that *acri arte* to which the poem written by Brunfels in his honor may have been alluding—were by his hand.

Every illustration in Brunfels' herbal was manifestly drawn from life, the artist seeking to capture both the form and characteristic elements of each plant, from the root system to the flowers, in accurate and highly legible drawings. Details such as the slightly wilted leaves of the borage (*Borago officinalis,* p. 113) and hellebore (*Helleborus niger,* p. 30), or the broken stems in the illustration of comfrey (*Symphytum officinale,* p. 53), betray the fact that the plants had been plucked from the earth some time before they were brought to the artist. He had an innate sensitivity for his subject matter, skillfully capturing the pliancy of the living plant from the upward curve of the stem to the downward arc of its leaves, as may be seen in the illustration of the 'Nenuphar' (European White Water Lily, *Nymphaea alba,* p. 37), with its elegant stems and tangled root filaments.

The plants have been drawn without any shading, as the publisher intended to have a certain number of copies—particularly those destined for illustrious recipients—colored by hand. It was obvious that the woodcut, while it offered the immense advantage of allowing multiple copies to be made, could not convey the lifelike colors that made the painted herbal such a useful aid to the herbalist. The artist nevertheless devoted great attention to capturing significant details, such as the fine veins of the leaves, the delicate hairs on the stems, and the complicated root system of the stinging nettle (*Urtica dioica,* p. 151). Most of the illustrations take up an entire page; only occasionally has the image of a plant been

C

HEPATICA.

D

Edel Leberkraut. Hyrtklee.

DE HEPATICA,

Rhapsodia XXXIIII.

NOMENCLATVRA

Græca, λειχὴν.
Latina, Hepatica, Hepataria, Iecoraria.
Germanica, **Leberkraut.**

HEPATARIAM, siue Hepaticam, omnes Lichen interpretantur eam quæ est apud Dioscoridem libro. α. cap. λειχὴν. Verba Dioscoridis sunt. Lichen, quæ saxis est familiaris, aliquibus Bryon appellatur, asperginosis petris adhæret.

Paraphrasis, BARBARI, super DIOSCORIDEM.

Lichen, siue Lechen, est muscus, ut hic, itemq; hærba alia nascens in petrosis: nomine utriq; inuento, quoniam contra Lichenas, remedijs omnibus anteponuntur. Folium unum ad radicem latum, caule uno paruo, loogis folijs dependentibus. Vtramq; recentiores herbarij Hepaticā appellant. Appuli hærbam, cuius unum ad radicem folium habere diximus, Iecorariam, siue Ficatellam cognominãt, folio succi pleno, & crasso:

Hepatica uni folia.
Ficatella.

OTTO BRUNFELS, *Herbarum vivæ eicones*, 1530. Liverleaf (*Hepatica*), page 190 inserted into the text and the elegance of the layout is always maintained, as can be seen in the illustration of the *Hepatica* on page 190 and the *Arum* that embellishes page 56.

Otto Brunfels has been described as 'the first great mind in modern botany' (Hunt 30), but the text of his herbal was not nearly as innovative as its visual component. Although his intention with *Herbarum vivæ eicones* was to present the flora of his native Germany, he still used as his point of departure the texts of Antiquity and in numerous cases failed in his endeavour to link a plant to one mentioned in the classical texts, since, in the infancy of botanical knowledge, he did not understand the variations of regional flora. Thus, about fifty species and varieties unknown to the ancient Greeks and Romans were described by him as 'herbae nudae,' or without name or identity. For each plant that he did identify, he provided the names in Greek, Latin, and German and, as in every botanical-medical herbal of the period, described its 'temperamentum' and 'vires et iuvamenta' (properties and uses). *Hepatica,* for example, was recommended for liver ailments and dog bites, while the *Geranium* could be used to treat kidney stones and heal wounds. In the Oak Spring Garden Library copy, French plant names have been written in with the heading 'Gallice' in red ink and the names in black.

Another singular feature of Brunfels' herbal is the religious faith that permeates it, beginning with the opening *Invocatio Divini Auxilii* (Invocation of Divine Aid) and the affirmation that the plant world must be considered a munificent gift offered by God to humanity. The first species discussed by Brunfels is the plantain, which he describes as most ordinary and unrefined but nonetheless a sign of the greatness of God, which expressed itself even in the most humble and insignificant aspects of nature.

In his introduction Brunfels also acknowledges the contribution of the Strasbourg printer Johann Schott, who had been producing books that were in his view particularly deserving of praise ever since the invention of the art of printing. In a poem the author eulogizes the expense, effort, and care (*sumptus, labor, studium*) which Schott had dedicated to the preparation of his text. Indeed, the printing of *Herbarum vivæ eicones* must have been a complicated undertaking; the first edition was published in Latin in three folio volumes that appeared in succession, the first volume in 1530, the second in 1532, and the third—an appendix that the author entitled *Corollarius*—posthumously in 1536. In 1532 a German translation of the work saw the light. The positive reception accorded *Herbarum vivæ eicones* led to the preparation of a second Latin edition whose three volumes came out in 1532, 1536, and 1540, respectively. According to many historians, the work was then republished in a single volume in 1539 with the title *Herbarium Othonis Brunfelsii tomis tribus.*

6. Leonhart Fuchs (1501–1566)

De historia stirpium commentarii insignes, maximis impensis et vigiliis elaborati, adiectis earundem vivis plusquam quingentis imaginibus, nunquam antea ad naturæ imitationem artificiosius effictis & expressis, Leonharto Fuchsio medico hac nostra ætate longè clarissimo, autore. Regiones peregrinas pleriq[ue], alij alias, sumptu ingenti, studio indefesso, nec sine discrimine uitæ non-nunquam, adierunt, ut simplicium materiae cognoscendae facultatem compararent sibi: eam tibi materiam uniuersam summo & Impensarum & temporis compendio, procul discrimine omni, tanquam in vivo iucundissimoq[ue]. uiridario, magna cum voluptate, hinc cognoscere licebit. Accessit ijs succincta admodum uocum difficilium & obscurarum passim in hoc opere occurrentium explicatio. Unà cum quadruplici Indice, quorum primus quidem stirpium nomenclaturas græcas, alter latinas, tertius officinis seplasiariorum & herbarijs usitatas, quartus germanicas continebit. [*Large printer's device*]. Cautum praeterea est invictissimi Caroli Imperatoris decreto, ne quis alius impunè usquam locorum hos de stirpium historia commentarios excudat, iuxta tenorem privilegij antè à nobis euulgati. Basileae, in officina Isingriniana, Anno Christi M.D.XLII.

2° 37.5 × 23 cm. α^6 β^8 A–3f^6 *i–xxviii* 1–33 35–60 *61* 61–896 [*4*] (255 as 254, 473 as 476, 476 as 473, 535 as 533, 596 as 569, 769 as 779) [928 pp.].

PLATES: Printer's device on title-page and last leaf; 509 plants; portrait of Fuchs on verso of title-page, and Heinrich Füllmaurer and Albrecht Meyer, the artists, and Veit Rudolph Speckle, the engraver, on page 897 (3f5), all hand-colored woodcuts.

BINDING: Contemporary calf over wooden boards; gilt tooling on spine.

REFERENCES: Anderson, pp. 137–147; *Botany in the Low Countries*, No. 9 and pp. 51, 90–91; Blunt & Raphael, pp. 123–128; Bush-Brown, pp. 95–109; Cambridge F1099; Cleveland 59; Dobat; Grolier Club, pp. 64–67; Hunt 48; Klebs 72; Lack 5, 6; Meyer; Murray, *German* 175; Nissen 658; Reeds, pp. 21, 32, 37; Rohde, pp. 82–83, 95; Stafleu & Cowan 1909; Voigt, p. 269; Wheelwright, pp. 108–109.

OTTO BRUNFELS' *Herbarum vivæ eicones* opened the way, about a decade later, for the work of another of the great founders of modern botany—Leonhart Fuchs. His *De historia stirpium*, which was published in 1542, is extraordinarily complex in conception and represents a landmark in the study of natural history in Europe. Indeed, according to some scholars the new scientific spirit that pervades it was analogous to that which informed the anatomical treatise of Andreas Vesalius (1514–1564) and the astronomical works of Nicolaus Copernicus (1473–1543), which saw the light one year later in 1543, all of them contributing to revolutionize man's understanding of himself, the world of nature, and the cosmos.

Here, as in Brunfels' herbal, the title is extremely significant: 'Outstanding commentaries on the history of plants, drawn up at great expense and effort, to which are added more than five hundred completely new images, portrayed with great art in imitation of nature.' The author underlines the fact that his text describes many plants originating from 'regiones peregrinæ' (distant lands) that were collected at 'great expense, with enormous fatigue and considerable danger' in order to provide the reader with immediate and simple access to knowledge of a 'materia universalis' presented in the form of a 'vivo iocundissimo viridario' (a living and most pleasant garden).

Leonhart Fuchs was born in 1501, the son of the burgomaster of the city of Wemding

LEONHART FUCHS, *De historia stirpium*, 1542. Portrait of Fuchs handling a species of *Veronica*, verso of title-page

in Bavaria. A precocious child, he was sent to study Latin and Greek at the School of St. Mary in the town of Erfurt in Thuringia, and in 1515 he enrolled in the faculty of arts of the university in Erfurt. After earning the title of *medicinæ doctor* from the university in Ingolstadt in 1524, he settled in Munich where he married, established a large family, and exercised his profession for some time. In 1526 we find him teaching medicine at the university in Ingolstadt. In 1528 he accepted an invitation to serve as physician to the Protestant Prince Georg, Margrave of Brandenburg, at his court in Ansbach, a sign of his growing reputation. This position left him ample time to devote to his scientific research and literary interests, and he was able to write several books. One of these was *Errata recentiorum medicorum* (Errors of modern doctors, 1530), his contribution to the debate regarding the merits of Arab medicine in comparison to the medical tradition of the ancient Greeks, Fuchs being a staunch advocate of the latter. A similar work appeared one year later *Compendiaria ... in medendi artem...*; this work was so popular that the author was continually asked to revise and update it for new editions. In 1532 he produced a translation of the work of Hippocrates on epidemic diseases, *Hippocratis medicorum omnium longe principis epidemiorum liber sextus*, with an extended commentary.

In 1533 Fuchs decided to take up teaching once again, but since he had in the meantime embraced the Lutheran faith, in 1535 he left the university in Ingolstadt, which was a bastion of Catholicism, in order to accept a position at the Protestant university in Tübingen. This university was celebrated for its humanistic studies and he found an influential supporter there in the celebrated humanist and reformer Philipp Melanchthon (1497–1560), who had known him since he was student in Ingolstadt. At Tübingen, Fuchs threw himself into the work of reform, both of the church and the medical profession. He served as rector of the university from 1538 to 1565 and, with the support of Joachim Camerarius the Elder (1500–1574), who had been a fellow student at Erfurt, formulated a new set of statutes for the university, reinforcing the study of astronomy, reducing the amount of time spent teaching Arab medicine, and introducing a course based on Andrea Vesalius' *De humani corporis fabrica*, the first modern text of human anatomy. He even acquired a skeleton for the professors to use during their anatomy lectures. During this period he published an amplified version of *Errata* entitled *Paradoxorum medicinæ*.

Fuchs also improved the course of botanical studies at the university by organizing regular botanizing expeditions into the countryside and mountains. Conducted during the summer months by botany professors, they allowed students to acquire knowledge of plants from life. Furthermore, there is evidence to suggest that Fuchs founded a botanical garden in the city of Tübingen. So great was the fame that he attained all over Europe that, after the Giardino dei Semplici of the university in Pisa was completed in 1553—the first botanical garden, together with the one in Padua, to be established in Europe—the grand duke Cosimo I de' Medici extended an invitation to him in 1554 to serve as its director and to teach botany at the university. Fuchs politely declined this prestigious offer—partly because of misgivings regarding the Catholic establishment south of the Alps and the possible conflicts that might arise—and the position was eventually taken up by the botanist Luca Ghini of Imola (1490–1556).

When *Herbarum vivæ eicones* was published in 1530, Fuchs was already at work on his own monumental herbal in Ansbach. Although Brunfels' book was immediately hailed as an important achievement, Fuchs was quick to note that it was not devoid of errors and furthermore that it made no mention of the many exotic plants that he in the meantime had had the opportunity to collect and study. In contrast to his colleague, Fuchs' greater knowledge of the classics allowed him to carry out a more critical analysis of the texts of the Greek and Roman authors and in particular of the nomenclature used by them. In 1538 he wrote to Duke Albert of Prussia (1490–1568):

> I have now completed a herbarium, but it has not yet been printed. In it, more than three hundred and fifty herbs are illustrated with their roots, stems, leaves, seeds and flowers—[it] will be a merry book to look at. Have brought it to completion, praise God, with great cost and work (Voigt, p. 269).

It was only in 1542, however, that *De historia stirpium* finally came off the presses of the famous publisher Michael Isingrin of Basel, a city renowned for the outstanding quality of its printed books. This work was followed one year later by a revised edition translated into German in order to make it accessible to a broader public, the *New Kreüterbuch*. Fuchs continued his studies with unabated intensity and immediately after the publication of these two books began another ambitious editorial project, the first part of which was completed between 1550 and 1551, containing more than three hundred new illustrations, in large part the work of the artist Jerg Ziegler. A second volume with four hundred illustrations was finished in 1557, followed by a third in 1563–4. Unfortunately, the author was not able to find a prince willing to sustain the costs of publishing this extensive work and it was lost to view until the eight-volume manuscript known as the *Codex Fuchs,* which is composed of 1,529 illustrations, was discovered in the National Library of Vienna and displayed at an exhibition in Vienna and Bonn in 2000 (Lack, pp. 36–45). Some of the drawings in this manuscript, such as the *Ornithogalum* and the pistacchio, had been sent to him by Luca Ghini, the botanist who accepted the post that he had refused in Pisa.

De historia stirpium is today considered to be Leonhart Fuchs' most significant work. In the elegantly written Latin introduction, after reminding the reader of the divine origin of all flora, the author enumerates the *veteres auctores* (ancient authors) who served as his principal sources, particularly the great authorities in *materia medica*. He acknowledges his

LEONHART FUCHS,
De historia stirpium,
1542. Herb Paris (*Paris quadrifolia*), page 87

profound debt to Dioscorides among the ancients and cites various contemporaries, including the humanists Ermolao Barbaro and Marcello Virgilio; the French naturalist Jean Ruelle (1474–1537); Valerio Cordo (1515–1544), a gifted botanist who died quite young but left behind him an important manuscript on Dioscorides; Otto Brunfels; and his friend Hieronymus Bock (known as Tragus, 1498–1554), the author of a widely read *Kreüterbuch*, the first edition of which was published without illustrations in Strasbourg in 1539.

In Fuchs' herbal the plants are arranged in alphabetical order by their Greek names, and each description is divided into seven sections that methodically cover names, kind, form (external appearance), habitat, time (of flowering and fruiting), nature and temperament (warm, cold, dry, etc., in accordance with the classical theory of the four humors), and finally power and effects. The author furnishes not only the plant's names in Greek, Latin, and German, but also the vernacular names by which it was generally known among apothecaries. For many plants, recipes for medicines and suggestions regarding other practical applications are given (for example, he observes that the fibers of the hop stem could be spun into thread or used to make paper). Moreover, Fuchs describes about forty new plants that are not mentioned by Brunfels, several of which he defined as 'new discoveries' because they were unknown to the ancients. Some of these were in fact such exotic plants as the pumpkin gourd, chili pepper, and maize from America, the latter of which Fuchs referred to as 'Turkish' because he thought that it came from the Near East.

Fuchs was the first botanist to systematically study the mutations in color of certain flowers such as Blueweed (*Echium vulgare*, p. 269) and the various stages in the development of the plant from the flower to the fruit. Indeed, sometimes the different stages are depicted in a single illustration, as in the case of the peach (*Prunus sylvestris,* p. 404). It is interesting to note that Fuchs was the first botanist to observe the galls on leaves caused by the gall wasp, which he meticulously reproduced in the illustration of the English Oak (*Quercus robur*, p. 229).

Fuchs—like Brunfels—assigned great importance to the role of the illustration in documenting his botanical studies, as is attested to in an extended passage in his introduction. Indeed, he was convinced that the visual image was much more exact and intelligible than the written description, no matter how eloquently couched ('res multo clarius exprimere, quam verbis ullis, etiam eloquentissimorum'). His description of the preparation of his illustrations reflects a new approach to the study of the natural world. As he declared, every one of the plants in his herbal was drawn from life as accurately as possible (*absolutissimae*) with its roots, leaves, flowers, seeds, and fruit. Many of the plants presented in this herbal were grown by Fuchs in his garden in Tübingen. He clearly took great pains to supervise the work of his artists, admonishing them not to indulge in the use of shading or any of the other expedients of painters, which might falsify the image and lead to a misinterpretation of the data presented. This work does contain errors, however. In the preface,

LEONHART FUCHS,
De historia stirpium,
1542. The Wild Basil or Hedge Basil (*Clinopodium vulgare*), page 896

Fuchs maintains that he has attempted not to ascribe bogus or unexamined properties to any plants depicted in his herbal; nonetheless, his confidence in old authorities produced many mistakes and, therefore, several plants are spuriously identified. Having never travelled to the Mediterranean region, Fuchs often confused the properties of German plants with those from the Mediterranean based on previous descriptions by Dioscorides. Some species are delineated twice with different names as a result of variations in the leaves, flowers, and fruit.

Who were the artists that collaborated with Fuchs on this important project, succeeding so well in respecting his precisely expressed *desiderata* that he acknowledged they were deserving of the highest honors for their accomplishment? Fuchs solicitously refers to them by name in his introduction, and we discover that they were talented artists (albeit not as celebrated as Hans Weiditz) working for Basel's many printing houses. Indeed, they may be considered genuine professionals, for each was specialized in a particular step in the complex procedure of producing woodcut illustrations. Albrecht Meyer was the *delineator* who made the drawing of each plant for Fuchs from life, often employing watercolors as well as pen and ink. The drawing was then transferred to the woodblock by Heinrich Füllmaurer, and finally Veit Rudolph Speckle of Strasbourg, the *sculptor*, carved the design into the wood, which was then ready to be inked and printed.

Fuchs does not merely name his artists and shower them with praise. He took the unusual step of having them portrayed as well, in order to underline the importance of their contribution to a project that he clearly regarded as a collaborative effort. Thus, in addition to the traditional full-page portrait of the author (clad in elegant doctoral robes and holding a flower) that follows the title-page, we find two other woodcuts (on a different page in different copies of the work—in the exemplar at the Oak Spring Garden Library they appear on page 897) portraying the artists engaged in their tasks. In the first illustration Meyer and Füllmaurer are seated before a table depicting a vase of flowers *dal vivo*, while in the second Speckle is pictured separately. All three are dressed in doctoral gowns, as if Fuchs wished to make plain the intellectual connotation of their art, raising it above the level of a purely manual endeavor.

While the illustrations in *De historia stirpium* do not always match the elegance and immediacy of Hans Weiditz's work, they are nevertheless characterized by a refined realism that makes them quite outstanding. In each case the entire plant is portrayed, the author's intention being that they should be colored by hand afterwards, sometimes as soon as they came off Isingrin's press. In a letter written in 1542, Fuchs informed his recipient that a hand-colored copy of his book cost the colossal sum of 15 gulden (Dobat, p. 21). A number of exemplars colored *ab antiquo* have come down to us, as a detailed study by Frederick Meyer shows.

The success of Fuchs' *De historia stirpium* was great and enduring. As we have already mentioned, within a year the author had prepared and published a German translation of the work, the *New Kreüterbuch*, in folio like the first edition. In 1545 Isingrin published two editions consisting of the illustrations alone, one in German (*Läbliche Abbildung und Contrafaytung aller Kreüter ...*) and one in Latin (*Primi de stirpium historia ...*), in an octavo format that was much more manageable in size (the prototype of the field guide), as well as a folio edition in Dutch (*Den Nieuwen Herbarius*, 1543). These were followed by editions prepared by other printers in Germany and France, while the woodcuts, especially those in the smaller octavo format, were copied in many herbals until the eighteenth century (see Nos. 8 and 10).

The Oak Spring Garden Library possesses an extraordinarily fine copy of *De historia stirpium* whose hand coloring—executed with particular skill and care by a talented artist—preserves all of its original freshness.

7. Pietro Andrea Mattioli (1500–1577)

Petri Andreæ Matthioli Senensis Medici, Commentarii in sex libros Pedacii Dioscoridis Anazarbei de Medica materia, iam denuo ab ipso autore recogniti, et locis plus mille aucti. Adiectis magnis, ac novis plantarum, ac animalium Iconibus, supra priores editiones longè pluribus, ad vivum delineatis. Accesserunt quoque ad margines Græci contextus quàm plurimi, ex antiquissimis codicibus desumpti, qui Dioscoridis ipsius deprauatam lectionem restitvunt. Cum locupletissimis indicibus, tum ad rem Herbariam, tum Medicamentariam pertinentibus. Cum privilegiis amplissimis, ut videre est statim post Præfationem ad Lectores. [*Large printer's device*]. Venetiis, Ex Officina Valgrisiana. M D LXV.

[*Colophon on 6G4*]: Venetiis, apud Vincentium Valgrisium. MDLXV.

2° 39.8 × 26.6 cm. ★^{6} 2★^{8} πA–M^6 A–6F^6 6G^4 6H^6 *i–clxxii* 1–1,459 [*13*] (717 as 617, 1,045 as 1,054, 1,328 as 1,228, 1,412 as 412) [1,644 pp.].

[*on 6H1 recto*]: De ratione distillandi aquas ex ominbus plantis, et quomodo genuini odores in ipsis aquis consevari possint.

PLATES: 1,007 plants, fishes, animals, farming scenes, and distilling apparatuses; portrait of Mattioli on πM6, all woodcuts by Giorgio Liberale and Wolfgang Meyerpeck printed on blue-grey paper with several illuminated in silver and gold.

BINDING: 16th-century white pigskin over wooden boards; tooled in blind; clasps and hasps.

PROVENANCE: Sydney Howard Vines (1849–1934), with inscription on front paste-down: 'belonged to Professor S H Vines, FRS d. 1934.' Sir Sydney Carlyle Cockerell (1867–1962), inscribed: 'Sydney C. Cockerell Cambridge 13 April 1923.' Bookplate of John Roland Abbey (1896–1969), inscribed: 'Bought from Sir Siydney Cockerell By J.R. Abbey, March 20th 1952. A letter to Abbey from Cockerell, dated March 25, 1952, is also on front paste-down.

REFERENCES: Anderson, pp. 163–172; Arber, pp. 95–96, 226; Blunt & Raphael, pp. 132–137; Blunt & Stearn, p. 74; Cambridge D665; Cleveland 93; Hunt 94; *An Oak Spring Sylva* 2; Lack 9 and 1; Nissen 1,305; Pritzel 5,985; Reeds, pp. 124, 127–128, 155–164; *Mattioli Woodblocks*; Tongiorgi Tomasi and Hirschauer; Wellcome 4,138; Wheelwright, p. 140.

One of the most prestigious volumes in the Oak Spring Garden Library is the *Commentarii* by Pietro Andrea Mattioli, one of two surviving copies of the 1565 Valgrisi edition printed on blue-gray paper and illustrated with woodcuts beautifully highlighted in silver and gold. While we know from an old manuscript inventory in the Sachsische Landesbibliothek Dresden that the second copy, now in its collection, originally came from 'the library of the elector of Saxony in Annaberg, 1574'—this was the cultured Augustus of Saxony (1526–1586) who ruled from 1553 to 1586—we have no such information regarding the volume in the Oak Spring Garden Library. Only its modern owners are recorded; it once belonged to Prof. Sydney Howard Vines, F.R.S., and then to John Roland Abbey before it was acquired in 1952 by Sir Sydney Carlyle Cockerell for his celebrated collection in Cambridge, England. In 1962 it was acquired for the collection of the Oak Spring Garden Library.

The *Commentarii* of Pietro Andrea Mattioli was the most renowned herbal produced in Europe in the modern age. The first edition appeared in Italian in 1544 to great acclaim, and during the course of his life the author amplified and corrected the work several times for new editions; a total of about forty-five would be published in various languages over a period of two centuries, the last in Venice by Niccolò Pezzana in 1744.

Pietro Andrea Mattioli was born in Siena in 1501, the son of a doctor with a flourishing practice in Venice. He enrolled as a student at the faculty of law of the University in Padua, but soon turned to the study of medicine and medical botany. He became a physician and practiced in Rome until 1527, when he joined the entourage of Cardinal Bernardo Clesio, the Bishop of Trento, who until his death remained Mattioli's patron and protector. In 1539 Mattioli published *Il Magno Palazzo di Trento*, a short verse work in octavo celebrating the restoration by the Cardinal of the splendid Palace of Buonconsiglio.

Mattioli's studies of *materia medica* took the philologically correct form of an examination of the classical sources, above all Dioscorides, based on his firm belief in the close relationship that bound theoretical research to the practical applications of botany. The conviction expressed in the *Proemio* (preface) to the 1548 Venice edition that 'without a correct knowledge of simples one cannot medicate in a rational manner' was grounded in the author's own experience, most of it gained during the period which he spent in the Val di Non in the region of Trentino as a practicing physician. There he had the opportunity to conduct extensive studies in a unique *teatro di natura* and to elaborate his ideas for the revival of a 'true doctrine and science of simples,' aimed at re-imposing order in what he referred to as its 'neglected garden.' In Trentino and afterwards in the region around Gorizia, the city where he served from 1542 to 1555 as public physician after the death of his patron, Cardinal Clesio, he traversed and explored 'with considerable toil' mountains, valleys and watercourses, learning not only to recognize the region's fauna, flora, and

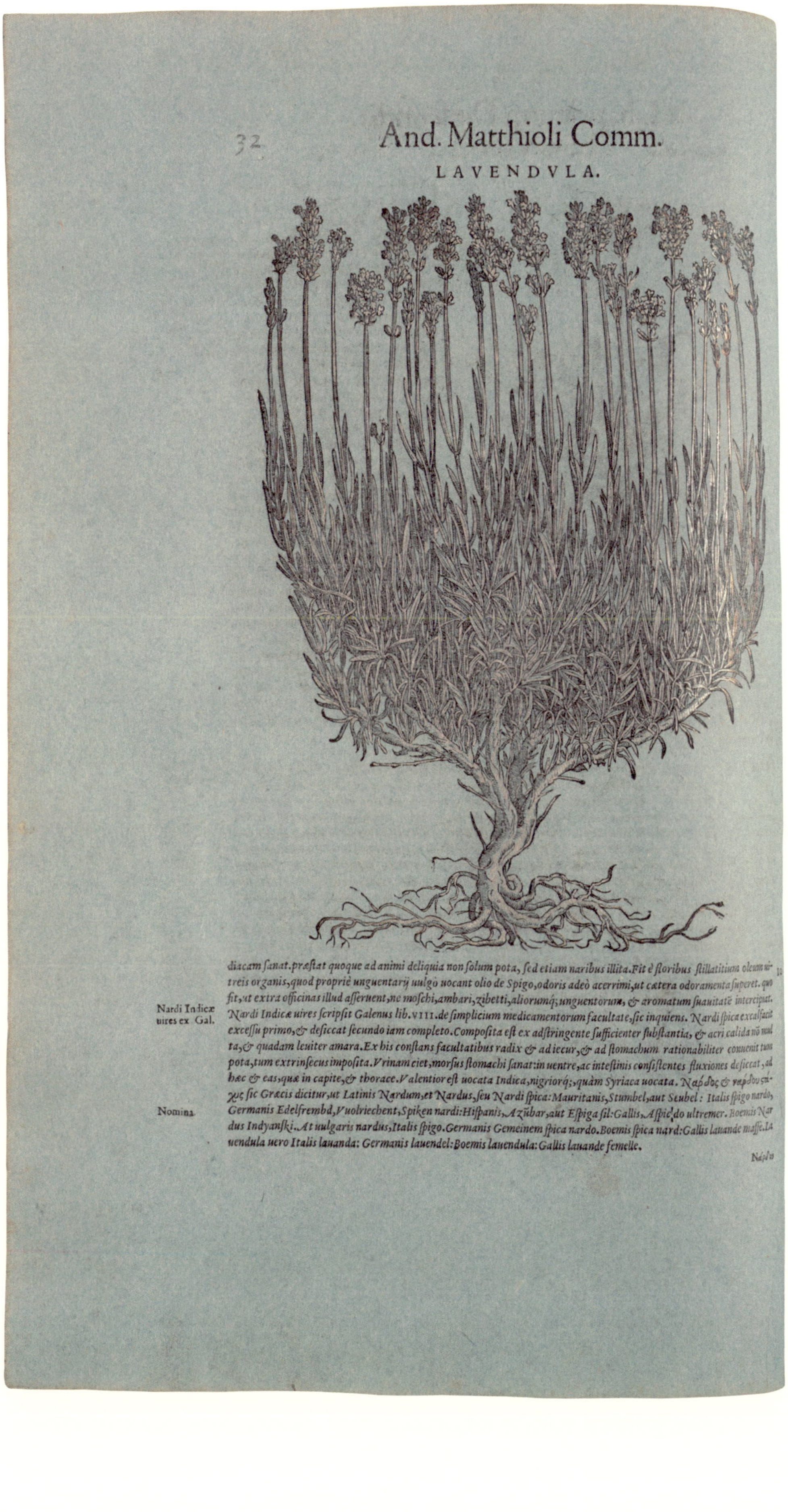

32

And. Matthioli Comm.

LAVENDVLA.

diacam ſanat. præſtat quoque ad animi deliquia non ſolum pota, ſed etiam naribus illita. Fit è floribus ſtillatitium oleum uitreis organis, quod propriè unguentarij uulgò uocant olio de Spigo, odoris adeò acerrimi, ut cætera odoramenta ſuperet. quo fit, ut extra officinas illud aſſeruent, ne moſchi, ambari, zibetti, aliorumq́; unguentorum, & aromatum ſuauitatē intercipiat.

Nardi Indicæ uires ex Gal.

Nardi Indicæ uires ſcripſit Galenus lib. VIII. de ſimplicium medicamentorum facultate, ſic inquiens. Nardi ſpica excalfacit exceſſu primo, & deſiccat ſecundo iam completo. Compoſita eſt ex adſtringente ſufficienter ſubſtantia, & acri calida nō multa, & quadam leuiter amara. Ex his conſtans facultatibus radix & ad iecur, & ad ſtomachum rationabiliter conuenit tum pota, tum extrinſecus impoſita. Vrinam ciet, morſus ſtomachi ſanat: in uentre, ac inteſtinis conſiſtentes fluxiones deſiccat, ad hæc & eas, quæ in capite, & thorace. Valentior eſt uocata Indica, nigriorq́;, quàm Syriaca uocata. Νάρδος & ναρδουστάχυς ſic Græcis dicitur, ut Latinis Nardum, et Nardus, ſeu Nardi ſpica: Mauritanis, Stumbel, aut Seubel: Italis ſpigo nardo,

Nomina.

Germanis Edelfrembd, Vuolriechent, Spiken nardi: Hiſpanis, Azūbar, aut Eſpiga ſil: Gallis, Aſpic do ultremer. Boemis Nardus Indyanſki. At uulgaris nardus, Italis ſpigo. Germanis Gemeinem ſpica nardo. Boemis ſpica nard: Gallis lauande maſſe. Lauendula uero Italis lauanda: Germanis lauendel: Boemis lauendula: Gallis lauande femelle.

Νάρδου

PIETRO ANDREA MATTIOLI, *Commentarii*, 1565. A species of Lavender (*Lavandula*), page 32

minerals, but also becoming familiar with the simple remedies used by the most modest of its herbalists. Greatly enriched by this experience, Mattioli prepared a vernacular Italian translation of five of the six books of Dioscorides' *De materia medica* with an extensive commentary entitled *Libri cinque della historia & materia medicinale*; this was published in 1544 by the Venetian printer Nicolò de Bascarini. In his work Mattioli presented the Greek physician's text completely updated and purged of the errors of interpretation that had accumulated over the centuries, incorporating the contributions made by Arab physicians and the knowledge acquired in Europe up to his own time, often benefiting from the suggestions and information furnished by his many friends and colleagues, with whom he maintained frequent correspondence.

Mattioli's ongoing work, considerably augmented by descriptions of the *novità* that had reached Europe following the geographic discoveries of explorers, was published in 1548 by the celebrated Venetian printer Vincenzo Valgrisi in a new edition, this time comprising all six books of *De materia medica*, with the title *Il Dioscoride dell'eccellente dottor medico P. Andrea Mattioli da Siena*, with reprints issued in 1550 and 1552. In 1554 Valgrisi published the first Latin edition, *Petri Andreae Matthioli Commentarii*, which included a significant modification: hitherto Mattioli's herbal had consisted exclusively of text, whereas this edition was accompanied by a series of small but realistic botanical illustrations. Another Latin edition was published in Lyon in the same year.

The fame of this work and Mattioli's exceptional professional qualifications led to his being invited to serve as the personal physician of the cultivated Hapsburgs Archduke Ferdinand of Further Austria and Tyrol and of his brother Maximilian II who would rule the Holy Roman Empire from 1564 to 1576. Mattioli spent the next twenty years at the imperial courts in Prague and Vienna. He finally retired and returned to Trento in 1569, although he continued to furnish the Prince Ferdinand and his wife Philippine with medical advice, visiting the court in Innsbruck from time to time. In 1578 the great botanist and physician died of the plague and was buried in the Cathedral of Trent, in a tomb that bears his portrait sculpted in bas-relief. Among Mattioli's children only Ferdinando followed in his footsteps and became a physician, serving at the court of the elector Augustus of Saxony, the owner of the Dresden copy of his father's *Commentarii*.

After his arrival in Prague, Mattioli published *Epistolarum medicinalium libri quinque* (1561), a collection of about eighty letters in Latin drawn from his correspondence with colleagues all over Europe and containing much erudite discussion and debate on various matters concerning medicine and botany. In the Bohemian capital of Prague the printer Jiří Melantrich acquired a five-year copyright to Mattioli's *Il Dioscoride*, and republished the work in 1562 in a Czech edition entitled *Herbartz* that was prepared by the humanist

CROCVM FLORENS.

CRocum quidem ubique terrarum cognoſcitur. Planta eſt comoſis folijs longis, anguſtis, capillaceis planis tamen non rotundis, tragopogoni modo, ſed anguſtioribus & numeroſioribus, humi accubantibus, denſis, ac mollibus. Floret poſt æquinoctium Autumni anteaquam folia proueniant, floribus Colchico Ephemero ſimilibus, purpureis, ac aſpectu iucundis. E medio rubra prodeunt ſtamina in craſſiuſculum cacumen deſinentia, quibus cum ex eadem ferè origine, quidam exeunt proceſſus exiguæ linguæ figura aurei coloris, ijs ſimilibus, qui in colchici flore habentur. Floret Crocum continuo ferè menſe. A floribus prodeunt ſubinde folia, quæ tota hyeme uireſcunt, ſpretis algoribus. Areſcunt hæc, ac prorſus deperduntur uere, & nunquam per æſtatem apparent. Radicem habet Crocum bulboſam, pluribus contectam tunicis, ipſiſque ſubflaueſcentibus, Gladioli modo. Eruuntur Bulbi quarto anno, uerno tantum tempore ſeruanturque æſtate tota in horreis & ſeruntur: deinde appetente autumno, in tenui ſolo; nam & ſtercora odêre, & ſolidam humum. Crocum in Italia, & præſertim in Hetruria, Mauritanorum modo uulgò uocatur Zaffarano: quanquam non deſint rura in agro noſtro Senenſi, quæ corrupto à Croco uocabulo ipſum appellant Gruogo. Cæterùm crocum Coryceum, & Lycium in Olympo monte decerptum (quod ſciam) ad nos non aducbitur. Quapropter quod ab Aquila Veſtinorum urbe affertur, primum ſibi locum uendicat

Croci conſideratio.

PIETRO ANDREA MATTIOLI, *Commentarii*, 1565. *Crocus* sp., possibly *Crocus biflorus*, page 69

Tadéas Hájek (1525–1600), who had studied medicine in Bologna and eventually became physician to the Hapsburg emperors Maximilian II and his successor Rudolph II (1552–1612). This edition, which had an immense influence on medical and pharmacological studies in central Europe, served as the basis for a second edition prepared by the naturalist Joachim Camerarius, son of the eminent German humanist of the same name, which appeared in 1596 and included more than 1,700 modifications and additions (see No. 42). In 1563 Melantrich simultaneously published a German edition of the work entitled *New Kreüterbuch* which was dedicated to the Emperor Ferdinand I (1503–1564) and his sons; the translation was carried out by a close collaborator of the author, the physician Georg Handsch.

The preparatory work for these two esteemed Prague editions, the Czech and the German, was not only intellectual and scientific, but considerable time and effort were devoted to the visual component as well. Just a decade earlier, in 1554, the first illustrated edition of *Commentarii* had been published with 562 small woodcuts, and in the intervening period Mattioli became increasingly convinced of the crucial support that illustrations could provide to a scientific text. We know his interest in botanical illustration began much earlier, for when he was working in Gorizia he commissioned skilled artists to portray the plants that he was studying from life, and in a letter sent in September 1553 to Ulisse Aldrovandi (1522–1605) he laments the fact that a serious illness had prevented 'his artist' from completing the drawing of a sesame plant, which he had been hoping to send to the Bolognese naturalist for his opinion. He was referring here to Giorgio Liberale of Udine, whom Giorgio Vasari described in the 1568 edition of *Le vite de' più eccellenti pittori, scultori, e architettori* as a proven artist and whom Mattioli praises as 'homo in artibus pingendi peritissimus' (a man most expert in the art of painting) in the *proem* to the 1565 edition of his commentary. The author also acknowledges the talents of the woodcut artist who translated Liberale's drawings into woodblock engravings, Wolfgang Meyerpech of Meissen. Today we know that two other artists collaborated on the Bohemian edition, the German Hans Minich and an anonymous engraver who signed his works 'GS' (Arber, p. 226), perhaps identifiable as the George Van Sichem who drew the illustration of the 'Morus' or mulberry on page 112. Together, these artists produced the more than nine hundred folio woodcuts of plants and the hundred illustrations of animals that appeared first in the Prague editions by Melantrich, and then in the two Venice editions by Valgrisi: the 1565 Latin edition and the 1568 Italian edition. At the end of these two editions Valgrisi inserted a short illustrated treatise on the distillation of plant essences and methods for conserving their fragrance: *De ratione distillandi aquas ex omnibus plantis.*

In addition to the botanical illustrations produced for him by the four artists, the Sienese naturalist incorporated images of other plants received from friends and colleagues with whom he was in constant correspondence. Of particular significance are the 'new' plants which may be found portrayed in the herbal, such as the tulip and the lilac (*Syringa vulgaris*). The former, which first appeared on page 1,237 of the 1565 edition, was a flowering species that had been sent from Constantinople to Vienna by the Flemish diplomat and man of letters Ogier Ghiselin de Busbecq (1522–1592). De Busbecq served as the ambassador of the Emperor Ferdinand I to the court of the Ottoman Sultan Süleyman I the Magnificent in Constantinople from 1554 to 1562 and from there not only introduced the tulip to Europe, but also discovered and brought to Vienna the precious sixth-century illustrated manuscript of Dioscorides' text known as the *Codex Vindobonensis* or Codex Juliana Anicia (see pp. xxix–xx and Lack 1).

As with Fuchs' *De historia stirpium*, many copies of the commentary on Dioscorides by Mattioli were colored by hand and, in the case of a limited number destined for illustrious recipients, commissioned from celebrated artists. One example is Gherardo Cibo, an artist much admired by Mattioli; various copies of *Commentarii* decorated by his hand have come down to us—one each of the 1548, 1558 and 1573 editions, and an extraordinarily fine miniated copy of the 1568 edition commissioned by the Duke of Urbino. Cibo colored the woodcuts with great care, sometimes embellishing them with landscape backgrounds. These four copies are conserved in the Biblioteca Angelica of Rome, while another copy of the 1568 edition decorated by Cibo can be found in the Biblioteca Alessandrina in Rome. Agnes Arber notes that the diplomat Sir Henry Wotton presented the wife of Charles I (1600–1648), Queen Henrietta Maria (1609–1669), with a copy of this work 'naturally colored' (pp. 95–96).

The exemplars illuminated in silver and gold deserve separate discussion. The author clearly had a predilection for this form of ornamentation. As we learn from a letter sent to Aldrovandi many years earlier, in 1554, Mattioli had 'kept in his house a Miniator for three months who colored and decorated [the work] all in gold and silver in such a way that in Venice it was retained the most rare thing of this kind that had ever been seen' (Tongiorgi Tomasi, 1997). Given the date of the letter, it is clear the author was referring to some work other than the two copies of *Commentarii* at Oak Spring Garden Library and at Dresden, which were decorated much later, the first most probably for an important member of the Hapsburg family and the second perhaps for the elector of Saxony, whose personal physician was Mattioli's son.

The *Commentarii* was an enormous success; both its text and illustrations were universally praised and its influence was profound and lasting. The French botanist Antoine Du Pinet (see No. 28), for example, followed Mattioli's model closely in his work *Historia Plantarum*, which was published in Lyon in 1561, while the miniature painter Giovanna Garzoni (1600–1670) copied some illustrations from *Commentarii* and drew inspiration from others when she compiled her own celebrated *Erbario Miniato*, today conserved at Dumbarton Oaks in Washington, D.C. (Tongiorgi Tomasi and Hirschauer).

Testifying to Mattioli's fame are the many portraits of him that have come down to us. Some appear in his books, including the *Epistolarum medicinalium* (1561), the 1562 Prague edition of *Commentarii*, and the 1565 and 1568 Venetian editions. In addition to the bas-relief carved on his tomb, at least four portrait paintings exist: one by Alessandro Bonvicino, called 'il Moretto,' at the Palazzo Rosso in Genoa; one anonymous work portraying the botanist at the age of sixty-seven, in the collection of the Botanical Garden of Pisa; another anonymous work in a private Italian collection (Tavoni, Tongiorgi Tomasi, and Tongiorgi, p. 65); and a portrait attributed to Giorgio Liberale in a private collection, which shows Mattioli proudly displaying a hand-colored copy of his work (Tosi, 1997).

7. Pietro Andrea Mattioli (1500–1577)

Three woodblocks of pear wood for *Commentarii in sex libros Pedacii Dioscorides Anazarbei de medica material*: (1)'Limonium' (sea lavender), 22 × 16 cm., p. 980, on verso, inscription on white paper label (not legible) and inked on block 'Ben Rvbrvm'; (2)'Iuncus' (rush), 21.9 × 15.4 cm., p. 1,036, on verso, inscribed on white paper label 'Juncus...' and engraved on block 'GIVNCO ·I·'; (3)'Aconitum V' (possibly *Geranium sylvaticum*, wood cranesbill), 22.1 × 16 cm., p. 1,088, on verso, inscription on white paper label (not legible) and inked on block 'Aconitum[?] Q...' On the back of the first block are impressions of type sorts laid on their sides, which were used as underlayments in order to optimize the printing impression.

These blocks were owned by the Duhamel du Monceau family until the middle of the last century; many of the engraved blocks for the large illustrations to the *Commentarii* were acquired by the celebrated French agronomist Henri Louis Duhamel du Monceau (1700–1782) and used to print the 150 illustrations for his two-volume *Traité des Arbres et Arbustes*, which was published in Paris in 1755 by Guerin and Delatour. The author recalls in his preface that: 'J'ai le bonheur de recouvrer presque toutes les planches de la belle édition latine du Matthiole de Valgrisi: les imprimeurs de mon Ouvrage ont fait graver avec soin celles qui y manquoient [I had the good fortune to recover almost all of the blocks of the fine Latin edition of Mattioli by Valgrisi; the printers of my Work engraved with great care those that were missing].' It must nonetheless be acknowledged that the eighteenth century re-printing of the woodcuts that had been prepared for the Sienese naturalist cannot match the precision, refinement, and subtle tonalities attained in the original editions.

Nearly one hundred of the woodblocks that were used to illustrate the works of Mattioli have appeared during the course of the last century on the antique market and they

PIETRO ANDREA MATTIOLI, *Commentarii*, 1565. Woodblock: Sea lavender (*Limonium* sp.), image on page 980

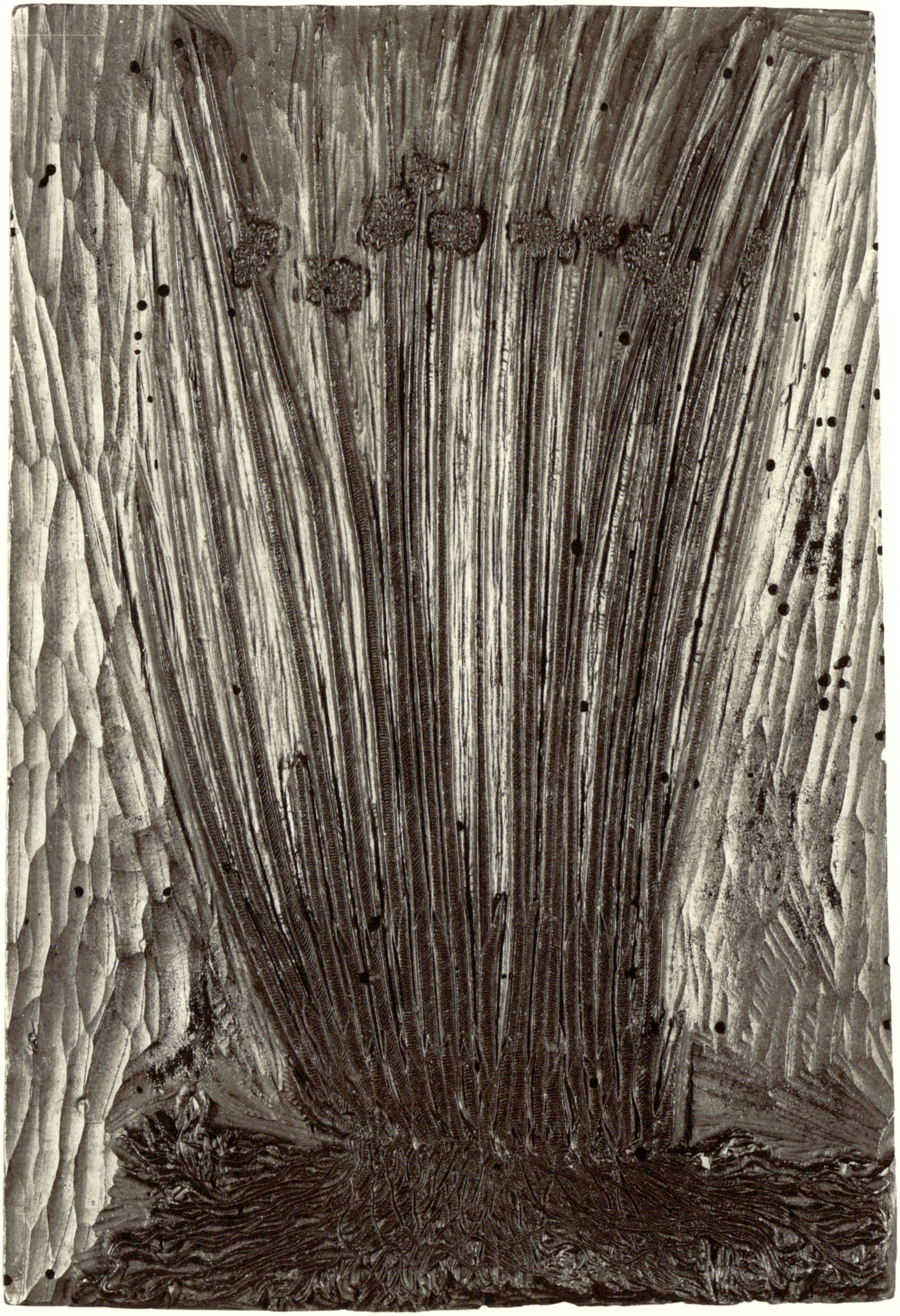

PIETRO
ANDREA MATTIOLI,
Commentarii, 1565.
Woodblock: Rush
(*Juncus* sp.), image
on page 1,036

PIETRO ANDREA MATTIOLI, *Commentarii*, 1565. Woodblock: Wood cranesbill (*Aconitum*) possibly *Geranium sylvaticum*, image on page 1,088

provide ample proof of the exceptional skills of Giorgio Liberale as a botanical artist and Wolfgang Meyerpech as an engraver. Most of these were prepared for the 1562 Prague edition, some for the Lyon edition of 1578, while the blocks in the Duhamel du Monceau collection were destined for the 1565 Venice edition. Many of these have found their way into collections in the United States—at the Oak Spring Garden Library, the Hunt Institute for Botanical Documentation at Carnegie Mellon University (Pittsburgh), the Department of Printing and Graphic Arts of Harvard University (Cambridge, Massachusetts), and the Landscape Architecture Studies Library at Dumbarton Oaks (Washington, D.C.).

These rare artifacts of the printer's craft—to which may be added those on botanical and zoological subjects conserved in the Ulisse Aldrovandi collection at the Biblioteca Universitaria in Bologna and others produced by the great Flemish botanists for the publisher Christophe Plantin of Antwerp (Plantin-Moretus Museum)—constitute evidence of fundamental importance for our understanding of the nature and significance of the woodcut illustration in the botanical texts of the sixteenth and seventeenth centuries.

In the three blocks at the Oak Spring Garden Library we can admire a technique that had reached its highest point in terms of refinement and virtuosity; the effects of shading in the woodcut of the '*Iuncus*' are no less delicately modulated than those obtainable by means of the more refined technique of metal engraving, whose use in botanical illustration was introduced just a short time later, around 1580. The realistic portrayal and sensitive *mise en page* of the plants in the obdurate medium of wood testify to the ability of the two artists who collaborated so closely with Mattioli and succeeded in rendering the necessary chiaroscuro values with unmatched clarity.

8. William Turner (*c.* 1510–1568)

[*Part I*]: The first and seconde partes of the herbal of William Turner Doctor in Phisick, lately oversene, corrected and enlarged with the thirde parte, lately gathered, and nowe set oute with the names of the herbes, in Greke Latin, English, Duche, Frenche, and in the Apothecaries and herbaries Latin, with the properties, degrees, and naturall places of the same. Here unto is ioyned also a Booke of the bath of Baeth in England, and of the vertues of the same with diverse other bathes, moste holsom and effectuall, both in Almanye and England, set furth by William Turner Doctor in Phisick. God save the Quene. [*Large Royal Arms device*]. Imprinted at Collen by Arnold Birckman. In the yeare of our Lorde M.D.LXVIII. Cum Gratia & Privilegio Reg.Maiest.

[*Part II*]: The seconde parte of Uuilliam Turners Herball, wherein are conteyned the names of herbes in Greke, Latine, Duche, Frenche, and in the Apothecaries Latin, and somtyme in Italiane, with the vertues of the same herbes with diverse confutationes of no small errours, that men of no small learning have committed in the intreating of herbes of late yeares. God save the Quene. [*Large Royal Arms device*]. Imprinted at Collen by Arnold Birckman In the yeare of our Lorde M.D.LXVIII. Cum Gratia & Privilegio Reg.Maiest.

[*Part III*]: The thirde parte of Uuilliam Turners Herball, wherein are conteined the herbes, trees, rootes and fruytes, whereof is no mention made of Dioscorides, Galene, Plinye, and other olde Authores. God save the Quene. [*Large Royal Arms device*]. Imprinted at Collen by Arnold Birckman, In the yeare of our Lorde M.D.LXVIII. Cum Gratia & Privilegio Reg.Maiest.

[*Bound with*]: A Booke of the natures and properties, as well of the bathes in England as of other bathes in Germanye and Italye, very necessarye for all sycke persones that can not be healed without the helpe of natural bathes, lately oversene and enlarged by William Turner Doctor in Phisick. God save the Quene. [*Large Royal Arms device*]. Imprinted at Collen by Arnold Birckman, In the yeare of our Lorde M.D.LXVIII. Cum Gratia & Privilegio Reg.Maiest.

[*and*]
A most excellent and perfecte homish apothecarye or homely physick booke, for all the grefes and diseases of the bodye. Translated out the Almaine Speche into English by Ihon Hollybush. [*Large printer's device* (*Brickman's earlier version*)]. Imprinted at Collen by Arnold Birckman, In the yeare of our Lord M.D.LXI.

2° 30 × 17.8 cm. Three parts in one volume. Part I: $\star^4$ A–B^6 C^8 D–T^6 *i–viii* 1–40 37–208 *209–223* [*5*] (139 as 136, 159 as 169) [240 pp.]. Part II: π^2 a^2 A–$2E^6$ $2F^4$ $2G^2$ fols. [*4*] 1–107 *108*–113 *114*–116 *117–118* 119–171 [*3*] (69 as 96, 121 as 133, 128 as 127, 131 as 132, 133 as 131, 134 as 132, 135 as 133, 136 as 134, 137 as 135, 138 as 136, 139 as 137, 140 as 138, 141 as 139, 142 as 140, 143 as 141, 144 as 140, 145 as 143, 146 as 144, 147 as 145, 148 as 144, 149 as 147, 150 as 148) [356 pp.]. Part III: $\star^4$ 3A–$3G^6$ $\star^4$ ($\star$1 + χ1) B–D^6 a–g^6 h^4 (-h4) pp. *i–viii* 1–14 *15*–81 *82–86* (45 as 46); fols. [*4*] 1–17 *18*; fols. *1*–45 (11 as 8, 25 as 24, 33 as 24, 35 as 24) [228 pp.].

PLATES: 386 woodcuts of 499 plants.

BINDING: Old calf tooled in blind; repaired and rebacked by Sutcliffe and Sangorski, 1968.

REFERENCES: *An Oak Spring Sylva*, pp. 8–13; Anderson, pp. 148–155; Arber, pp. 119–128; Bush-Brown, pp. 110–123; Chapman & Tweddle; Cleveland 99; Henrey I.18–26 and Nos. 366–368; Hunt 65; *Mattioli Woodblocks*; Nissen 2,013; Pritzel 9,570; Rohde, pp. 75–97; Wheelwright, pp. 104, 118, 186.

WILLIAM TURNER, who has been called 'The Father of British Botany,' was born in Morpeth, Northumberland, around 1510. He enrolled as a student of Pembroke Hall, Cambridge, in 1526 and, after receiving his baccalaureate, became a fellow of the same college in 1531. Later, however, he would take the opportunity in the dedication to the 1568 edition of *A New Herball* to express his profound dissatisfaction with the way in which botany was taught in England. As he wrote, 'Beyng yet felow of Pembroke hall in Cambridge,' he found the professors unable to teach their students the Greek and Latin names of many species, and that furthermore some plants did not even have names in English, a fact which clearly demonstrated 'the ignorance of the simples at that tyme.' As a result there were countless errors of nomenclature in English herbals, and here we may perhaps detect a veiled reference to the 1526 edition of *The Grete Herball* (see No. 4), and to works by the most illustrious European botanists of the time from Pietro Andrea Mattioli (see No. 7) to Leonhart Fuchs (see No. 6) and Hieronymus Bock (better known by his Latin name, Tragus).

Therefore Turner devoted his first book—*Libellus de re herbaria novus*, published in 1538—to the plants of his native region that had not yet been described by botanists. In 1548, an amplified edition in English, *The Names of Herbes in Greke, Latin, Englishe, Duche and Frenche...*, was prepared for the benefit, as he wrote, of apothecaries who had little knowledge of the classical languages. His *opus magnum, A New Herball,* would also be published in the vernacular.

Turner was an ardent Nonconformist, preaching his faith all over England. The country was in the throes of religious strife and dynastic conflicts, alternating between Protestant and Catholic sovereigns, and those who voiced their opposition too openly paid a heavy price. Turner was imprisoned several times, saw his books destroyed, and was finally forced to seek exile on the Continent. He made the best use that he could of this time, travelling around Italy, Switzerland, and the Low Countries to study the native flora and, above all, he endeavored to cultivate the company of Europe's most eminent botanists. Thus, he found the opportunity to discuss botany with Luca Ghini and Konrad Gesner, and to enter into correspondence with Fuchs. With the encouragement of Ghini, whom he refers to with great respect as 'maister' in the preface to *The New Herball*, Turner studied medicine and earned degrees from the universities of Ferrara and Bologna. It is probable that Turner also learned from Ghini how to create a *hortus siccus* or dried herbal by conserving plants between sheets of paper; he, in fact, mentions one such collection assembled by 'Maister John Falconer,' another English naturalist who had studied in Italy and had some relationship with Luca Ghini (Arber, p. 140).

When Henry VIII died in 1547, Turner returned to England, where he was appointed physician and chaplain to the Lord Protector Edward Seymour, Duke of Somerset. In 1551 the first part of *A New Herball*, which the author dedicated to his patron, Lord Edward, was published in London at the printing shop of Steven Mierdman, a Fleming who in his turn had escaped to England for religious reasons. This work was written with the aim of providing a useful and accurate revision of the botanical nomenclature and in it Turner launches a broadside against his colleagues Fuchs (see No. 6), Mattioli (see No. 7), and Rembert Dodoens (see No. 10), accusing them of countless errors and affirming that while in the past he had learned much from them, now they had as much to learn from him. Unfortunately, in the climate of religious intolerance that prevailed during the reign of Mary Tudor, a large number of the copies of this work were destroyed and Turner had to depart for the Continent once again. There he continued to study the flora of Europe and to prepare the sequel to his herbal.

Religious faith and religious conflicts affected the lives and careers of many botanists, from Fuchs (see No. 6) and Brunfels (see No. 5) to Garcia da Orta (see No. 17). Turner bitterly lamented the interruptions to his studies, for he was prevented from concentrating on his research and extending it, as he would have liked, to all three kingdoms of nature. While he succeeded in producing a short work on ornithology, *Avium præcipuarum, quarum apud Plinium et Aristotelem mentio est, brevis et succincta historia* (Cologne, 1544), in the dedication addressed to Elizabeth I that opens *A New Herball*, Turner expresses his regret that he did not have sufficient time to devote to his ichthyological studies and to the writing of 'a booke of the names and natures of fishes that are within youre Mayesties dominion.'

Of Bulbine or yelowe Leke.

Bulbus siluestris.

Pliny maketh mention of a kind of Bulbus / which he sayeth / is called of the Grecianes Bulbine / and hath leues lyke Leekes / and a heade or knoppe. The herbe whiche I take to Bulbine / groweth amonge the corne / and hath floures about the beginninge of Aprill. It hath long smal blades lyke Leekes / but much smaller and sharper / The blades that come from the roote / are verye small & longe / but they that come oute of the top of the stalke / & growe hygher then the stalkes / are broder. The floures growe out of the top of the stalke / and they are yellowe. The vessel that holdeth seede is thresquare / the stalke is small / and not a span longe / The roote is rounde lyke an oynion with a dunne huske / and somethynge redish within / and a littel bitter and clammy. It groweth muche in Germanye aboute Bon and Colon. Plinye writeth that the propertye of thys herbe is to heale woundes / and close the it Bulbus / whereof this is one kinde / hath manye other good properties. But I will not gyue no more vnto the herbe then my autor doth / except I had experience that he had not sene or proued. Some in Duch land cal this herbe Hundts vllich / and it maye be called in Englishe Corne leeke / or wilde leeke. As for the gardin Bulbus / and the other called Vomitorius / although I haue longe and diligently sought for them these twentye yeares / both in

I

WILLIAM TURNER, *A New Herball*, 1568. Yellow Star of Bethlehem, a species of field *Gagea*, page 97

Finally, with Elizabeth, the daughter of Henry VIII, securely on the throne, Turner was able to return permanently to England and finish the second part of his herbal. This was published in Cologne in 1562 by the printer Arnold Birckman, who made close copies of the woodblocks that were used in the 1545 octavo edition of Fuchs' *De historia stirpium* published in Basel (see No. 6).

The third and final part of *A New Herball* was completed in London just a few months before the author's death and was published in 1568 by Birckman together with the first two parts in a magnificent volume dedicated to Queen Elizabeth, which bore the title *The first and second partes of the Herbal of William Turner, Doctor in Phisick, lately oversene, corrected and enlarged with the Thirde parte....* To this the author added a treatise on the hot mineral springs of the city of Bath, *A Booke of the natures and properties, as well of the bathes in England as of other bathes,* which had already appeared in print in 1562, while the printer added a translation of Hieronymus Braunschwieg's 'Thesaurus pauperum,' entitled *A most excellent ... homish apothecarye* (1561).

A New Herball presents more than two hundred thirty native English plants that had never been described before. It is lavishly illustrated with more than five hundred woodcut illustrations; furthermore, the initial letter to each chapter is elegantly decorated with plant motifs and human, monstrous, and emblematic figures. The plants are generally depicted with their flowers, fruit and roots, sometimes arranged side by side two to a page, and are labelled with their English and Latin names.

Most of the illustrations are not original, about four hundred of them having already appeared in Fuchs' *De historia stirpium*. The text represents a genuine contribution to the botanical studies of the period, however, because the descriptions of the plants were almost all based on direct observation. The plants are conventionally arranged in alphabetical order with no attempt at classification on the basis of their morphology or other botanical criteria. Turner was an expert botanist who appears to have cultivated an extensive garden of his own near Kew and, although well acquainted with the texts of classical Antiquity, he did not allow himself to be unduly influenced by them, relying instead on his own intuition and acute powers of observation. For example, he was one of the first naturalists to challenge the traditional belief in the anthropomorphic nature of the mandrake root (Arber, p. 123). In his herbal he provides the names of each plant, a detailed description of its appearance and a discussion of its properties. He includes information drawn from the classical authorities, but always in a critical spirit, and on various occasions refutes their assertions.

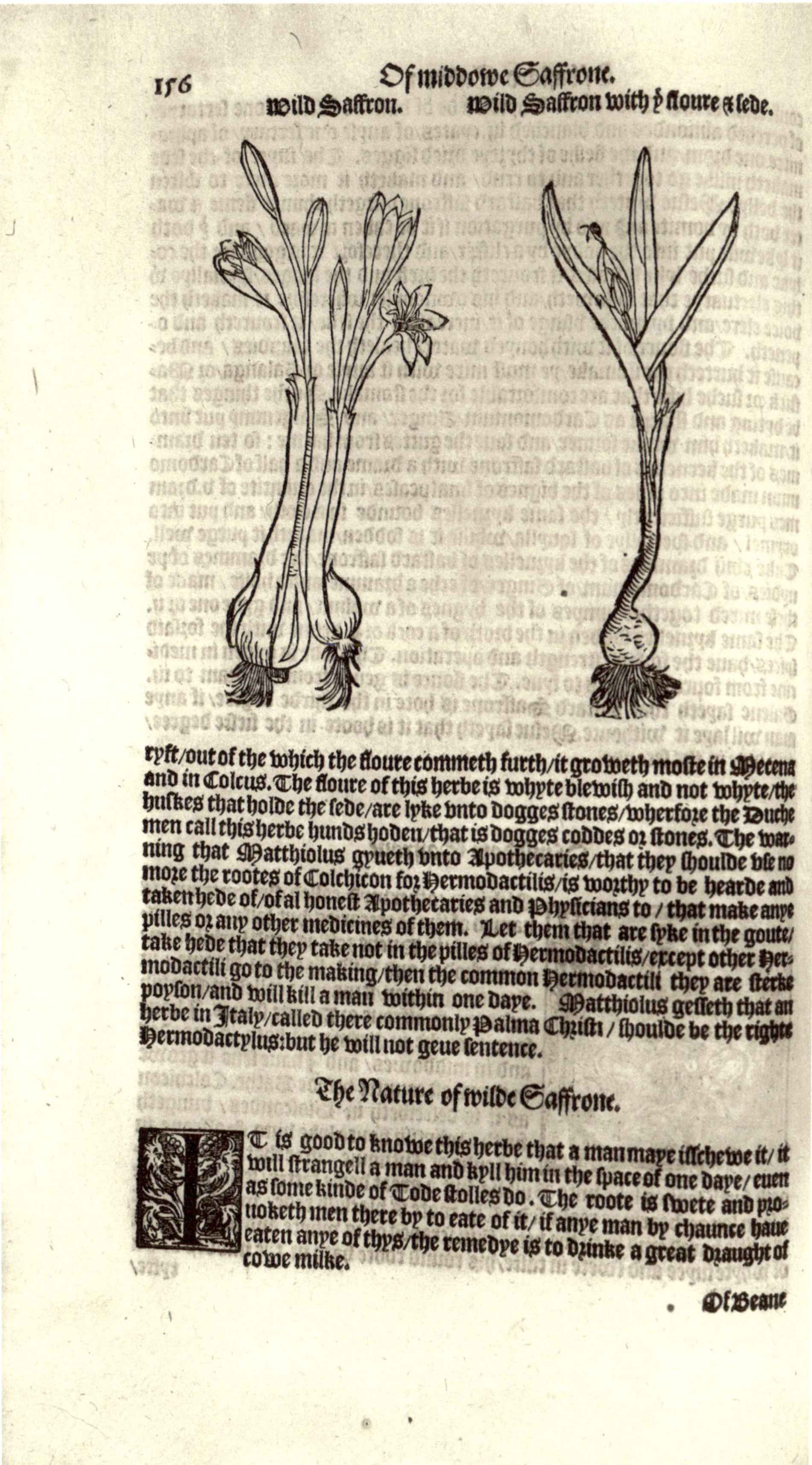

Wild Saffron. Wild Saffron with ẏ floure & sede.

ryst/out of the which the floure commeth furth/it groweth moste in Mecena and in Colcus. The floure of this herbe is whyte blewish and not whyte/the huskes that holde the sede/are lyke unto dogges stones/wherfore the Duche men call this herbe hunds hoden/that is dogges coddes or stones. The warning that Matthiolus gyueth unto Apothecaries/that they shoulde use no more the rootes of Colchicon for Hermodactilis/is worthy to be hearde and taken hede of/of al honest Apothecaries and Physicians to / that make anye pilles or any other medicines of them. Let them that are syke in the goute/take hede that they take not in the pilles of Hermodactilis/except other Hermodactili go to the making/then the common Hermodactili they are sterke poyson/and will kill a man within one daye. Matthiolus gesseth that an herbe in Italy/called there commonly Palma Christi / shoulde be the righte Hermodactylus: but he will not geue sentence.

The Nature of wilde Saffrone.

IT is good to knowe this herbe that a man maye isschewe it/ it will strangell a man and kyll him in the space of one daye/ euen as some kinde of Tode stolles do. The roote is swete and prouoketh men there by to eate of it/ if anye man by chaunce haue eaten anye of thys/the remedye is to drinke a great draught of cowe milke.

Of Beane

WILLIAM TURNER,
A New Herball, 1568.
'Autumn crocus,' Wild saffron (*Colchicum autumnale*), left, in flower and right, with fruit, page 156

9. Carolus Clusius (1526–1609)

Caroli Clusii Atrebat [*i.e., Atrbatensis or Atrebatæ*]. Rariorum aliquot stirpium per Hispanias observatarum Historia, libris duobus expressa: Ad Maximilianum II. Imperatorem. [*Printer's device*]. Antverpiae, Ex officina Christophori Plantini, Architypographi Regij: M.D.LXXVI.

8° 17.6 × 10.6 cm. A–Z⁸ a–l⁸ *1–2* 3–529 [*15*] [544 pp.].

PLATES: 230 woodcuts of plants.

BINDING: Old vellum with lettered title on spine.

REFERENCES: Arber, pp. 86–88; Blunt & Stearn, pp. [?]; *Botany in the Low Countries*, pp. 42–43, 50, 62, 114–115, 132; Egmond, Hoftijzer, and Visser; Garbari, Tongiorgi Tomasi, and Tosi; Hopper; Hunt 125; Cleveland 113; Menten; Nissen 370; Pritzel 1,756; Stafleu & Cowan 1,145; Swan, 2006.

JULES CHARLES DE L'ÉCLUSE, generally known by his Latinized name Carolus Clusius, was an eminent Flemish botanist who deserves credit for having introduced to the gardens of Europe many unknown plants, in particular species from the Near and Far East. He also contributed to the advance of horticulture, cultivating and encouraging the dissemination of a large number of bulbous and tuberous plants from Asia, including the crown imperial and new varieties of the iris, hyacinth, anemone, ranunculus, narcissus, lily, and above all the tulip. So great was his reputation that the Flemish neo-Stoical philosopher Justus Lipsius (1547–1606) and Marie de Brimeu, Princess of Chimay and Duchess of Aarschot, both referred to him as 'le père de tous les beaux jardins de ce pays.'

Clusius was born in 1526 in Arras (Atrecht), in the region of Flanders. After receiving his early education at the chapter school of Sint-Vaast between 1540 and 1542 and then at the Houckaert Latin School in Ghent (1543–1546), he studied in numerous European universities, acquiring a culture that embraced the classical and modern languages, law, medicine, botany, philosophy, history, cartography, and even numismatics and epigraphy. In addition to his botanical texts, he produced a translation from Latin into French of the humanist Donato Acciaiuoli's work *Les vies d'Hannibal et Scipion l'Africain* (Paris, 1567), and two maps, one of 'Gallia Narbonensis' (southern France) for the atlas of the famous cartographer Abraham Ortelius (1527–1598) and one of the Spanish peninsula which was published at the famous printing works of Christophe Plantin in Antwerp.

Clusius earned a degree in law from the university in Louvain and then continued his peripatetic studies at the university in Wittenberg, where his interests turned to medicine and botany. We know that in 1551 he was in Montpellier following the lectures of the celebrated naturalist Guillaume Rondelet (1507–1556), where he studied botany and zoology. In this way Clusius discovered his true calling; henceforth he would dedicate himself above all to the study of botany.

Rather than limiting his research to the identification of the plants described by the *veteres auctores*, Clusius was a 'militant botanist' who traversed Europe in search of new plants, and not only those with medicinal properties. He showed an unusual flare for the identification of unknown species and described them with exceptional precision, an achievement that was all the more remarkable if one considers the limited terminology at his disposal. It appears that he added no less than six hundred new species to the known flora of Europe and his pioneering work would prove invaluable to the botanists who succeeded him. As Wilfrid Blunt observed: 'His description and the associated illustrations ... help to typify the species of later authors' (Blunt and Stearn, p. 64).

Two of Clusius' earliest destinations were Spain and Portugal, and he published the results of his botanizing expeditions in 1576 in *Rariorum aliquot stirpium per Hispanias observatorum historia*. This treatise was dedicated to Maximilian II (1527–1576), Holy Roman Emperor, who in 1573 had invited Clusius to come to Vienna and serve as *prefectus* of the imperial botanical garden. The Flemish botanist already possessed some experience in this area, having laid out and tended the splendid garden of Jean de Brançion at Malines until 1568. He remained at the Hapsburg court for about fourteen years, taking advantage of this period to study the flora that was completely new to him, that of the Austro-Hungarian alpine region. He published his findings in 1583 in a work that also contained the descriptions of exotic plants from other continents—for example, a series of 'faseoli' or leguminous plants from the Americas and the horse chestnut (*Aesculus hyppocastanus*), whose seeds were first sent to Europe in 1581, probably by the Count of Weissenfeld, David Ungnad, who succeeded Ogier de Busbecq (see No. 7) as ambassador to Constantinople. When his *opus magnum*, *Rariorum plantarum historia*, was published in 1601, Clusius included in it his two early works on the flora of the Iberian peninsula and the Alps, as well as what may be considered the first printed monograph on mycology, *Fungorum ... Historia*.

In addition to these original works, Carolus Clusius rendered an invaluable service to the botanical sciences by translating many important texts by other authors, among them the *Cruijdeboeck* by his friend Rembert Dodoens (see No. 10) from Flemish into French in 1557; *Coloquios dos simples e drogas* by Garcia da Orta (see No. 17) from Portuguese to Latin in 1567 (*Aromatum et simplicium ... historia*); and in 1574 the works of Cristobal Acosta (*c.* 1525–*c.* 1594) and Nicolas Monardes (see No. 16) from Spanish to Latin.

After retiring from his position at the court of Maximilian II, Clusius spent a period of time in England and then returned to Vienna in 1588, where he received two tubers of the potato from the governor of Mons, Philippe de Sivry. A watercolor of this exotic plant from the Americas bearing the inscription: 'Taratoufli à Philipp° de Sivry acceptum, Viennae 26 Januarij 1588' can be seen at the Plantin-Moretus Museum in Antwerp (*Botany in the Low Countries*, p. 119; Blunt and Raphael, p. 147).

In 1593, although old and infirm, he accepted an invitation to lay out an herbal garden for the university of Leiden. Following the model of the botanical gardens of Pisa and

Padua, Clusius drew up a ground plan based on botanical rather than medical-pharmacological criteria. He was also responsible for the 'ostensiones simplicium,' or lecturing to students on plants *dal vivo*. Clusius died in Leiden in 1609 before he had finished correcting his last work, *Curae posteriores*; this and a treatise on the geography of Belgium were published posthumously in 1611 and 1619, respectively.

Carolus Clusius' fame as a botanist reached every part of Europe; he won the respect of the greatest naturalists of the period and corresponded regularly with many of them, from Joachim Camerarius the Younger (see No. 42) and Ulisse Aldrovandi to Nicolas Claude Fabri de Peiresc (1580–1637). He was also acquainted with many intellectuals from Melanchthon to Ogier de Busbecq. Through his extensive contacts he continually received new botanical specimens from all over the world; de Busbecq sent him bulbs and seeds regularly from Constantinople and Sir Francis Drake gave him plants from the Americas (see No. 14). The renown of the botanist is further attested to by the large number of his portraits that could be found in circulation, one of the most widely disseminated being an engraving by Jacques de Gheyn the Younger depicting him at the age of seventy-five; there is also a fine painting conserved *ab antiquo* at the Botanical Garden of Pisa (Garbari, Tongiorgi Tomasi, and Tosi, pl. 30, p. XIII).

Clusius assigned great importance to the botanical illustration as a means of documenting his work; he sometimes made his own drawings, and to illustrate his texts employed a professional artist, Pieter van der Borcht (1540–1608). Van der Borcht had already demonstrated his talent in his work for Dodoens (see No. 10) and beginning in 1564 worked on a regular basis for the publisher Christophe Plantin of Antwerp. Another artist who collaborated with Clusius, particularly during the latter part of his life, was Jacques de Gheyn the Younger (1565–1629), the delineator of his portrait; a series of drawings in gouache by de Gheyn, many portraying species described by Clusius, is conserved at the Custodia Foundation in Paris (Hopper and Swan, 2006). A large number of botanical illustrations prepared for Clusius can also be found among the *Libri Picturati*, an extraordinary collection of more than 1,860 illustrations of plants and 260 illustrations of animals that once belonged to the Prüssiche Staatsbibliothek in Berlin but were then lost, and which have recently been rediscovered in the Jagiellonian Library in Cracow.

Rariorum aliquot stirpium per Hispanias observatarum Historia, an exemplar of which may be found in the Oak Spring Garden Library, was therefore the first botanical work written by Clusius. Published by Plantin in 1576, it presents the results of herborizing expeditions undertaken in 1564 and 1565 by Clusius in Spain and Portugal in the company of Jacob Fugger, son of the Augsburg banker Anton Fugger. It constitutes an early example of what today would be referred to as a 'flora,' that is, a work describing the plants indigenous to a specific region. During the course of his expeditions the botanist gathered about two

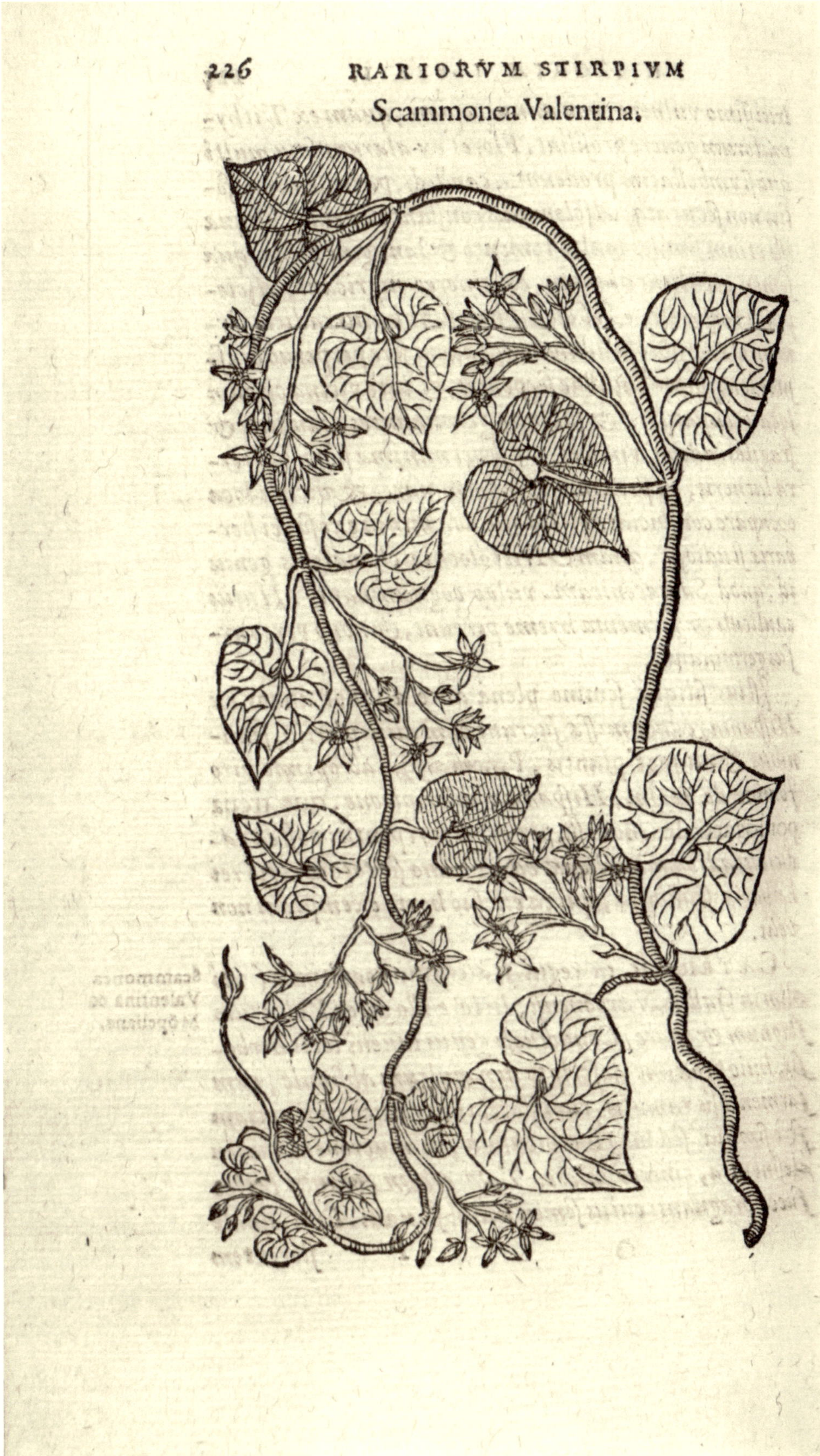

CAROLUS CLUSIUS, *Rariorum aliquot stirpium*, 1576. 'Scammonea Valentina,' Swallow wort (*Cynanchum acutum*), page 226

hundred hitherto unknown plants, which he describes in careful detail and illustrates in his text. In the introduction we learn that he drew sketches of the specimens himself in charcoal and red pencil, but that the drawings for the book itself were executed in September and October 1568 by Pieter van der Borcht in Malines on the basis of the dried specimens that he had brought back with him. One year later both the text and illustrations were ready, but due to a misunderstanding or perhaps to a lack of funds, publication was delayed and thirty-nine illustrations intended for Clusius' work were first used by Plantin in Rembert Dodoens' *Purgantium … historiae* (1574).

Rariorum aliquot stirpium is divided into two volumes, the first dedicated to woody plants and vines and the second to herbaceous plants and shrubs. Clusius describes each species accurately, including its color, scent, and habitat; assigns it a name, often adopting a binomial nomenclature which would in large part be maintained by Linnaeus; and classifies the plant based on its external characteristics following the system developed by Dodoens. The dragon tree (*Draecena draco*), which Clusius saw for the first time in Lisbon in 1566, is just one of the unusual species described; a watercolor drawing of it may be found in *Libri picturati* (18, fol. 1847, *Botany in the Low Countries,* p. 79).

At the end of *Rariorum aliquot stirpium* is an appendix in which the author describes a number of plants and bulbs from the Orient, obtained in large part through the good offices of de Busbecq. Of particular interest is the section devoted to the 'Tulipa' (page 509).

Rariorum aliquot stirpium was published by the celebrated printer Christophe Plantin (1514–1588), at whose works in Antwerp many of the most important scientific texts of the period were produced, including the herbals and botanical treatises of Clusius, Matthias de L'Obel, and Rembert Dodoens. The books by these three botanists were in fact closely correlated, because the authors were on extremely good terms with one another and engaged in the reciprocal exchange of information. Plantin initiated his activity around the year 1576, establishing a tradition of scholarly publishing that was continued by his sons-in-law Jean Moretus the Elder in Antwerp and Frans van Ravelingen (Franciscus Raphelengius) in Leiden, and by their descendants for another eight generations. The texts were printed at the sign of the Golden Compasses, whose emblem included the motto 'Labor & Constantia' (Labor and Constancy), and whenever possible were illustrated. To this end Plantin employed a permanent staff of artists; in addition to van der Borcht, the engravers Arnold Nicolai (*c.* 1525–1590), Gerard Janssen van Kampen (*c.* 1540–1590), Antoine van Leest (*c.* 1545–1592) and Cornelis Müller are known to have worked for him. Since the creation of a new set of illustrations was prohibitively expensive, Plantin did not hesitate to use the same images in more than one work, a circumstance that can frequently be observed in the botanical texts printed by him. Indeed, *Plantarum seu Stirpium … Eicones* by L'Obel, published in 1581, represents a veritable compendium of the illustrations from various books written by the three botanists.

A visit to the Plantin-Moretus Museum in Antwerp, whose exhibits include a unique collection of more than four thousand engraved wood-blocks, provides a glimpse into the workings of a sixteenth-century printing shop, and the intellectual and technical endeavor that lay behind the publication of every book by this remarkable printing house.

10. Rembert Dodoens (1517–1585)

[*Within woodcut border*]: A Niewe Herball, or Historie of Plantes: wherin is contayned the vvhole discourse and perfect description of all sortes of Herbes and Plantes: their divers & sundry kindes: their straunge Figures, Fashions, and Shapes: their Names, Natures, Operations, and Vertues: and that not onely of those whiche are here growying in this our Countrie of Englande, but of all others also of forrayne Realmes, commonly used in Physicke. First set foorth in the Doutche or Almaigne tongue, by that learned D. Rembert Dodoens, Physition to the Emperour: And nowe first translated out of French into English, by Henry Lyte Esquyer. At London by me Gerard Dewes, dwelling in Pawles Churchyarde at the signe of the Swanne. 1578.

[*Colophon on 3Y4 verso*]: [*large printer's device*] Imprinted at Antwerpe, by me Henry Loë Bookeprinter, and are to be solde at London in Povvles Churchyarde, by Gerard Devves.

2° 29.2 × 18.5 cm. $\star^6$ $2\star^6$ A–$3G^6$ 3H–$3I^4$ 3K–$3X^6$ $3Y^4$ *i–xxiv* 1–356 359–362 361–779 [*25*] (150 as 151, 151 as 152, 162 as 163, 163 as 164, 357 as 359, 358 as 360, 359 as 361, 360 as 362, 773 as 775) [828 pp.].

PLATES: Woodcut border surrounding title-page with Henry Lyte's coat-of-arms (the translator) on verso, portrait of Dodoens on verso of ⋆6, woodcut printer's device above colophon, plus 640 woodcuts of 873 plants and one of insects.

BINDING: Eighteenth-century quarter calf and marbled boards.

PROVENANCE: 'T. Shaftsbury' inscribed on front pastedown.

REFERENCES: Anderson, pp.173–180; *Botany in the Low Countries*, No. 34 and pp. 98, 102, 105–106; Blunt & Raphael, pp. 153, 156, 163–165; Cleveland 118; Hunt 132; Meerbeeck; Nissen 516; Pritzel 2,345; Stafleu & Cowan 1,488; *An Oak Spring Flora*, pp. 154–158.

REMBERT DODOENS was born in 1517, the illegitimate son of Denys Dodoens, physician of the city of Malines (Mechlin) in Flanders. After receiving his degree in medicine from the University of Louvain, Dodoens continued his studies in Italy, France, and Germany. In 1541 he became, like his father before him, town physician of Malines, where he practiced and also found time to pursue his interests in medicine and cosmography. Stimulated by the work of the Byzantine physician Paulus Aegineta, he wrote a short treatise on the subject of fevers, *Pauli Aeginetae de febribus* ..., which was printed in Cologne in 1546. In 1548 his *Cosmographica in astronomiam et geographiam isagoge* appeared, published in Antwerp by the city's most important printer, Jan van der Loe, whose important scientific work was taken up, when he died in 1563, by Christophe Plantin.

In this period the horizon of the botanical sciences was rapidly expanding following the publication in Basel of Leonhart Fuchs' *De historia stirpium* in 1542 with its innovative text and magnificent illustrations (see No. 6). In addition to a German translation, Michael Isingrin prepared a Dutch edition in 1543 (perhaps translated by the twenty-five-year-old Dodoens). In any event, Fuchs' publication inspired Dodoens to plan the compilation of his own herbal, to which work he devoted the rest of his life, beginning with his first major publication, the translation and adaptation of Fuchs' herbal as the *Cruijdeboeck* of 1554, and ending with his masterwork *Stirpium historiæ pemptades sex* (see No. 12), which contains more than 1,300 illustrations and was printed by Plantin in 1583, just two years before the botanist's death. Dodoens' work was itself quite new, for he attempted to arrange his plants systematically rather than in mere alphabetical order. The plants are divided into categories based on their morphology and uses (fragrant herbs, medicinal herbs, flowers, seeds, roots, etc.) and the characteristics of the various species and varieties have been clearly distinguished. The author furnishes not only the Greek and Latin names of the plants, but also their vernacular names in German, Bohemian, French, and English. As befitted the serious physician, he lists the medical-pharmaceutical properties of each plant, although credulously describing some rather unlikely nostrums for various ills. Nonetheless, as his studies progressed, Dodoens' approach gradually changed from that of the physician whose knowledge was based almost exclusively on his reading of the classical texts, to that of the naturalist who examined plants from life, gathering data and drawing his conclusions based on direct observation (see Meerbeeck, p. 103).

Jan van der Loe obtained the privilege to update and re-issue Fuchs' herbal, and in 1552 he published Dodoens *De frugum historia*, a treatise on cereals and leguminous species, the first small step to this end. Then—as was sometimes done to generate interest and help cover the printer's costs—in anticipation of the appearance of Dodoens' herbal, the illustrations alone were published: those of the first part (*Trium priorum de stirpium historia … imagines*) in 1553, and those of the second part (*Posteriorum trium … historia*) in 1554. Finally, later in the year 1554 the completed *Cruijdeboeck* appeared, and, in 1568, in response to the growing interest in horticulture and plants destined for cultivation in the gardens of wealthy amateur botanists, he published a brief treatise on flowers, particularly those species prized for their fragrance, *Florum et coronariarum … historia* (see *An Oak Spring Flora*, pp. 154–158).

In 1574, with the backing of his friend the Flemish botanist Carolus Clusius, Dodoens was appointed personal physician to Emperor Maximilian II, whom he attended at his court in Vienna and then Prague. The emperor died in 1576 but he continued to serve his son and successor, Emperor Rudolph II, until the year 1582 when he was offered the chair of botany at the university in Leiden; he would teach there until his death in 1585.

As the author noted in the preface to the *Cruijdeboeck*, at the insistence of his publisher

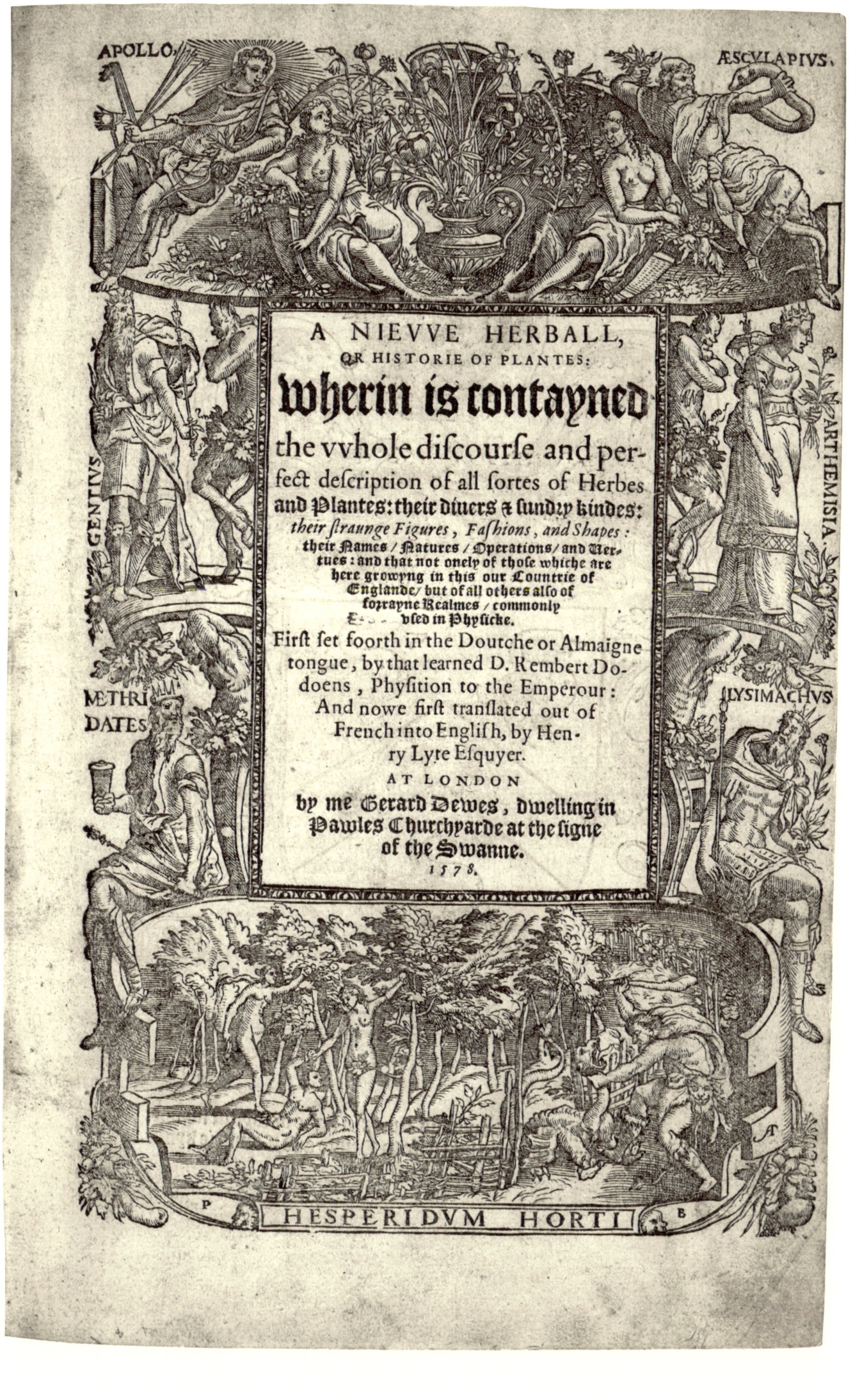
A NIEVVE HERBALL,
OR HISTORIE OF PLANTES:
wherin is contayned
the vvhole discourse and perfect description of all sortes of Herbes and Plantes: their diuers & sundry kindes: *their straunge Figures, Fashions, and Shapes:* their Names / Natures / Operations / and Vertues: and that not onely of those whiche are here growyng in this our Countrie of Englande / but of all others also of forrayne Realmes / commonly vsed in Physicke.
First set foorth in the Doutche or Almaigne tongue, by that learned D. Rembert Dodoens, Physition to the Emperour: And nowe first translated out of French into English, by Henry Lyte Esquyer.
AT LONDON
by me Gerard Dewes, dwelling in Pawles Churchyarde at the signe of the Swanne.
1578.

REMBERT DODOENS,
A Niewe Herball, 1578.
Title-page

Jan van der Loe, the book was written in Dutch rather than Latin in order to attract the largest possible number of readers. Dedicated to Maria of Hungary, the sister of Charles V and ruler of the Low Countries, the herbal boasts an ornate title-page border that is signed near the bottom 'P.B.,' the initials of the artist Pieter van der Borcht, as well as with the monogram '*A*' of his engraver, Arnold Nicolai. The engraving is reminiscent of the one on the title-page of Otto Brunfels' *Herbarum vivæ eicones* (see No. 5), and it too portrays several scenes and personages. Again we see beneath the space containing the title, the Garden of the Hesperides with Hercules battling its guardian dragon. In the upper part of the page Apollo and Asclepius are portrayed, while on either side of the title are various historical-mythological figures connected with medicine, such as Gentius, King of the Illyrians; Mithradates VI Eupator, an expert on poisons; Artemisia, Queen of Caria; and the Macedonian general Lysimachus.

A portrait of the author at the age of thirty-five, perhaps based on a drawing by van der Borcht, appears a few pages later. The botanist is shown wearing a doctoral gown and holding a flowering plant, with his family coat-of-arms—consisting of two stars and a crescent moon—in the background. This engraving first appeared in the work *Trium priorum … imagines* in 1553, and the woodblock from which it was printed is conserved in the Plantin-Moretus Museum in Antwerp.

In the *Cruijdeboeck* many plants native to northern Europe are described for the first time. The herbal is embellished with an impressive set of illustrations comprising over seven hundred woodcuts, many of which were based on drawings made from life. Others were either copied from previously published works, such as Mattioli's *Commentarii,* or printed from woodblocks used in the 1545 octavo edition of Fuchs' herbal. Dodoens' book was an immediate success all over Europe, thanks in part to an excellent French translation from the Dutch prepared by Clusius, *Histoire des Plantes*; this was published in 1557 by Jan van der Loe.

An English edition of Dodoen's work—*A Niewe Herball or Historie of Plants*—was published in 1578, more than twenty years after the French translation by Clusius, on which it was based. It was issued in London by Gerard Dewes 'at the signe of the Swanne,' as is announced on the title-page, but printed by Henrick van der Loe, Antwerp. The translation was the work of an amateur botanist, Henry Lyte (1529–1607) (see No. 12), whose coat-of-arms appears on the verso of this page. The woodcut border on the title-page and the portrait of the author have both been included unaltered from the original edition, except that the Spanish coat of arms at the head has been replaced by a vase of flowers. *A Niewe Herball* opens with a dedication to Queen Elizabeth, a preface addressed to 'the friendly and indifferent Reader,' and some verses written in honor of the author Dodoens and, more particularly, the translator Lyte.

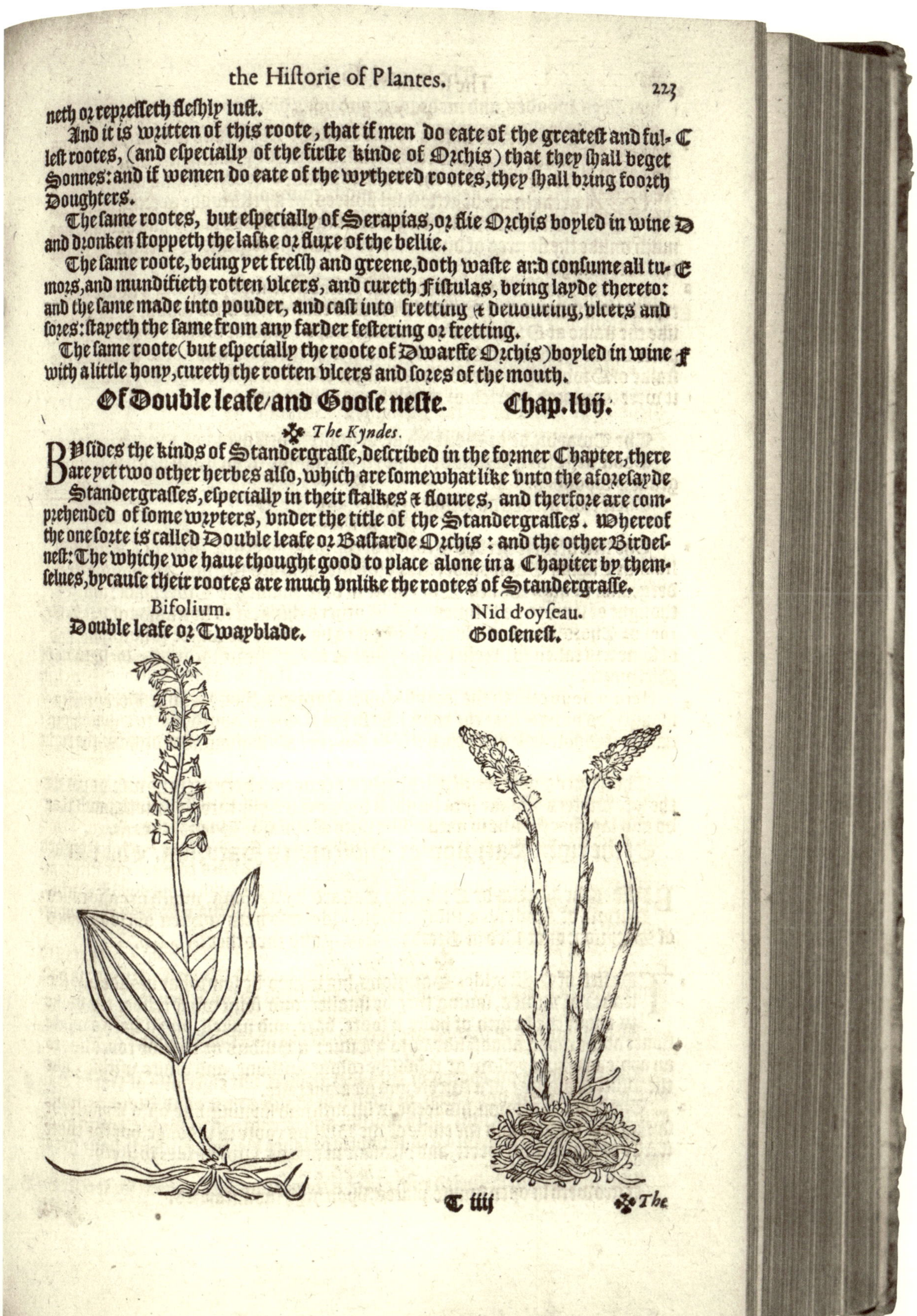

neth or repreſſeth fleſhly luſt.

And it is written of this roote, that if men do eate of the greateſt and fulleſt rootes, (and eſpecially of the firſte kinde of Orchis) that they ſhall beget Sonnes: and if wemen do eate of the wythered rootes, they ſhall bring foorth Doughters. C

The ſame rootes, but eſpecially of Serapias, or flie Orchis boyled in wine and dronken ſtoppeth the laſke or fluxe of the bellie. D

The ſame roote, being yet freſſh and greene, doth waſte and conſume all tumors, and mundifieth rotten vlcers, and cureth Fiſtulas, being layde thereto: and the ſame made into pouder, and caſt into fretting & deuouring, vlcers and ſores: ſtayeth the ſame from any farder feſtering or fretting. E

The ſame roote (but eſpecially the roote of Dwarffe Orchis) boyled in wine with a little hony, cureth the rotten vlcers and ſores of the mouth. F

Of Double leafe/and Gooſe neſte. Chap. lvij.

The Kyndes.

Byſides the kinds of Standergraſſe, deſcribed in the former Chapter, there are yet two other herbes alſo, which are ſomewhat like vnto the aforeſayde Standergraſſes, eſpecially in their ſtalkes & floures, and therfore are comprehended of ſome wryters, vnder the title of the Standergraſſes. Whereof the one ſorte is called Double leafe or Baſtarde Orchis: and the other Birdeſneſt: The whiche we haue thought good to place alone in a Chapiter by themſelues, bycauſe their rootes are much vnlike the rootes of Standergraſſe.

Bifolium.
Double leafe or Twayblade.

Nid d'oyſeau.
Gooſeneſt.

The

REMBERT DODOENS, *A Niewe Herball*, 1578. European common twayblade (*Listera ovata*) and the Bird's nest orchid (*Neottia nidus-avis*), page 223

A Niewe Herball constitutes a fine example of the printer's art; its 640 small woodcuts have been arranged with care, sometimes two or three to a page (none of the illustrations are in folio), in such a way as to create a pleasing balance between the text and images. The plants are almost always portrayed with their roots and flowers to aid the herbalist in their identification. Many of the illustrations were printed from the wood-blocks (still in excellent condition) that were prepared for Fuchs' 1545 octavo edition of *De historia stirpium* (see No. 6). Acquired by Jan van der Loe, the blocks were passed down to Henrick, who carried on the business after his father's death in 1563 until 1566, when they were sold to Christophe Plantin.

Also included are several illustrations that had already appeared in Dodoen's treatise *Florum et Coronariarum … historia*, such as the sunflower, here called the 'Indian Sunne or Golde Floure of Perrowe (*Chrysanthemum peruuianum*)' and described by the author as follows: '[it] groweth in Weste India, the whiche is called America, and in the Countrey of Perrowe: and being sowen in Spayne, it groweth to the length of foure and twentie foote' (p. 191).

The work by Dodoens was republished several times in England: in 1586 by Niman Newton, in 1595 by E. Bollifant, in 1606 by Simon Stafford, and, unillustrated, in 1619 by Edward Griffin.

11. Fabio Colonna (1567–1650)

ΦΥΤΟΒΑΣΑΝΟC sive Plantarum aliquot historia in qua describuntur diversi Generis Plantæ veriores, ac magis facie, viribúsque respondentes antiquorum Theophrasti, Dioscoridis, Plinij, Galeni, aliorúmque delineationibus, ab alijs hucusque non animaduersæ. Fabio Columna auctore. Accessit etiam piscium aliquot, Plantarúmque nouarum Historia eodem auctore. Ad illustrissimum et reverendissimum D. S. R. E. Card. M. Antonium Columnam. [*Printer's device*]. Ex Officina Horatij Saluiani. [*rule*] Neapoli, M. D. XC II. Apud Io. Jacobum Carlinum, & Antonium Pacem.

4° 20 × 13.9 cm. ¶–2¶[4] A–P[4] a–d[4] 2a[4] *i–xvi* 1–120 1–32 [*8*] [176 pp.].

PLATES: 37 etchings with letterpress borders, thirty of plants and seven of fishes.

BINDING: Modern dark green morocco.

REFERENCES: Anderson, pp. 210–217; Barker 1999; Blunt & Raphael, pp. 99–101, 172–173; Gabari and Tongiorgi Tomasi; Hunt 165; Nissen 386; Tognoni.

FABIO COLONNA was born in Naples in 1567. His father Girolamo was a descendent of the Neapolitan branch of the aristocratic Roman family of the Colonnas. Educated in various disciplines (Greek, philosophy, mathematics, botany, drawing, and music), he enrolled in the faculty of jurisprudence at the university in Naples and graduated in *utroque iure* (ecclesiastic and civil law) in 1589. He studied philosophy under Giovanbattista della Porta (see No. 51), who encouraged him to pursue his interest in natural history.

Frail in health and affected by a grave form of epilepsy, Colonna turned to the texts of the physicians of Antiquity, from Hippocrates to Galen, in search of a cure for his malady, analyzing the composition of the pharmacological remedies recommended by Pliny, Dioscorides, and Theophrastus. He did not limit his research to books, also venturing into the countryside and woods around Naples in search of simples, comparing the plants he found to those described by Dioscorides, and discovering errors and omissions which he recorded in a commentary on *De materia medica* that, unfortunately, was never published and has since been lost. He also had a talent for drawing and made illustrations of the plants that he had gathered. After careful trial and error, the novice botanist determined that *Valeriana officinalis* (Valerian), which he identified as the diuretic *phu* described by Dioscorides in *De materia medica* and which he illustrates on page 114 of *Phytobasanos,* alleviated the symptoms of his epilepsy.

Colonna's studies led him to become a convinced supporter of the 'orti dei semplici' (botanical gardens) that in this period were being established in numerous Italian cities—Pisa, Padua, Bologna and Rome—and he entered into correspondence with celebrated naturalists such as Bartolomeo Maranta (d. 1570) and Giovan Vincenzo Pinelli (1535–1601). In his tireless quest for knowledge he visited the natural science collections of Giovanbattista della Porta and Ferrante Imperato (*c.* 1550–1625) in Naples; the latter constituted a veritable museum whose fame reached every corner of Europe after the publication of the catalogue *Dell'istoria naturale libri XVIII* in 1599. On various occasions in his books Colonna mentions specimens that he had observed in Imperato's collection.

Fabio Colonna's first published work was *Phytobasanos*; it appeared in 1592 and to the author constituted merely the first stage in his ongoing botanical studies. Immediately after its publication he decided to undertake a series of botanizing expeditions in order to broaden his knowledge of the flora of Italy. By staying with relatives in various parts of Italy he was able to study the plant life of the mountainous area of Matese in the Campania region, the flora of the zone near the city of Cerignola in Puglia, and finally the plants that grew around Zagarolo near Rome, where he was appointed a captain of the city.

Colonna suffered a serious illness in 1560 and after his recovery took up the practice of law in Rome. Five years later we find him in Naples once again, where he had gone to examine a famous herbal in the monastery of San Giovanni in Carbonara. He published the results of his latest studies in a new work, *Minus cognitarum stripium* ... *Ἔκφρασις* (Exposition of lesser known plants), which proved to be an extremely costly and time-consuming project. Its publication was delayed by the fact that Colonna spent several years searching for a printer capable of reproducing its 156 illustrations in accordance with his specifications, and, when it was finally printed in Rome by Guglielmo Faciotti in 1606 at the author's expense, it was not very well received. The undertaking, especially given the large number of illustrations, took a heavy toll on the author's purse.

In 1610 Colonna had the good fortune to make the acquaintance of Federico Cesi, the founder of the first scientific academy in Europe, the Accademia dei Lincei. Cesi, who was very interested in botanical studies, had come to Naples in order to meet Ferrante Imperato and Giovanbattista della Porta and to persuade the latter to establish a branch of his academy there. A cordial relationship sprang up between Colonna and the Roman nobleman, who held his new friend in the highest esteem and introduced him to Galileo Galilei, another member of the academy. Colonna had occasion to advise the scientist to act with the greatest circumspection when he became embroiled with the Catholic Church over the theological questions raised by his discoveries. The affiliate academy was established in Naples in 1612 and Colonna became one of its first members, succeeding della Porta as its director when the latter died in 1615. He was a keen and diligent academician, participating actively in the editing of the botanical and mineralogical sections of *Rerum Medicarum Novae Hispaniae Thesaurus*, also known as *Tesoro Messicano*, an ambitious project begun by 'Prince' Cesi and continued by the academy after his premature death, its aim being to edit and update the work on the natural history of the New World by the Spanish physician Francisco Hernández (1515–1587) (see No. 16).

His profound learning and inquiring mind led Colonna to pursue research in other, sometimes disparate fields, from malacology and fossilized marine animals to music; he wrote a short treatise on the purple dye obtainable from certain shells and even invented a complicated musical instrument with fifty strings similar to a harpsichord. Born in frail health, Colonna made for himself a reputation as an outstanding naturalist during his lifetime. He died in Naples in 1650 at the very respectable age of eighty-three.

Phytobasanos was published at Colonna's expense in Naples in 1592 at the address of Orazio Salviani by Giovanni Giacomo Carlino and Antonio Pace, two of the city's most important printer/booksellers. The text, written in Latin and divided into two parts devoted to plants and marine life, marked the debut of the Neapolitan botanist in scientific circles. The thirty-eight-year-old author is portrayed holding a columbine, in an oval frame surrounded by the family coat-of-arms and various emblems. The Greek word *phytobasanos* means 'plant touchstone,' the touchstone being the stone slab on which goldsmiths rubbed precious metals to assay their quality and value.

To understand the genesis of this work dedicated, as Colonna wrote, to the pleasure and utility of the study of the three kingdoms of nature, it is enlightening to read the 'avviso al lettore' in which he outlines his aims. The author recounts that he had been

FABIO COLONNA, *Phytobasanos*, 1592. Two Primeroses (*Primula acaulis* and *Primula officinalis*), page 21

FABIO COLONNA,
Phytobasanos, 1592.
A member of the *Umbelliferae* family, possibly a species of *Pimpinella* or of *Oenanthe*, page 76

motivated to embark on the study of simples by the need to find a cure for his epileptic attacks. After consulting a series of physicians and pharmacists who proved to have little or no knowledge of medicinal plants, he decided to carry out his own research and become his own doctor, and the modest success that crowned his efforts encouraged him to undertake a more systematic study of the plant, animal, and mineral kingdoms in search of medicines and their active elements, based on the tradition handed down from the most famous physicians of Antiquity.

In his herbal every plant is provided with a complete description divided into various sections that sometimes vary in order but which always opens with a learned discussion of the etymology of the plant's name. This is followed by a detailed and accurate description of its parts (flower, stem, and root) and continues with notes regarding its habitat, and the period of its flowering. Considerable space is devoted to the *vires* or virtues, a compendium of the plant's medicinal properties drawn from the classical authorities including Galen, Dioscorides, Pliny, and the seventh-century Byzantine physician Paulus Aegineta or, if not mentioned by them, based on his own knowledge and contemporary sources. The scientific activity of Colonna did not end with the painstaking transmission of the *historia* of the plants as described by the *veteres auctores* (ancient authors). In symphony with the general work of revision being carried forward by naturalists all over Europe during the sixteenth century, he advocated and applied a more rigorous approach to the classification of simples, where theoretical study was combined with practical experimentation.

The illustrations to *Phytobasanos*—thirty-seven of which depict plants complete with their roots and often accompanied by details of the flower and seeds, seven of which portray fishes—are of particular importance and interest. The author prepared them with great care, based upon his personal experience with botanical illustration as a young man, and it is not improbable that many of the preliminary drawings were by his own hand, even if contemporary sources do not make any specific mention of this. Significantly, Colonna chose to employ the medium of copper-plate etching, which allowed his engraver to create *chiaroscuro* effects and achieve results that were not only more refined artistically but also more convincing scientifically than the traditional woodcut illustration.

Each botanical illustration consists of an etching surrounded by an elegant border printed separately from letterpress ornaments. The author's deep interest in the art of naturalistic illustration is expressed in the preface to *Phytobasanos*, in a passage that has been open to misinterpretation. Various scholars have in the past used it to demonstrate the importance placed by Colonna on the scientific illustration, but they have gone further and read into his words a declaration that he himself participated in the *delineatio* (drawing) of the illustrations. A more attentive reading of the passage and the recent discovery of an important manuscript by Colonna now allow us to correct the misconception that he was the draftsman-engraver of the illustrations to his own work.

It is, in fact, unlikely that in the declaration 'icones ipsis plantis nova quadam arte à me excogitata effinx' (I personally have prepared the images of plants with a new technique by me invented) the author was referring to copper-plate printing, for this technique was already well known to Italian artists and printers and had been used to produce the illustrations to many books on natural history, particularly after the remarkable success of the ichthyological plates in *Aquatilium animalium historiae* by Ippolito Salviani. This work had appeared in Rome almost half a century earlier and must certainly have been known to Colonna, while the first botanical book to use the technique was the rare Pietro di Nobilis *Herbal*, Rome, *c.* 1580. We can only conclude that the author was alluding to a new process specifically designed to portray botanical specimens. This matter can now be disentangled thanks to the discovery of two voluminous manuscripts by Colonna's hand, which were displayed in 1999 at the Royal Oak Foundation and The Grolier Club in New York as part of the exhibit curated by Nicolas Barker, *Treasures from the Libraries of National Trust Country Houses*. These manuscripts constitute, as the title *Icones ipsis plantis ad vivum expressae quoad fieri potuit nova quaedam arte excogitata ab ipso auctore* (Images of plants expressed from life according to a new method invented by the author himself) reveals, an herbal produced by the process today known as nature printing (see Barker, pp. 90–91). This complex technique involves sprinkling a plant, either freshly picked or dried, with organic pigments and then making a direct impression of it on a sheet of paper. Only a handful of examples of the use of this technique in the early sixteenth century are known, one being the herbal compiled by the Florentine pharmacist and perfumer Zenobio Pacini (*c.* 1520), today conserved in the Bibliothèque Nationale in Paris.

With nature printing the scientist could be certain that a botanical illustration was the mirror image of the plant itself, and avoid relying on professional artists who had no knowledge of botany or scientific research and were all too prone to indulge in representations that were more artistic than realistic. By means of this technique Colonna had a visual record systematically made of all the plants he had gathered during his many botanizing expeditions. It is evident that the smaller or more complex parts of a plant could not always be captured in a direct impression, in which case the illustration was retouched and the details drawn by hand. In the herbal *Icones ipsis plantis*, which consists of no less than 570 sheets of nature prints, various illustrations include parts outlined in ink with the finer details sketched in; then watercolor was added to recreate the colors of the flowers and the play of light and shadow on the leaves and petals. The technique of nature printing, which was adopted so enthusiastically by Colonna and to which he makes specific reference in the preface to *Phytobasanos*, became quite popular among botanists engaged in the compiling of their own herbals, as is demonstrated by various manuscripts conserved at the Oak Spring Garden Library (see Nos. 57, 59, and 60), for it combined many of the qualities of the 'orto secco' and the 'orto dipinto.'

Colonna made a fundamental contribution to modern science for, although he began in his youth with an investigation of the medicinal properties of plants, he arrived at a quintessentially modern conception of botany. With his natural bent for direct observation and experimentation, he progressively detached himself from the study of the classical texts to arrive at the direct analysis and comparison of the characters of different plants, focusing above all on the morphology of the flower and seed. He identified many new species and defined several genera; indeed Joseph Pitton de Tournefort acknowledged that the idea of the 'generum ratio,' or genus, originated with him, and Linnaeus described him as 'the first among all botanists.'

12. JOHN GERARD (1545–1612)

[*Engraved title-page signed* 'Io: Payne sculps:'] [*Within cartouche*]: The Herball or Generall Historie of Plantes. Gathered by John Gerarde of London Master in Chirurgerie[.] Very much Enlarged and Amended by Thomas Johnson Citizen and Apothecarye of London. [*within cartouche*]: London Printed by Adam Islip Joice Norton and Richard Whitakers Anno 1636.

2° 34.2 × 22 cm. $¶^{8}$ (-¶1) 2¶–$3¶^{6}$ A–B^{8} C–$6V^{6}$ $6X^{4}$ 6Y–$7B^{6}$ (-7B6) *i–xxxviii* 1–32 29–229 *230*–1,257 *1,258*–1,589 *1,590*–1,630 *1,631–1,632* [*46*] (20 as 18, 31 as 29, 32 as 30, 309 as 312, 312 as 309, 358 as 258, 359 as 259, 370 as 400, 371 as 401, 372 as 402, 391 as 491, 398 as 498, 403 as 379, 410 as 386, 445 as 449, 446 as 450, 487 as 587, 488 as 588, 493 as 593, 494 as 594, 681 as 761, 682 as 782, 683 as 783, 684 as 764, 795 as 797, 819 as 719, 904 as 890, 1,154 as 1,054, 1,207 as 1,217, 1,269 as 1,272, 1,272 as 1,269, 1,349 as 1,449, 1,351 as 1,451, 1,352 as 1,452, 1,353 as 1,453, 1,354 as 1,454, 1,355 as 1,455, 1,356 as 1,456, 1,357 as 1,457, 1,358 as 1,458, 1,360 as 1,460, 1,420 as 1,396, 1,457 as 1,459, 1,480 as 1,478, 1,581 as 1,481) [1,720 pp.].

PLATES: Engraved title-page, 2,776 woodcuts of plants.

BINDING: Contemporary calf; recently re-backed in dark brown morocco.

PROVENANCE: 'George Lyte his Booke' inscribed on dedication.

REFERENCES: Anderson, pp. 218–226; Arber, pp. 129–135; *Botany in the Low Countries*, No. 63 and pp. 80, 124; Blunt & Raphael, pp. 150–151, 154–155, 164–169; Bush-Brown, pp. 143–159; Henrey I.36–54 and Nos. 154–156; Hunt 230; Nissen 698; Pritzel 3,282; Rohde, pp. 98–119; Stafleu & Cowan [?].

JOHN GERARD'S *Herball*, if not one of the monuments of the English language, as Frank Anderson wrote, is certainly one of its 'great delights' (p. 218), because of its splendid Elizabethan prose and poetic imagery. It also constitutes an important landmark in the history of the botanical sciences in England and Europe.

John Gerard was born in 1545 Nantwich, Cheshire, to a Lancashire family, as we learn from the coat-of-arms that appears in the first edition of his herbal. After studying at the school in nearby Willaston (Wisterson), he trained under a barber-surgeon in London and then served for many years as a ship's doctor, probably on a merchantman. This gave him the opportunity, as he later wrote, to visit various foreign lands, including 'Moscow, Denmark, Swenia [Sweden], Poland'; it is possible that he also sailed around the Mediterranean basin. When Gerard returned to England in 1577, he immersed himself in the study of the native flora. We then find him in London, supervising the gardens of a fervent collector of rare plants, William Cecil, First Baron of Burghley (1521–1598), at his residence in the Strand and his country estate of Theobalds in Hertfordshire. Gerard himself kept his own garden at his residence Holborn where he cultivated hundreds of plants gathered during herborizing expeditions or purchased at dear price for—as George Baker (1540–1612), the surgeon of Queen Elizabeth, wrote in a commendatory letter that appears in the first edition of *The Herball*—Gerard sought 'all the rare simples which by any means he could attaine unto.'

In 1596 John Gerard published a modest quarto volume of about twenty pages listing over one hundred plants—among them the exotic potato—that could be found in his garden; *Catalogus arborum, fruticum ac plantarum tum indigenarum, quam exoticarum in horto Johannis Gerardi ... nascentium*, is of considerable interest because it represents one of the earliest printed catalogues of a private botanical collection. A copy of this very rare work that once belonged to the celebrated naturalist Sir Hans Sloane (1660–1753) is conserved in the British Museum. The second edition of the catalogue, published in 1599, was dedicated by the author to the great seaman and explorer of the New World, Sir Walter Raleigh (*c.* 1552–1618).

In 1595 Gerard was elected a member of the powerful Court of Assistants of the Barber-Surgeons' Company, which exercised control over the surgical profession in the city of London; within this organization he served in positions of increasing importance, culminating in his appointment in 1608 as Master of the Company. It is interesting to note that around the year 1595 he also drew up a petition addressed to the University of Cambridge, which his patron Lord Burghley apparently intended to endorse as well, urging the authorities to plant a botanical garden and offering his services as 'herbarist'; disappointingly, this proposal was ignored. Later, however, in 1586 when the College of Physicians decided to build a physic garden in London, Gerard was appointed as its curator.

In 1597 Gerard's *The Herball or Generall Historie of Plantes* appeared, an imposing volume in folio, written in English and printed at the works of Edmund Bollifant for the London bookseller and royal printer, John Norton. The genesis of this herbal, it must be said, was somewhat dubious and would form the subject of controversy for many years to come. It appears that in 1583 Norton commissioned an English translation of Dodoens' *Stirpium historiæ pemptades sex* (see No. 10) from a certain Dr. Priest. When Priest unexpectedly died, Norton seems to have asked Gerard to complete the project. He did so, in the process borrowing heavily from the work of the celebrated Flemish botanist Matthias de L'Obel (1538–

¶ *The Vertues.*

The leaues of water Plantain, as ſome Authors report, are good to be laid vpon the legs of ſuch as are troubled with the dropſie, and hath the ſame propertie that the land Plantain hath. A

‡ *Dioſcorides* and *Galen* commend the ſeed hereof giuen in wine, againſt fluxes, dyſenteries, the ſpitting of bloud, and ouermuch flowing of womens terms. B

Pliny ſaith, the leaues are good againſt burns. ‡ C

CHAP. 97. *Of land Plantaine.*

media L. S. P.

1 *Plantago latifolium.*
Broad leaued Plantaine.

2 *Plantago incana.*
Hoary Plantaine.

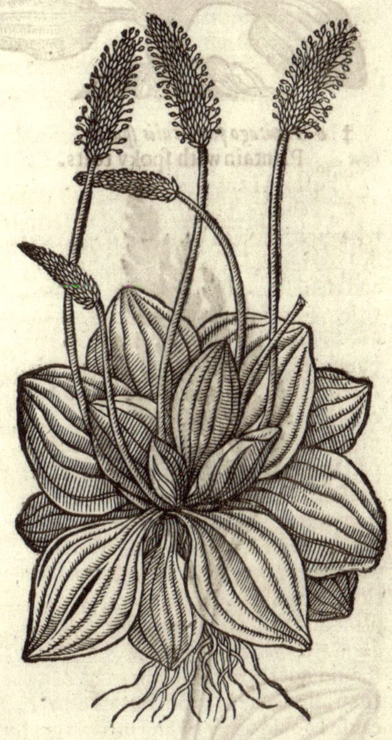

¶ *The Deſcription.*

1 AS the Greeks haue called ſome kinds of herbs Serpents tongue, Dogs tongue, and Ox tongue; ſo haue they termed a kinde of Plantain *Arnogloſſon*, which is as if you ſhould ſay Lambs tongue, well known to all, by reaſon of the great commoditie and plenty of it growing euery where; and therefore it is needleſſe to ſpend time about them. The greatnes and faſhion of the leaues hath been the cauſe of the varieties and diuerſities of their names.

2 The ſecond is like the firſt, and differeth in that, that this Plantaine hath greater but ſhorter ſpikes or knaps; and the leaues are of an hoary or ouerworne green colour: the ſtalks are likewiſe hoary and hairy.

3 The ſmall Plantain hath many tender leaues ribbed like vnto the great Plantain, and is very like in each reſpect vnto it, ſauing that it is altogether leſſer.

4 The ſpiked roſe Plantaine hath very few leaues, narrower than thoſe of the ſecond kinde of Plantain, ſharper at the ends, and further growing one from another. It beareth a very double floure vpon a ſhort ſtem like a roſe, of a greeniſh colour tending to yellowneſſe. The ſeed groweth vpon a ſpiky tuft aboue the higheſt part of the plant; notwithſtanding it is but very low in reſpect of the other Plantains aboue mentioned.

5 The

JOHN GERARD, *The Herball, or Generall Historie of Plantes*, 1636. Two plantains, Great Plantain (*Plantago major*) and Hoary Plantain (*Plantago media*), page 419

1616), who was then residing in England and who provided him with considerable material help as well. Gerard published the resulting work under his own name (Arber, p. 129), making only a passing reference to his sources in the preface addressed to his 'courteous and well wiling readers.' *The Herball* was dedicated to Lord Burghley, and to attest to its *bona fides* the author included numerous commendatory letters written by friends including Lancelot Browne, the physician of the Queen (d. 1605); the royal surgeon George Baker; and the botanist de L'Obel himself. Whatever its origin *The Herball* was a work of great scientific and literary appeal, due in large part to the brilliant pen of its author. In its compilation Gerard adhered to the most authoritative of models, Dioscorides' *De materia medica*, providing for each plant a complete 'description' with its 'names,' the places where it grew, and finally its 'virtues' or uses, primarily in medicine but on occasion for other purposes as well.

As always in this early period in the history of printing, the problem of how to illustrate the text posed itself. Norton had managed to obtain the woodblocks used for the voluminous herbal of Jacob Dietrich of Bergzabern (*c.* 1520–1590), a celebrated naturalist better known by his Latinized name, Tabernaemontanus, who had studied under Brunfels (see No. 5). The text of Tabernaemontanus' *Neuw Kreuterbuch* was published in Frankfurt in 1588 and 1591, while its illustrations, copied in large part from Hieronymus Bock (also known as Tragus), Fuchs, Mattioli, and Dodoens, and L'Obel, were printed in 1590 in an extremely successful album entitled *Eicones plantarum*.

To Gerard fell the (not always simple) task of matching these images to the plants described in his text and commissioning any that were lacking to be drawn and engraved *ex novo*. L'Obel, resident in London at the time, was engaged by the publisher to check the accuracy of the work, which went to press at the insistence of Gerard after over one thousand revisions had been made. As Jean Bauhin (1541–1612) observed in *Pinax Theatri Botanici* (1623), most of the illustrations to *The Herball* originally appeared in the *Neuw Kreuterbuch*, a few were borrowed from L'Obel, while only sixteen were completely original. The illustrations, labelled with their names in Latin and English, are arranged generally two to a page in alternation with the text, or else four to a page without any text. One of the most noteworthy inclusions is the first printed illustration of the potato (*Solanum tuberosum*, p. 927), which is drawn twice. On the left we can see the plant with its flowers, and on the right the root with its characteristic tubers, which Gerard describes in scrupulous detail: 'Some of them are round as a ball, some are oval or egge-fashion, some longer, some others shorter.' The author called the plant 'Virginian,' because he thought it was native to that region and had been brought to Europe by settlers from Sir Walter Raleigh's colony, whereas the potato was actually discovered in Central or South America by the Spaniards. He furthermore notes that the plant was already being cultivated in Europe by Clusius and that it was not to be confused with the better known 'batatas' or sweet potato.

round ſlender footſtalkes, whereon grow very faire and pleaſant floures, made of one entire whole leafe,which is folded or plaited in ſuch ſtrange ſort,that it ſeemes to be a floure made of fiue ſundry ſmal leaues,which canot eaſily be perceiued,except the ſame be pulled open. The whole floure is of a light purple colour,ſtriped downe the middle of euery fold or welt with a light ſhew of yellownеſſe,as if purple and yellow were mixed together. In the middle of the floure thruſteth forth a thicke flat pointall yellow as gold,with a ſmall ſharp green pricke or point in the midſt thereof. The fruit ſucceeds the floures,round as a ball,of the bigneſſe of a little Bulleſſe or wilde plumme, green at the firſt,and blacke when it is ripe,wherein is contained ſmall white ſeed leſſer than thoſe of Muſtard: the root is thicke,fat,and tuberous,not much differing either in ſhape,colour,or taſte, from the common Potato's,ſauing that the roots hereof are not ſo great nor long,ſome of them are as round as a ball,ſome oual or egge-faſhion, ſome longer, and others ſhorter; the which knobby roots are faſtned vnto the ſtalks with an infinite number of threddy ſtrings.

Battata Virginiana ſiue Virginianorum, & Pappus,
Virginian Potatoes.

¶ *The Place.*

It groweth naturally in America,where it was firſt diſcouered,as reporteth *Cluſius*, ſince which time I haue receiued roots hereof from Virginia,otherwiſe called Norembega,which grow & proſper in my garden as in their owne natiue country.

¶ *The Time.*

The leaues thruſt forth of the ground in the beginning of May; the flours bud forth in Auguſt, the fruit is ripe in September.

¶ *The Names.*

The Indians call this plant *Pappus*,meaning the roots; by which name alſo the common Potatoes are called in thoſe Indian countries. Wee haue it's proper name mentioned in the title. Becauſe it hath not only the ſhape and proportion of Potato's, but alſo the pleaſant taſte and vertues of the ſame,we may call it in Engliſh,Potatoes of America or Virginia.

‡ *Cluſius* queſtions whether it be not the *Arachidna* of *Theophraſtus*. *Bauhine* hath referred it to the Nightſhades,and calls it *Solanum tuberoſum eſculentum*; and largely figures and deſcribes it in his *Prodromus*,*pag*.89. ‡

The

JOHN GERARD,
The Herball, or Generall Historie of Plantes, 1636.
Virginian Potatoes
(*Solanum tuberosum*),
page 927

The Herball also contains the descriptions of numerous fungi, mosses, and marine organisms such as sponges and corals, which at the time were believed to be plants, as well as a chapter on the mythical 'goose tree' or 'barnacle tree.' It was long believed that the barnacle goose or *Branta leucopsis* was born from a shell that grew on this tree, a legend inspired by a marine crustacean that could be found attached to floating tree trunks.

Many errors made by Gerard in his text and in the identification of the illustrations were corrected by the apothecary and botanist Thomas Johnson of London, who prepared an expanded edition of *The Herball* that first appeared in 1633 and was re-published in 1636. Johnson's painstaking revision of Gerard's herbal constituted in itself a valuable contribution to botany and to the art of the printed book. He added the descriptions of many new plants with illustrations, some of them borrowed from the botanical texts published by Christophe Plantin in Antwerp (see No. 9), in this way bringing the total number of engravings to 2,776. Any new passages were carefully marked with special symbols so that the reader could distinguish them from the original text.

The title-page to the 1636 edition of *The Herball* is particularly attractive. It was executed by John Payne (1607–1647), one of the most talented engravers of the period, to replace the original page by William Rogers (Rogers was also the author of the portrait of Gerard that appears in the 1597 edition). In the upper part of Payne's work we see a luxuriant garden with the goddesses Ceres and Pomona on either side. Below them are the fathers of botany, Theophrastus and Dioscorides, while in the lower section two imposing vases filled with flowers surround a portrait of Gerard, who is shown facing in the opposite direction to that of the portrait by Rogers. The vase on the left is crowned with a bunch of bananas as a tribute to Johnson, who prepared a drawing for a new woodcut of the plant, a sample of which he had obtained from the Bermudas.

One of the most significant additions made by Johnson was his chapter on the 'Maracot' or 'Grandilla' as it was called at the time (actually the passion-flower). He includes a full page illustration (p. 1592) and refers the reader to Monardes (see No. 16) for more information on this exotic species. The conscientious botanist also corrects Gerard in his chapter on the 'goose tree,' affirming that no such plant existed and citing in support of his conclusion Fabio Colonna's observations in *Phytobasanos* (see No. 11). In a long preface Johnson traces the history of the botanical sciences, analyzing the contributions of celebrated figures from the mythical King Solomon to William Turner (see No. 8) and contemporaries such as John Parkinson (1567–1650), the London pharmacist and author of *Paradisi in Sole* (see *An Oak Spring Flora*, No. 40, pp. 158–161). He closes with some critical remarks on John Gerard and the origins of his herbal.

A beautiful copy in excellent condition of Thomas Johnson's edition of Gerard's herbal is conserved at the Oak Spring Garden Library.

13. Adriaan van de Spiegel (1578–1625) *and* Adolph van Voorst (1597–1663)

[*Engraved title-page; within cartouche*]: Adriani Spigelii Philos. ac Medici Patauini Isagoges in rem herbariam, 1633 Isagoges in rem herbariam Libri duo. [*within compartment at foot*]: Lugduni Batavorum. Ex officina Elzeviriana. Anno 1633.

12° 10.8 × 5.3 cm. A–S[8] *1–2* 3–15 *16–20* 21–222 *223–224* 225–272 [*16*] [288 pp.].

PLATES: Engraved title-page.

BINDING: Contemporary calf; gilt tooled borders on covers; gilt tooling on spine.

PROVENANCE: Bookplate of La Bibliothèque de Laussat.

REFERENCES: Arber, p. 142; Hunt 184 (1606 edition); Pritzel 8,827.

The second edition of Spiegel's *Isagoges* was published by Elzevir in a diminutive duodecimo volume together with a catalogue of the plants in the Botanical Garden of Leiden. The Elzevirs were an eminent Dutch family of printers and booksellers. Louis Elzevir (1540–1617) published his first book in Leiden in 1583, and his successors would continue with this activity until the end of the eighteenth century. Members of the family established printing shops in Amsterdam, Utrecht, The Hague, and other European cities, where they became renowned for their beautifully printed, reasonably priced books. These were often produced in a relatively small format with a clear, legible typeface, in what might be considered an early version of the modern 'pocket edition' that could be easily and quickly consulted.

The slender, elegant volume by Spiegel is not illustrated, but it opens with a handsome title-page depicting the goddess Flora, who presents an oval cartouche decorated with a floral garland and inscribed with the name of the author and the title of the work. In the background, a gardener holding a rake, a spade, and a flower pot can be glimpsed, while arranged at the bottom of the page are distilling vessels, a mortar and pestle, a lancet, jars containing medicinal preparations, and an herbal that has been opened to a page illustrating a plant.

Adriaan Spiegel was born into a medical family in Brussels in 1578; both his father and grandfather were surgeons. He attended the universities in Louvain and Leiden before deciding in 1601 to complete his studies at the illustrious faculty of medicine in Padua. There he adopted the Latinized form of his name—Spigelius—and joined the 'Natio Germanica,' the university's association for students from northern Europe. Padua was especially renowned for its school of anatomy and Spiegel was able to study under the celebrated professors Girolamo Fabrici d'Acquapendente (1533–1619, Acquapendente was the name of the town in Lazio where Fabrici was born, see No. 42) and Giulio Casserio (1561–1616). He seems to have graduated in the year 1603, and in 1606 was appointed physician of the 'Natio Germanica.' Padua was also famous for its academic botanical garden, the first—together with the botanical garden of Pisa—to be founded in Italy, and there Spiegel could devote himself to the study of medicinal plants.

During this period he probably also served as assistant to the by-now elderly Fabrici in his private practice, for he accompanied the physician on a trip to Florence, where Fabrici had been summoned to treat Grand Duke Ferdinand I de' Medici, and on another trip to Venice. After failing in 1607 to win a post on the faculty of medicine in Padua, Spiegel returned to Belgium in 1612. He remained there for only a short time, however, before departing for Germany where he served as 'medicus primarius' in Moravia and Bohemia. He finally achieved his life's ambition when in 1616 the Senate of Venice appointed him professor of anatomy and surgery at Padua, the 'studio' of the Republic of Venice. A respected and honored member of the faculty—in 1623 he was elected 'Cavaliere di San Marco' (Knight of the Order of Saint Mark)—Spiegel would remain in Padua until his death in 1625.

In addition to *Isagoges*, Spiegel wrote various works on the subject of medicine: *De lumbrico lato liber...*, which was printed in Padua by Pasquati in 1618; *De semitertiana libri quatuor*, the first extensive account of malaria, which was published by de Bry in Frankfurt in 1624; *Catastrophe anatomiae publicae in Lycaeo patavino feliciter absolutae* (Padua, 1624); and *De formato foetu liber singularis...*, which was published posthumously in Padua in 1626 by Spiegel's son-in-law, Liberale Crema. One of his most significant books was *De humani corporis fabrica libri X* (Venice, 1627), whose title pays homage to the famous anatomical treatise by Andrea Vesalius published in 1542. It describes with great accuracy the muscles of the human body, and is splendidly illustrated with engravings that had originally been prepared by the artists Francesco Valesio (*c.* 1560–after 1643) and Odoardo Fialetti (1573–1638) for Giulio Casserio's unfinished 'Theatrum anatomicum,' whose plates were published as *Tabulae anatomicae LXXIIX* (Venice, 1627). A new edition of the complete works of Spiegel was published in Amsterdam in 1645.

Isagoges, which is in the nature of a general treatise on botany, was printed in Padua in 1606 by Magietto & Pasquali and was dedicated by the author to the 'Natio Germanica,' to which he had such close links. It is quite significant because it was the first botanical text to provide detailed instructions on how to create a collection of dried plants (see Arber, p. 142). This technique, which is still used by botanists today, was perhaps invented by the Italian botanist Luca Ghini of Imola (1490–1556), who served as the first director of the 'Giardino dei Semplici' of Pisa after the position had been turned down by his German colleague Leonhart Fuchs (see No. 6). The dried herbal was an invaluable tool because it offered a means of conserving plants so that they could be studied when live specimens were not available. Many of Ghini's students would go on to create their own collections

ADRIAAN VAN DE SPIEGEL, *Isagoges*, 1633. Title-page

of dried plants, beginning with the Italians Andrea Cesalpino (1519–1603), whose dried herbal is conserved in the Botanical Museum's library at the University of Florence, and Gherardo Cibo (1512–1600), whose collection may be found at the Biblioteca Angelica in Rome. Botanists in other countries engaged in similar activities, for numerous citations mention the collections of the English naturalist John Falconer (1553), the Swiss Felix Platter (1554), the German Caspar Ratzenberger (1556), and the Frenchman Jean Girault (1558). During the seventeenth century the dried herbal became one of the most widespread tools for the study of plants (see Nos. 56 and 61).

Isagoges consists of two books, the second of which is dedicated to the uses and properties of plants and contains little that is new. The first book begins with a general discussion of plants, their form, similarities and differences, and continues with the description of various species, but the chapters of greatest interest are the last three (LVI, LVII, LVIII). In Chapter LVI the author notes that the botanist must engage in the constant, detailed study of plants, their form and their varieties if he wishes to remember and be able to recognize them. In Chapter LVII he underlines the importance of the botanical illustration and describes a method to help those who are not gifted at drawing. He suggested pressing the plant or a part of it between two sheets of paper so that it would leave an outline that could then be retraced and colored, in a procedure that is today referred to as 'nature printing' (see Nos. 57, 58, 59). Since it was not always possible to apply this technique (if, for example, the plant had stiff branches or thorns), the author recommended in his final chapter that the botanist prepare what he called 'Horti Hyemales' (winter gardens) or 'orti secchi' (dried gardens), in the following manner. Each plant should be arranged on a sheet of rough paper, left to dry, and then placed between two sheets of white paper ('inter cartae candidae duo folia ponantur'). When a sufficient number of specimens have been prepared, the sheets could be arranged in a pile and a moderate weight applied ('libro huic superponatur pondus') for three or four hours in order to squeeze out the last remaining traces of moisture. At this point a heavier weight could be placed on the pile so as to press the specimens until they are quite flat. The dried plant should then be attached to a fresh sheet of paper using a special type of glue, the recipe for which he duly provides. Finally, the dried herbal should be carefully stored in a dry place.

Bound together with *Isogages* was a catalogue of the plants that were being cultivated in the botanical garden of Leiden in the year 1633. This list had been drawn up by Adolph van Voorst (also known by his Latinized name, Vorstius), a professor of medicine and botany at the university in Leiden. Van Voorst began his studies in botany and classical and oriental languages in Leiden, before transferring to the university in Padua where he received his degree. He returned to Holland as part of the retinue of Marcus Antonius Maurocenus, the Venetian envoy to the States of Holland. It may be noted that in this period Leiden's garden of simples, which had been constructed in 1587, was one of the most celebrated in Europe. It boasted an extensive collection of exotic plants that was started by the great Flemish botanist Carolus Clusius (see Nos. 9 and 14) when he was appointed curator in 1593. Clusius was succeeded by Aelius Everard van Voorst, the father of Adolph, who initiated a plant collecting project with the Dutch East India Company that greatly enriched the garden's collection of oriental plants. Many species native to North and South America were also added at this time.

Adolph took over from his father as director of the botanical garden and professor at the university in 1625. He spent many years writing a commentary on the work of Theophrastus, although this project was never completed. In 1628 he published the first catalogue of the botanical garden of Leiden, which included many plants from Canada and Virginia. By the time the second catalogue was compiled in 1633 the collection had grown significantly and included new plants such as the common spiderwort (*Tradescantia virginiana*) and the double syringa (*Philadelphus coronarius*). A total of 1,107 plants are listed, to which is appended a further catalogue of 289 wild plants that could be found growing in the countryside around the city of Leiden. This work was reprinted in 1636, 1649, and 1658. Beginning in the 1620s many plants gathered by explorers around the Cape of Good Hope reached Leiden, and from there—like the exotic species that had arrived earlier from the Near East, the Far East and the Americas—were distributed to botanists all over Europe after the publication in 1687 of *Horti Academici Lugduno Batavi catalogus* by the professor of botany Paul Hermann (1640–1695).

III

HERBALS AND PLANTS FROM DISTANT LANDS

CARLOUS CLUSIUS,
Aliquot notæ in Garciæ,
1582. Coconut palm
(*Cocos nucifera*),
page 10

14. Carolus Clusius (1526–1609)

Caroli Clusii Atreb. [*i.e., Atrebatensis or Atrebatæ*] Aliquot Notæ in Garciæ Aromatum Historiam. Eiusdem Descriptiones nonnullarum Stirpium, & aliarum exoticarum rerum, qu[æ] à Generoso viro Francisco Drake Equite Anglo, & his obseruatæ sunt, qui eum in longa illa Nauigatione, qua proximis annis uniuersum orbem circumiuit, comitati sunt: & quorundam peregrinorum fructuum quos Londini ab amicis accepit. [*Printer's device*] Antuerpiæ, Ex officina Christophori Plantini, M. D. LXXXII.

[*Colophon, C6 verso*]: Antuerpiae Excudebat Christophorus Plantinus, Architypographus, Anno CIƆ. IƆ. LXXXII. Mense Feb.

8° 16.2 × 10.2 cm. A–C⁸ *1–2* 3–43 [5] [48 pp.].

PLATES: 15 woodcuts of exotic plants in text.

BINDING: Modern tan morocco with gilt tooled borders on covers and gold tooling on spine.

REFERENCES: *Botany in the Low Countries*, No. 47 and pp. 42–43, 109; Cambridge C2240; Egmond, Hoftijzer, and Visser; Hunt 140; Nissen 373; Pritzel 1,737; Sabin 13,800; Stafleu & Cowan 1,146.

This rare work by Carolus Clusius was dedicated to Balthasar de Batthyany, a professional soldier whose wide-ranging interests included a profound love of plants and books. Its modest length belies its extraordinary importance, for *Aliquot notæ in Garciæ* provided the Flemish botanist with the occasion to make important observations on a number of exotic plants and their uses.

The first part consists of a brief supplement to the third edition of *Aromatum et simplicium … historia* (1579), Clusius' translation of Garcia da Orto's herbal of Indian medicinal plants, which was first published by Christophe Plantin in 1567 (see No. 17). In the second part of the work the botanist describes several specimens that he had the good fortune to obtain from Sir Francis Drake (1545–1595), the brilliant English seaman who circumnavigated the globe. Indeed, *Aliquot notæ in Garciæ* contains the first lengthy account of Drake's historic voyage, as is reflected in its title: 'Some notes on the history of aromatic [plants] of Garcia [da Orta] by Charles de L'Écluse [Carolus Clusius] of Arras, and accounts by the same [author] of observations made of some plants and other exotic things by the noble Englishman, Sir Francis Drake, and by those who accompanied him on the long journey he made of late years around the world; and also foreign products which [the author] received from friends in London.'

In December 1577, with the patronage of Queen Elizabeth and accompanied by the naturalist Lawrence Eliot, Drake embarked on a voyage around the world with a fleet of five small ships (of which only Drake's *The Golden Hind* was destined to return). Their itinerary took them across the Atlantic, through the perilous Strait of Magellan, and then up along the coasts of Chile and Peru to California. From there Drake crossed the Pacific, stopping in the Moluccas Islands and Java, passed round the Cape of Good Hope, and fi-

CARLOUS CLUSIUS,
Aliquot notæ in Garciæ,
1582. Cacao seeds
(*Theobroma cacao*),
page 29

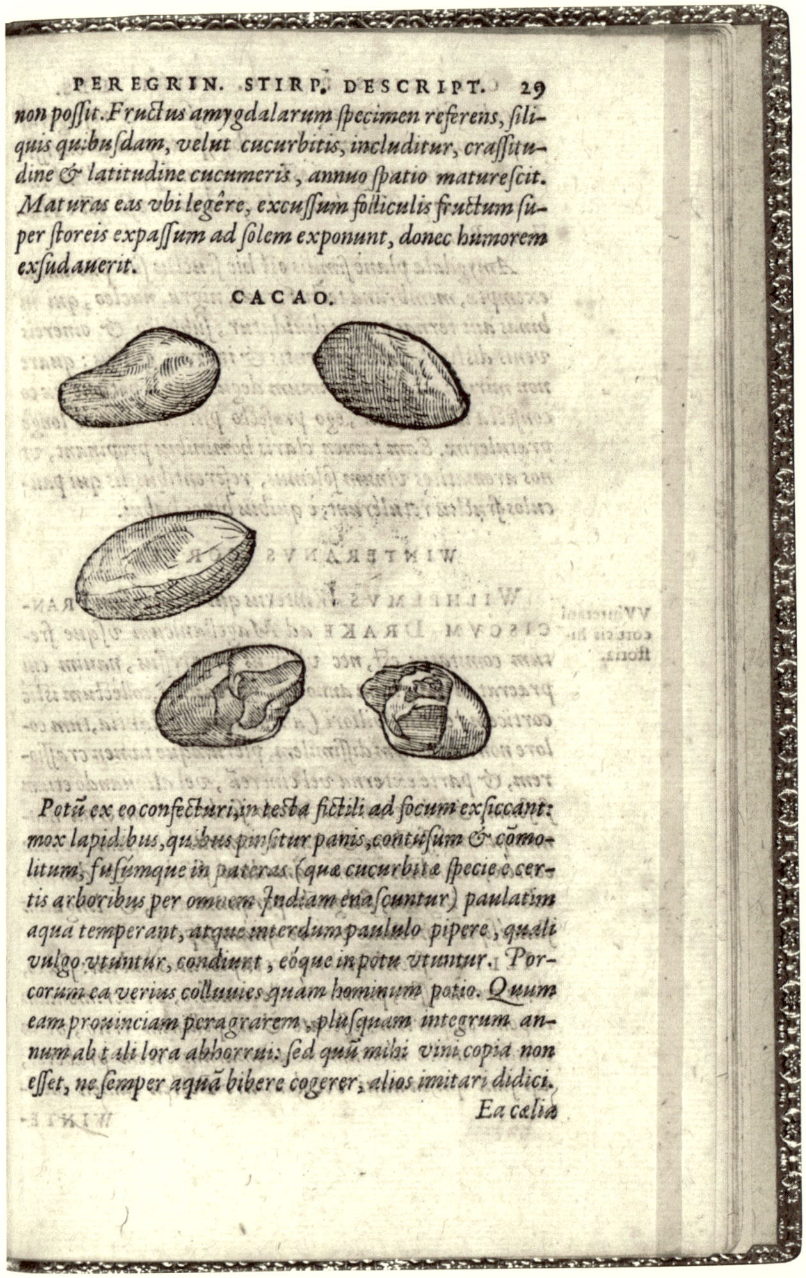

PEREGRIN. STIRP. DESCRIPT. 29

non possit. Fructus amygdalarum specimen referens, siliquis quibusdam, velut cucurbitis, includitur, crassitudine & latitudine cucumeris, annuo spatio maturescit. Maturas eas vbi legêre, excussum folliculis fructum super storeis expassum ad solem exponunt, donec humorem exsudauerit.

CACAO.

Potū ex eo confecturi, in testa fictili ad focum exsiccant: mox lapidibus, quibus pinsitur panis, contusum & comolitum, fusumque in pateras (quæ cucurbitæ specie è certis arboribus per omnem Indiam enascuntur) paulatim aqua temperant, atque interdum paululo pipere, quali vulgo vtuntur, condiunt, eóque in potu vtuntur. Porcorum ea verius colluuies quam hominum potio. Quum eam prouinciam peragrarem, plusquam integrum annum ab tali lora abhorrui: sed quū mihi vini copia non esset, ne semper aquā bibere cogerer, alios imitari didici.

Ea cælia

nally sailed triumphantly into Plymouth Harbor on 26 September 1580, with just fifty-six of his original crew of 100 but laden with treasure taken from the Spaniards.

It was probably during his sojourn in London after leaving the Hapsburg court that Carolus Clusius met Drake and his company and received the gift of various botanical specimens, including plants, fruits, and pieces of bark, which they had brought back with them from their long voyage. Of particular interest to the naturalist were the seeds of the cacao plant and the 'batatas' or sweet potato. While in London Clusius also made the acquaintance of Hugh Morgan, the personal physician to Queen Elizabeth, from whom he received a great curiosity—two pieces of bread made with flour from the sago palm, a tree native to the East Indies.

Among the plants cited in the first part of *Aliquot notæ in Garciæ* are the coconut palm and the elegant 'Jasminum mexicanum' (*Mirabilis jalapa*), which the author declares that he saw for the first time in 1580. Joachim Camerarius sent him some seeds from Rome, where the plant had been acclimatized and was known as 'Jasminum Indicum.' This exotic species was much prized by wealthy collectors, who cultivated it in their gardens and commissioned botanical painters such as Jacopo Ligozzi (1547–1627) to portray its beautiful, fragrant flowers.

One of the most important plants brought back by Drake and described by Clusius was the cacao (illustrated on page 29), of which 'great use was made in America.' The botanist notes that the fruit was similar in shape to an almond and was utilized by the natives to prepare a drink that had a 'bitter and unpleasant' taste.

Aliquot notæ in Garciæ is illustrated with fifteen woodcuts; some take up an entire page while others are smaller and have been inserted into the text.

15. Prospero Alpini (1553–1617)

Prosperi Alpini de plantis Aegypti liber. In quo non pauci, qui circa herbarum materiam irrepserunt, errores, deprehenduntur, quorum causa hactenus multa medicamenta ad usum medicinæ admodum expetenda, plerisque medicorum, non sine artis iactura, occulta, atque obsoleta iacuerunt. A Ioannem Maurocenum Antonij Filium Patricium Venetum Clarissimum. Accessit etiam liber de Balsamo aliàs editus. [*Printer's device*]. Venetiis, M. D. XCII. [*rule*] Apud Franciscum de Franciscis Senensem.

[*Bound with*]: Prosperi Alpini de balsamo, dialogus. In quo verissima balsami plantæ, opobalsami, carpobalsami, & xilobalsami cognitio, plerisque antiquorum atque iuniorum medicorum occulta, nunc elucescit. Ad illustrissimos et sapientissimos Patauinæ Academiæ Curatores. [*Printer's device*]. Venetiis, M. D. XCII. [*rule*] Apud Franciscum de Franciscis Senensem.

4° 21.4 × 14.9 cm. a^4 A–X^4 a–b^4 fols. [*4*] 1–40 37–57 *58–60* 61–80 [*8*] (72 as 92) [192 pp.].

PLATES: Forty-nine plants in text, head- and tail-pieces, all woodcuts.

BINDING: Old speckled calf with gilt tooled borders; rebacked.

PROVENANCE: 'Feb. 7th 1748/9. Richard Richardson Esqr. of North Bierley in Love gave this Book to Thomas Knowlton,' inscribed on front free endpaper.

REFERENCES: Arber, pp. 100–102; Blunt & Raphael, p. 141, 144; Cleveland 136; *DBI*, II.529–530; Mortimer, *Italian* II.17; Hunt 164; Nissen 20; Osler 1,799; Pritzel 111.

PROSPERO ALPINI, the son of a well-known physician, was born in Marostica near the northern Italian city of Vicenza in 1553 and died in 1617 in Padua, where he was buried in the Basilica of Saint Antonio. In 1574 Alpini began his studies at the illustrious University of Padua, where he treated patients during a terrible plague that ravaged the city. His professors included men of science of great renown, including the anatomist Girolamo Fabrici d'Acquapendente and the physician Gerolamo Mercuriale. The teacher who had the greatest influence on him, however, was the German Melchior Wieland, whose name was Italianized to Melchiorre Guilandino. Guilandino served as the second *prefetto* (director) of the Botanical Garden of Padua and travelled to every part of the Mediterranean in search of plants for the garden's collection. He initiated Alpini into the practice of botanizing and influenced him with his enthusiasm for scientific exploration in faraway lands.

In 1578 Alpini completed his studies in philosophy and medicine and began to practice, but his restless, inquiring mind was not satisfied with this circumscribed activity. He therefore accepted with alacrity the offer of the patrician Giorgio Emo to join his entourage as his personal physician, and accompanied him to Cairo where Emo had been appointed Consul by the Republic of Venice. This sojourn provided Alpini with the opportunity to study *dal vivo* the flora of a country that was entirely unknown to him.

The Venetians' journey by ship was long and full of incident; they were forced to make various detours along the way, one of the most important for Alpini being a stop in Crete, then considered to be a veritable botanical paradise. On this island he was able to gather many plants and make various original observations. Alpini stayed in Egypt for three years, until 1584, and this experience proved crucial to his formation as a scientist. His research on new species was aimed not only at discovering their therapeutic properties, but he also studied their morphology and habitat, thus making a significant contribution to the modern science of botany.

Prospero Alpini published four works during his lifetime, which constituted the tangible fruit of his travels and botanical studies. The first was *De medicina Aegyptiorum*, which was written in the form of a dialogue taking place between himself and Guilandino in the Botanical Garden of Padua. This was published in Venice in 1591, and was immediately followed by a short work on the balsam plant, *De balsamo dialogus* (1591). One year later his most influential work, *De plantis Aegypti*, appeared. He spent the remainder of his life working on *De plantis exoticis*, which was published posthumously in 1629 by his son. Significantly, in these works he did not limit himself to a discussion of the botanical and pharmacological aspects of the species under consideration, but included reflections of an ethnological and archaeological nature that testify to his acute powers of observation and wide-ranging curiosity.

PROSPERO ALPINI, *De plantis Aegypti liber*, 1592. 'Papyrus Burdi' (*Cyperus papyrus*), page 43. the first image of the plant to be published in Europe

When Alpini returned to Italy he resumed his practice and for ten years, from 1584 to 1594, led the busy life of a highly successful physician. In the town of Bassano near Vicenza one of his patients was the well-known painter Leandro Da Ponte, surnamed Bassano, who executed a fine portrait of the physician, a copy of which may be seen at the Botanical Garden of Padua, while the original is conserved in the Staatsgalerie of Stuttgart. He then transferred his practice to Genoa, where he served as a physician at the court of Giovanni Andrea Doria. In 1595 Alpini was appointed professor of medicinal plants at the University of Padua, and in 1603 he became director of the city's botanical garden, where he introduced and cultivated many Egyptian plants.

In *De plantis Aegypti* the physician describes fifty-six indigenous species, including trees and shrubs that were cultivated and used for medicinal purposes in the regions that he had visited. The work is preceded by an introduction in which the author and his professor Guilandino discuss the reasons why it was important to acclimatize these African plants in Europe. Among the exotic species described are the banana (*Musa* sp.), the papyrus (*Cyperus papyrus*), the tamarind (*Tamarindus indica*), the date palm (*Phoenix dactylifera*), the Arabian or Sambac jasmine (*Jasminum sambac*), and the baobab tree (*Adansonia digitata*); the artist also depicts the baobab's edible fruit, cut open to show its pulp and seeds. The author devotes ample space to the coffee plant (*Coffea arabica*), which he called bun or buna, praising the therapeutic properties of the drink that could be made from its roasted seeds. Alpini was the first botanist to describe this species, and probably never suspected the vogue that it would soon enjoy all over Europe.

The forty-nine woodcuts that illustrate the text were in all certainty executed from life, and are models of clarity, accuracy, and elegance. Among these the illustration of the papyrus—which, Alpini noted, grows along the banks of the Nile and 'presents one or two very erect stalks'—is especially captivating. It also represents the first image of the plant to be published in Europe.

A new edition of *De plantis Aegypti*—considerably revised and augmented by the physician Johannes Vesling who, like Alpini, spent a period botanizing in Egypt and eventually became a professor of botany at Padua—was published in 1638.

The Oak Spring Garden Library's exemplar of *De plantis Aegypti* has been bound together with a copy of Alpini's dialogue on the balsam, an aromatic plant that fascinated naturalists since Antiquity. The name 'balsam' was in fact assigned to a variety of plants containing an aromatic resin that was believed by the ancients to have marvelous healing powers. Alpini had already described this plant—which he called opobalsamum—in *De medicina Aegyptiorum*, and had furthermore brought back with him from Egypt a specimen of what was perhaps *Commiphora opobalsamum,* which he subsequently planted. This treatise gave rise to a vast amount of literature written either in support of or refuting

Alpini's affirmations regarding the identity and nature of the plant. The debate was only concluded in 1764 when Wilhelmus Le Moine, a student of Linnaeus, published a dissertation confirming the theories propounded by Alpini (see Immaeus, *Opobalsamum declaratum* ... [Upsala 1764]).

To preface his description Alpini presents once again a learned debate, held this time between himself, the Egyptian physician Abdella, and the Jewish physician Abdachim in a garden in Cairo. The treatise is embellished with a fine woodcut of the plant, and in his text the author has included some pertinent observations made by the French explorer Pierre Belon during his travels across the Mediterranean and published in his work *Les observations de plusieurs singularitez et choses memorables, trouvées en Grece, Asie, Iudée, Egypte, Arabie, et autres pays estranges* (Paris, 1553).

16. Nicolas Monardes (1493–1588)

[*Part I: within border of typographical ornaments*]: Joyfull Newes Out of the New-found VVorlde. Wherein are declared, the rare and singuler vertues of diuers Herbs, Trees, Plantes, Oyles & Stones, with their applications, as well to the use of Phisicke, as of Chirurgery: which being well applied, bring such present remedie for all diseases, as may seeme altogether incredible: notwithstanding by practice found out to be true. Also the portrature of the said Hearbs, verie aptly described: Englished by John Frampton Marchant. Newly corrected as by conference with the olde copies may appeare. Wherunto are added three other bookes treating of the Bezaar Stone, the herb Escuerconera, the properties of Iron and Steele in Medicine, and the benefit of Snow. London, Printed by E. Allde, by the assigne of Bonham Norton. 1596.

[*Part II: woodcut headpiece*]: The Second part of this Booke is of the things that are brought from our Occidentall Indias, which serve for the use of Medicine, wherein is treated of the Tabaco, and of the Sassafras, and of the Carlo Sancto, and of many other hearbes & plants, seedes and licoures, that newly are brought from those partes, of great vertues and mervellous effectes. VVritten by Doctor Monardus Phisition of Sevill. [*woodcut tailpiece*].

[*Part III: woodcut headpiece*]: The Third parte of the Medicinall Historie, which treateth of the thinges that are brought from our Occidentall Indias, serving for the use of Medicine. Wherein there is mention made of many things Medicinall, that have great secretes and vertues. ¶ Nowe newely set foorth by the sayde Doctor Monardus, after that he had made the first and second partes [*typographical ornaments*].

[*Bound with, on leaf 2E3 recto, within border of rules and typographical ornaments incorporating the initials E and A*]: A Booke which treateth of two medicines most excellent against all venome, which are the Bezaar stone, & the hearbe Escuerconera. Wherein are declared their marvellous effectes & great vertues, with the manner howe to cure the sayd venoms & the order which is to be used for to be preserved from them. Where shall be seene greate secretes in medicine and many experiences. Newly compyled by Doctor Monardes of Sevill. 1574. Translated out of Spanish into English, by John Frampton. 1580.

[*Bound with, on leaf 2M3 recto, with woodcut headpiece*]: The Dialogue of Yron, which treateth of the greatnesse thereof, and how it is the most excellent metall of all others, and the thing most necessarie for service of man: and of the greate medicinall vertues which it hath. An Eccho for the Doctor Monardus[,] Phisition of Seville. [*woodcut device*] In Sevill in the house of Alonso Escrinano.

[*Bound with, on leaf 2S4 recto, with woodcut headpiece*]: The Boke which treateth of the Snow, and of the properties & vertues thereof: And of the maner that should be used to make the drinke cold therwith, & of the other waies wherewith drinke is to he made cold: Whereof is shewed partly, in the latter part of the second Dialogue of Iron. With other curiosities which will give contentment by other ancient thinges worthy to bee knowen, which in this treatise shall be declared. Written by Doctor Monardes Phisition of Sevill. 1574.

4° 18.3 × 13.4 cm. π^4 (-π2) A–2Y⁴ fols. [*3*] 1–109 *110–111* 112–163 173 172 173 166 175–187 (66 as 69, 98 as 103, 106 as 115, 132 as 133, 157 as 153, 159 as 157) [366 pp.].

PLATES: Twelve woodcuts in text.

BINDING: Full red morocco with gilt tooling on covers and spine.

PROVENANCE: Bookplate of Paul Mellon.

REFERENCES: Anderson, pp. 206–209; Blunt & Raphael, pp. 148–149; Bush-Brown, pp. 124–142; Cleveland 140; Hunt 173; Pritzel 6,366; Sabin 49,946.

NICOLAS MONARDES was born in Seville in 1493 and, after studying medicine at the university in Alcalá de Henares, set up practice in Seville where he spent the remainder of his life. Although he never visited the New World, he dedicated his life to studying the flora of these recently discovered lands, in particular their medicinal properties.

In Spain there was widespread interest in the flora and pharmacopia of the Americas, especially during the reign of Philip II (1556–1598); the king went so far as to send his personal physician Francisco Hernández (1514–1587) to Mexico in 1569 to study the remedies that were prepared by the natives from indigenous plants and animals. Seven years later, having exhausted his funds, Hernández returned to Spain, bringing back with him about twenty manuscript volumes filled with notes and drawings. These were deposited in the library of the Escorial Palace, but most unfortunately the originals were eventually lost in the catastrophic fire of 1671. Meanwhile, however, the king asked his new physician, the Italian Nardo Antonio Recchi, to edit the material that his predecessor had gathered, and this led to the publication, several years later in Mexico, of *Quatro libros de la naturaleza y virtudes de las plantas y animales que estan recevidos en el uso de la medicina en la Nueva España* (1615). Recchi's compilation, after much elaboration, would eventually be published by the Accademia dei Lincei in 1651 as *Rerum Medicarum Novae Hispaniae Thesaurus*, also known as *Tesoro Messicano* (see No. 11).

Even though Monardes never ventured beyond Seville and thus could not conduct his botanical studies *in situ*, he was the author of an important work on the plant life of the Americas entitled *Dos libros, el uno que trata de todas las cosas que traen de nuestras Indias Occidentales … al uso de la Medicina …* This was published in his native city by Hernando Diaz; the first part appeared in 1565 and was republished in 1569, the second part appeared in 1571, and the third was published together with the first two parts in a complete edition in 1574 which was reprinted in 1580.

Dos libros … was quite successful because it was one of the few texts then available that focused exclusively on the flora of the New World, and it was soon translated into several languages. Clusius prepared an abridgement of the first two volumes in a Latin

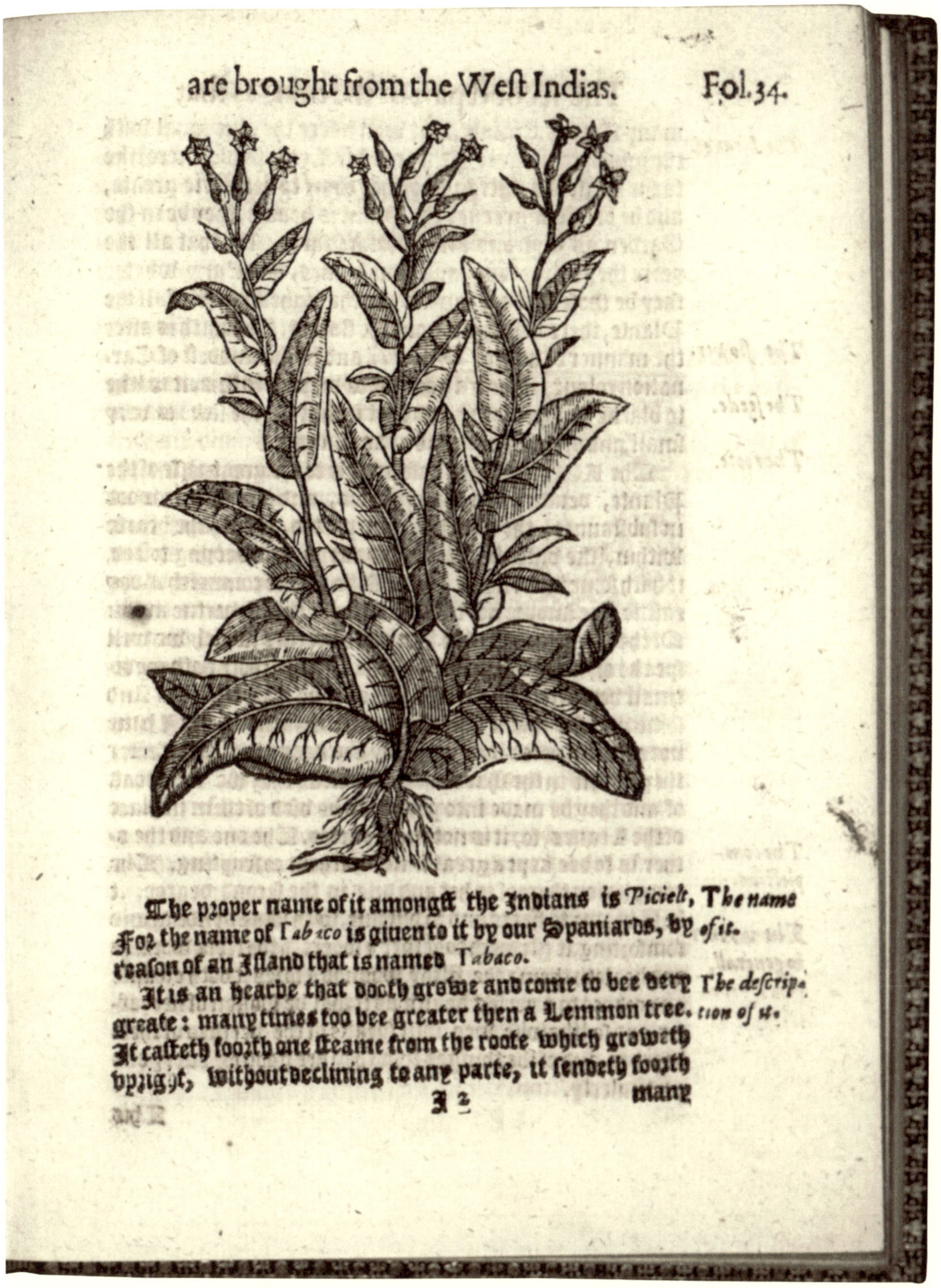

are brought from the West Indias. Fol.34.

The proper name of it amongst the Indians is *Picielt*, For the name of *Tabaco* is giuen to it by our Spaniardes, by reason of an Island that is named *Tabaco*. *The name of it.*

It is an hearbe that dooth growe and come to bee very greate: many times too bee greater then a Lemmon tree. It casteth foorth one steame from the roote which groweth vpright, without declining to any parte, it sendeth foorth many *The description of it.*

I 2

NICOLAS MONARDES, *Joyfull Newes Out of the New-Found Worlde*, 1596. Tobacco (*Nicotiana rustica*), folio 34

taking,for that it hath neither ſmell nor ſauour, it maketh his worke without paines. It is hot in my opinion, in ẏ firſt degree. I know not what maner of thing the tree is, wherout they take it, for hee which brought it, knoweth not ſo much himſelfe.

Of the Armadilio.

This beaſtes portraiture I tooke out of an other naturally made, which was in the Counting houſe of *Gonſalo de Molina*, a Gentleman of this Citie, in the which there is greate quantitie of Bookes of diuers Authours, and the faſhion & forme of many kindes of beaſtes, and birdes, and other curious things, brought from the Oriental Indias, as alſo from the Occidentall, and from other parts of the worlde. And great varietie of coynes and ſtones of antiquitie, and differences of armes, which with great curioſitie, & with a noble minde he hath cauſed to be brought thither.

They

NICOLAS MONARDES, *Joyfull Newes Out of the New-Found Worlde*, 1596. A member of the family Armadillo (*Dasypodidae*), folio 73

edition that was published by Plantin in 1574. In 1575 and 1582 the work appeared in Italian. Three English editions were published in London, the first in 1577 by William Norton including an issue with a alternate title reading: *The Three Bookes written in the Spanishe tonge, by the famous Phisition D. Monardes ....* The second expanded edition was published by Norton in 1580 and the third—a virtual reprint of the 1580 edition—was issued in 1596 by Edward Allde and Bonham Norton. The Oak Spring Garden Library possesses a copy of the last.

Joyfull Newes is composed of three parts. To these have been bound *A Booke which treateth of two medicines most excellent against all venome, which are the Bezaar stone, & the hearbe Escuerconera*, in which the author discusses the bezoar stone (the calculi of various animals discovered in Peru, which were believed to constitute miraculous remedies against poisoning), *The Dialogue of Yron*, and *The Boke which treateth of the Snow*. Monardes' work was translated by John Frampton, a merchant who had regular contacts with Spain, the country that was the first to receive most of the new plants and remedies arriving from the Americas. The English edition lacks the portrait of the author at the age of fifty-seven, which appears on the title-page of the 1569 Spanish edition. *Joyfull Newes* is illustrated with ten small woodcuts—nine depicting plants and one an armadillo, whose bones were believed to be an effective remedy against headache.

One of the most important chapters is dedicated to the tobacco plant and 'his great vertues' (pp. 33–45); according to Monardes it could be used to eliminate worms and to relieve pain, in particular headache, toothache, and stomach ache. The text is accompanied by a beautiful engraving showing the plant with its flowers and large, turgid leaves. This was one of the earliest illustrations of the tobacco plant to be published in Europe; it first appeared in 1570 in Pierre Pena and Matthias de L'Obel's *Stirpium adversaria nova* and then in de L'Obel's *Plantarum seu stirpium historia* (1576) and *Kruydtbœck* (1581). Frampton used it in place of the more rudimentary woodcut that appeared in the original Spanish edition of volume two published in 1571. Monardes named the plant *Nicotiana* in honor of his friend Jean Nicot who, as the French Ambassador to Lisbon, was the first to bring the plant to France (see No. 24).

Monardes' work includes chapters on 'the notable Hearbe of Sunne' or sunflower; the 'Gujacan wood' from the island of San Domingo; the 'long peper,' a variety with 'a sharper taste, than the pepper which is brought from the Oriental Indias,' the latter is accompanied by a modest illustration. He goes on to describe the uses made by the natives of the coca plant; the 'Batatas, a common fruite of those Countries'; and the Peruvian 'Granadilla' or passion-flower with its extravagant blossom, in which one may find 'the sign of the Passion of our Lord.'

17. Garcia da Orta (*c.* 1502–1568)

Dell'Historia dei Semplici Aromati, et altre cose che vengono portate dall'Indie Orientali pertinenti all'uso della Medicina. Di don Garzia dall'Horto Medico Portughese, con alcune brevi Annotationi di Carlo Clusio. Parte Prima, divisa in quattro Libri. Et due altri Libri parimente di quelle cose che si portano dall'Indie Occidentali; Con un Trattato della Neue & del bever fresco. Di Nicolo Monardes Medico di Siviglia. Hora tradotti dalle loro lingue nella nostra Italiana da Messer Annibale Briganti, Marrucino da Cività di Chieti, Dottore & Medico Eccellentissimo. Con Privilegio. [*Printer's device*] In Venetia, M. DC. XVI.

[*Colophon*]: Stampato in Venetia nella Stamperia di Giovanni Salis, M. DC. XVI.

8° 15.5 × 9.9 cm. a–b⁸ A–2K⁸ (-2K8) *i–xxxii* 1–154 *155–179* *180*–321 *322*–463 *464*–525 *526* (336 as 326) [558 pp.].

PLATES: 25 woodcuts in text.

BINDING: Contemporary vellum; 'Dall'hOrta D. Garzia' inscribed on the spine.

REFERENCES: Hunt 142, (1582 edition); Nissen 949; Pritzel 4,316.

GARCIA (ABRAHAM) DA ORTA (or, variously, Herto, Huerta, ab Orto, Dall'Horto) was born in Castelo de Vide, Portugal, at the turn of the sixteenth century. He was the son of a Jewish merchant who had been forced to leave Spain in 1492 as a result of religious persecution under the Catholic monarchs Ferdinand and Isabella. In 1497 the family, together with thousands of others, converted to Christianity on the order of King Manuel I of Portugal (1495–1521). After studying at the university in Alcalá de Henares, where Nicolas Monardes had also been a student, and then at the university in Salamanca, in 1530 da Orta became a professor of logic at the university in Coimbra, where he sought to pass on to his students his profound interest in the natural sciences, as well as an understanding of the importance of direct observation and experimentation. In 1533 he was elected a member of the Senate of the university.

The Old Continent, however, was too narrow and restrictive for Garcia da Orta; kindled by the accounts of travellers, and perhaps also fearful of a new wave of persecution, in 1534 da Orta left Portugal and travelled east, finally settling in Goa, the capital of the Portuguese Indies. There he served as personal physician to Martim Afonso de Sousa, a prominent naval officer who was appointed Viceroy of Portuguese India by the king of Portugal after de Sousa's return from Brazil, where he had colonized the coasts and established large sugar cane plantations.

Da Orta would stay in Goa for the remainder of his life, leading the peaceful existence of a physician and merchant in spices, medicines, and precious stones. He studied the texts of the Hindu and Arab authorities, which offered a *materia medica* completely unknown in the West, and travelled throughout the Portuguese possessions, as far as Ceylon. In Goa he established a garden for the cultivation of medicinal plants, and participated in the first autopsy conducted in the territory, which took place in 1543 during a cholera epidemic. Da Orta died in 1568 and was buried in his adopted land; in 1580, however, he was

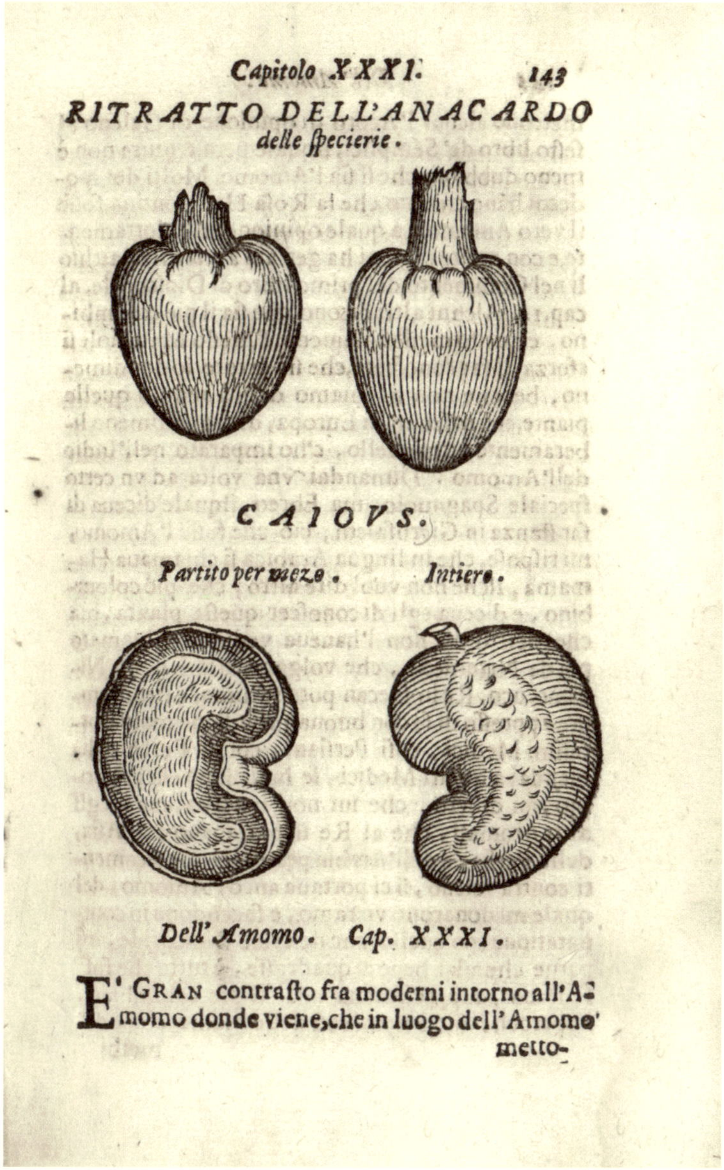

Capitolo XXXI. 143

RITRATTO DELL'ANACARDO
delle ſpecierie.

CAIOVS.

Partito per mezo. Intiero.

Dell'Amomo. Cap. XXXI.

E' Gran contraſto fra moderni intorno all'A-
momo donde viene, che in luogo dell'Amomo
metto-

GARCIA DE ORTA,
Dell'Historia dei Semplici Aromati, 1616. Cashew (*Anacardium occidentale*), page 143

GARCIA DE ORTA, *Dell'Historia dei Semplici Aromati*, 1616. Sassafras (possibly *Sassafras albidum*), page 407

Capitolo 11. 407

& il modo, come lo haueuano da vsare, & che cosi fecero, e si risanarono di molti mali. Il che certo apporta merauiglia, che vn solo rimedio facesse cosi

Cc 4 mera-

condemned posthumously for 'the crime of Judaism' by a tribunal of the Inquisition, which ordered that his body be exhumed and burned, and his ashes scattered in the sea.

Garcia da Orta published various works in Goa, the third and best known of which was *Coloquios dos simples, e drogas he cousas medicinais da India....* Originally written in Latin, and published by the printer Johannes de Endem in Portuguese in 1563, this unillustrated text describes various plants, minerals, and animal parts (presented in alphabetical order from 'ambra' to 'zerumbet') that were used by the natives to prepare medicinal remedies. Miscellaneous observations regarding their applications in cooking and even in the game of chess are included. However, to make his work more appealing, da Orta adopted a literary style then very much in vogue, structuring his text in the form of a series of spirited dialogues conducted between himself and various personages, beginning with the imaginary figure of Dr. Ruano, his alter ego and the embodiment of traditional orthodoxy based on the classical and Arab authorities.

Coloquios dos simples was extremely well received in Europe, but its contents were subsequently abridged and distorted by its translators, beginning with Carolus Clusius. The Flemish botanist had in fact carefully studied da Orta's herbal during the period he spent botanizing in Spain and Portugal between 1564 and 1565 (see No. 9), and he prepared a Latin edition loosely based on the original entitled *Aromatum et simplicium aliquot medicamentorum apud Indos nascentium historia.* This work, embellished with sixteen woodcuts by the artist Pieter van der Borcht and the engraver Arnold Nicolai, was published by Plantin in 1563. A translation into Castilian by Juan Fragoso was printed in Madrid in 1572, and Clusius published an expanded version of *Aromatum et simplicium ...* in 1574 with a third edition issued in 1579. An Italian edition appeared in Venice in 1576, followed by others in 1582, 1589, and 1616. The Oak Spring Garden Library has a fine copy of the 1616 Italian edition, printed and bound together with Monardes' *Joyfull Newes* (see No. 16) by the publisher Giovanni Salis.

Little of da Orta's original work survived in these translated versions; for example, the plants are no longer presented in alphabetical order and the author's witty dialogues have been entirely omitted. This is somewhat compensated for by the detailed annotations, set off in italic type, that Clusius added to the end of each chapter. The most interesting chapters describe what were at the time quite exotic species such as the cinnamon (accompanied by the illustration of a branch and leaf of the plant), sandalwood, cashew, tamarind, mango, rhubarb, and pepper.

18. Jean Robin *the elder* (1550–1629)

[*Within woodcut border*]: L'Histoire des Plantes Aromatiques qui croissent en l'Inde, tant Occidentale qu'Orientale. Reveu & augmente de plusieurs Plantes venues des Indes lesquelles ont esté prises et cultivées aux Iardin de Mr Robin Arboriste du Roy. (*⁎*) A Paris. [*within border cartouche*]: Chez Guillaume Macé, au mont S. Hilaire, à l'enseigne de la Pyramide. 1619.

12° 11.5 × 7.3 cm. π^4 2T–2X^8 *649* 650–704 [56 pp.].

PLATES: 51 woodcuts, 50 of plants, seeds, and roots, and one of a mummy (*Asphaltum Mumie*).

BINDING: Modern salmon paper boards.

PROVENANCE: Bookplate of Michael Zinman.

[*Within woodcut border*]: Histoire des Plantes, nouvellement trouvées en l'Isle Virgine, & autres lieux, lesquelles ont esté prises & cultivées au Iardin de Monsieur Robin Arboriste du Roy. Non encore veuës n'y Imprimées par cy devant. Dedie à Monsieur Morand. A Paris. [*within border cartouche*]: Chez Guillaume Macé au mont S. Hilaire, à l'enseigne de la Pyramide. 1620.

12° 11.7 × 7.3 cm. $ã^8$ *1–2* 3–16 [16 pp.].

PLATES: 14 woodcuts of plants in text.

BINDING: Dark blue morocco by Delanoé Jeune; all edges gilt; '2998' on small circular paper label on front cover.

REFERENCES: *An Oak Spring Flora*, p. 8, Vallet entry; Pritzel 7,672.

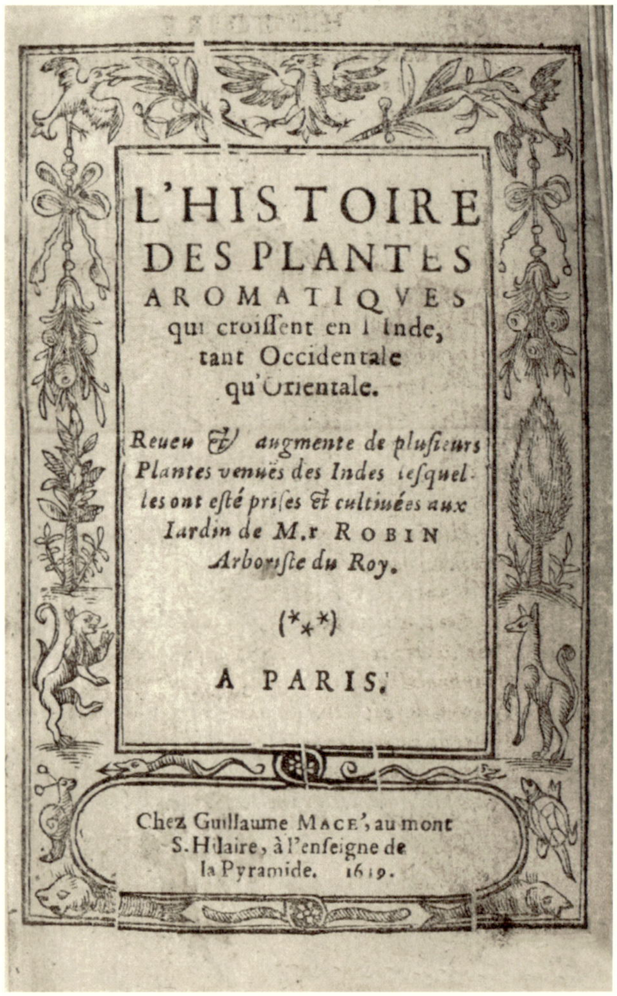
L'HISTOIRE
DES PLANTES
AROMATIQVES
qui croissent en l'Inde,
tant Occidentale
qu'Orientale.
Reueu & augmente de plusieurs Plantes venuës des Indes lesquelles ont esté prises & cultiuées aux Iardin de M.r ROBIN *Arboriste du Roy.*
(*⁎*)
A PARIS.
Chez Guillaume MACE', au mont S. Hilaire, à l'enseigne de la Pyramide. 1619.

JEAN ROBIN, *L'Histoire des Plantes*, 1619. Title-page

Jean robin was appointed Royal Gardener (*arboriste*) in 1590 by Henri IV, the first Bourbon king of France, who ruled from 1589 to 1610. The position of royal gardener required a profound knowledge not only of horticulture but also of botany and Jean Robin was fully prepared in this regard, having served as curator of the garden of medicinal plants that the Faculty of Médicine had established at the easternmost end of the Ile de la Cité. This garden was, however, rather modest in size and the royal subsidy barely covered the expenses of its maintenance, as Guy de la Brosse wrote sixteen years later when the decision was finally made to proceed with the construction of a new Jardin des Plantes Medicinales (see No. 43).

The royal gardens were officially public institutions but were little frequented by the physicians of the city and Jean Robin and his son Vespasien managed them as if they were private estates, holding themselves answerable to no one but the king himself. During the reign of Henri IV the Tuilleries Gardens and later the Garden of Luxembourg became magnificent showcases where the most singular plants and most beautiful flowers were cultivated, not only to enhance the prestige of the king but also for the pleasure of Marie de' Medici, whom he had married in 1600 and who was nostalgic for the luxuriant gardens of her native Tuscany.

In 1601 Jean Robin published a small volume illustrating 1,300 plants from the garden, *Catalogus stirpium tam indigenarum quam exoticarum quæ Lutetiæ coluntur* (see *An Oak Spring Flora*, p. 40). It appears, however, that he was extremely jealous of his collection and reluctant to share his knowledge or give away specimens to other colleagues. This idiosyncrasy of his became so well-known that Guy Patin (1601–1672), a celebrated professor of medicine at the Collège de France, ironically referred to Robin as 'Heunucus Hesperidum' (the Eunuch of the Hesperides) in his *Lettres*, which are full of witty commentary on the high society of his day. A portrait of Jean Robin at the age of fifty-seven may be found in *Le Jardin Du Roy Très Chréstien Henri IV*, a beautiful florilegium dating from the early seventeenth century that was compiled by the eclectic artist and 'brodeur du Roi' Pierre Vallet and dedicated to the queen (see *An Oak Spring Flora*, No. 83, pp. 38–41).

Jean Robin had a son—Jean Robin the Younger, also known as Vespasien (1579–1662)—who became a professional gardener like his father and in addition embarked on a series of expeditions to gather plants for the royal gardens, traversing England, Germany, and Spain, and then venturing as far as the Barbary Coast in his search for new plants. In recognition of his expertise he was appointed 'sous démonstrateur' of the Botanical Garden. In 1603 the Robins published a short work describing thirteen different exotic plants brought back from foreign countries by Jean the younger, *Exoticæ quædam Plantæ a Johanne Robino Juniore ex Guinea et Hispaniæ delatæ*.

Two short treatises published by Jean Robin the Elder in 1619 and 1620 may be found in the Oak Spring Garden Library. Each was printed in duodecimo format, and opens with a pleasing title-page embellished with a woodcut border. These works had no scientific pretensions, nor are their contents particularly original; indeed, they constitute little more than modest gardening catalogues. The first, *L'Histoire des Plantes Aromatiques qui croissent en l'Inde, tant Occidentale qu'Orientale: Revue & augmente de plusieurs Plantes venue des Indes lesquelles ont été prises et cultivées aux jardin de Mr. Robin arboriste du Roy*, consists of woodcut illustrations of fifty exotic plants accompanied by brief botanical descriptions. Among the species to be found are the 'dragon's blood palm' (*Calamus draco*, which derived its common name from the red resin contained in its fruit—whose illustration is copied from Clusius, see No. 9), the coconut palm, the cinnamon tree, and the tobacco plant, which was

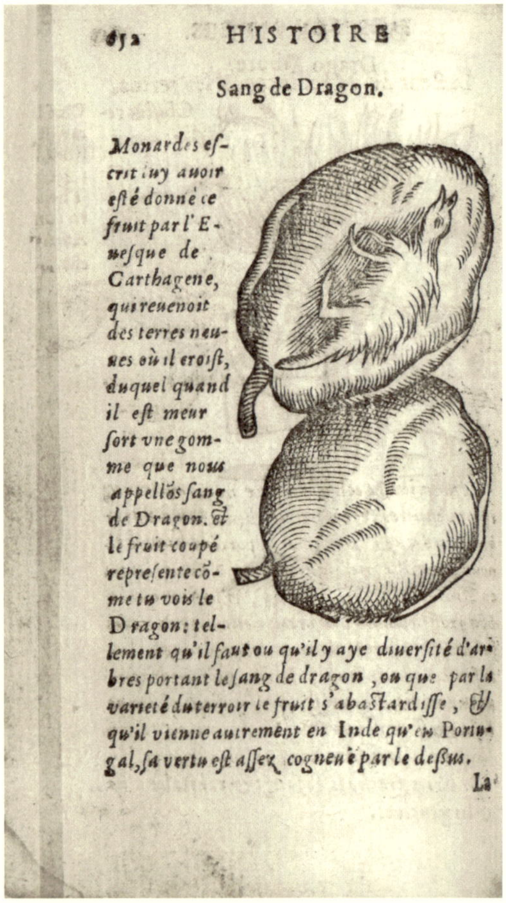
652 HISTOIRE

Sang de Dragon.

Monardes escrit luy auoir esté donné ce fruit par l'Euesque de Carthagene, qui reuenoit des terres neuues où il croist, duquel quand il est meur sort vne gomme que nous appellõs sang de Dragon. & le fruit coupé represente cõme tu vois le Dragon: tellement qu'il faut ou qu'il y aye diuersité d'arbres portant le sang de dragon, ou que par la varieté du terroir le fruit s'abastardisse, & qu'il vienne autrement en Inde qu'en Portugal, sa vertu est assez cogneuë par le dessus.

La

DES PLANTES. 15

Canna Indica flore rubro.

Cette plante a esté apportee des Indes, elle vient haute de quatre pieds & plus, les feuilles sont fort grandes & larges, elle porte ses fleurs en forme d'espic lesquelles sont extremement viues en couleur, elle fleurit en ce pays en Septembre & se meurt l'yuer.

Satirium

JEAN ROBIN, *L'Histoire des Plantes*, 1619. Fruit of the 'Sang de Dragon' (*Calamus draco*), page 652

Histoire des Plantes, 1620. 'Canna Indica flore rubro,' Indian shot (*Canna indica*), page 15

referred to variously in French as 'petum,' 'nicotiace' or 'herbe de la reine' (see Nos. 16 and 24). The page numbering begins with 650, which demonstrates that the work, although furnished with its own title-page, actually constituted an extract from a longer work, specifically Geofroy Linocier's *Histoire des Plantes* (1620), an herbal whose text was based on the works of Fuchs, Mattioli, and various Flemish botanists, and whose illustrations were copied from the same sources.

The second treatise, which is entitled *Histoire des Plantes, nouvellement trouvées en l'Isle Virgine …*, contains instead a most interesting 'Avertissement au lecteur' in which the author declares himself most pleased to offer: 'ce que la nature aurait poussé de plus beau, sur les tapisseries de la terre … Attendant que je te pourray contenter de quelques autre grande particulliereté selon ton merite …', from which it may be concluded that he was finally making available to the public some of the rare plants in the royal gardens. The treatise is accompanied by fourteen small woodcuts of exotic plants—most of them, it must be said, by now quite common all over Europe—such as *Fritillaria meleagris*, *Fritillaria imperalis*, *Canna indica*, the passion flower, and the Indian narcissus.

In 1624 the Robins published *Enchiridion isagogicum ad facilem notitiam stirpium, tam indigenarum, quàm exoticarum. Hae coluntur in horto D.D. Ioannis & Vespasiani Robin, botanicorum regiorum*, a completely updated edition of the 1601 catalogue, which lists no less than 1,800 plants. It is known that in 1635 Vespasien introduced a plant that can now be found growing all over Europe, *Robinia pseudoacacia*.

The king's gardener was by now so well known that Paul Contant, the pharmacist of Poitiers, dedicated a brief poem to him at the beginning of his work *Le Jardin et Cabinet Poetique* (see No. 31). Jacques Cornut (1606–1651) published a treatise on the flora of North America—*Canadensium Plantarum Historia* (1635)—even though he had never visited the New World, and it is probable that he was able to base some of his botanical descriptions on specimens to be found in the gardens of the Robin family and that of another celebrated family of gardeners, René and Pierre Morin.

19. Georg Wolfgang Wedel (1645–1721)

[*Title-page in red & black*]: Georgii Wolffgangi Wedelii, Med. Doctoris, Professoris Publici, et Medici Ducalis Saxonici, Opiologia ad mentem Academiæ Naturæ Curiosorum. [*Engraved vignette*]. Jenæ, Sumptibus Johannis Bielkii Bibliop. Typis Viduæ Samuelis Krebsii. [*rule of dashes 9.6 cm.*] Anno M.DC.LXXXII.

[*Bound with*]: Michaelis Ettmülleri, Phil. & Med. D. & Professoris Publici in Academiâ Lipsiensi De Virtute Opii Diaphoretica Dissertatio. [*full measure rule of dashes*] Lipsiæ et Jenæ, Sumptibus Joh. Bielkii. Literis Krebsianis.

4° 19.5 × 15.5 cm. a–b⁴ A–Z⁴ 2A⁴ A–F⁴ *i–xvi 1* 2–170 (142 as 42) *i–xxii 1–3* 4–5 *6*–13 *14*–48 [256 pp.].

PLATES: Engraved and etched vignette on title-page, signed bottom center: 'I. Linck f.'

BINDING: Half vellum and marbled boards.

REFERENCES: Cleveland A2; Pritzel 10,054.

GEORG WOLFGANG WEDEL, the son of a Protestant minister, Johann Georg Wedel, was born in Golssen, Germany, in 1645. He attended the Kursächsische School in Pforte and then, having been awarded a scholarship from the Elector of Saxony, was able to enroll as a student at the University of Jena in 1662. He earned his degree in medicine in 1667 after following courses in medicine and philosophy, and served as district physician of the town of Gotha from 1667 to 1672. In 1673 he was appointed to a professorship at the University of Jena where, during the course of a long academic career, he served several times as dean and rector. His teachings in anatomy, surgery, and botany were followed by a large number of students, and from 1719 until his death he also taught chemistry, medical theory, and applied medicine.

Wedel's fame spread well beyond the university; he received many awards and honors and became affiliated with such prestigious scientific institutions as the Berlin Academy, the 'Accademia Leopoldina,' and the 'Academia Naturae Curiosorum' of Jena. He was appointed personal physician to the Duke of Weimar in 1679 and to the Prince of Saxony in 1685. He was in addition a prolific correspondent and many of the letters that he wrote to physicians and scientists all over Europe have been conserved.

Wedel was an extremely learned, versatile, and eclectic physician, as is demonstrated by the endless stream of works that issued from his pen; his writings on certain subjects proved to be quite influential. He wrote many short treatises on plants, in which he discussed not only their medical properties but also their symbolic associations. In addition to these works on specialized topics he produced a scrupulous translation of the Bible from Greek into German.

Wedel was also drawn to the study of astrology and alchemy, but soon abandoned these archaic pursuits to take up chemistry instead. Intrigued by the possibilities that the emerging science offered for the discovery of new remedies, he plunged into the study of the earliest theoretical concepts and began conducting his own experiments. He became a convert to the medical doctrine of iatrochemistry conceived by Theophrastus Bombastus von Hohenheim, better known as Paracelsus (1493–1541). The ideas of the Swiss physician and occultist, which were disseminated by the physicians Georg Bauer (who Italianized his name to Agricola) (1494–1555), Andreas Libavius (d. 1616), and Ole Borch of Denmark (1626–1690), won many adherents during the course of the seventeenth century and eventually evolved into the modern science of chemistry. As many scholars including the English historian Frances Yates (*Giordano Bruno and the Hermetic Tradition*, 1964) have noted, in this period hermetic and Paracelsian theories regarding natural phenomenon held widespread credence among men of science, and indeed laid the groundwork for an entirely new conception of science.

According to the theory of iatrochemistry, life, health, and disease were the result of chemical balances and imbalances, and one should therefore be able to trace many physiological and pathological phenomena back to the actions of chemical processes, including fermentation and effervescence. Unlike their predecessors the alchemists, iatrochemists were not interested in turning base metals into gold, but rather in the production of effective pharmacological remedies. These concepts emerge clearly in Wedel's treatise on opium, a substance that offered the ideal occasion for experimentation, since various derivatives of its active ingredients could be obtained by chemical manipulation. The work *Opiologia ad mentem academiæ naturæ curiosorum* in the Oak Spring Garden Library represents an expanded edition of the original treatise published by Wedel in 1674. Many such monographs were written in this period; as knowledge grew, based on more sophisticated

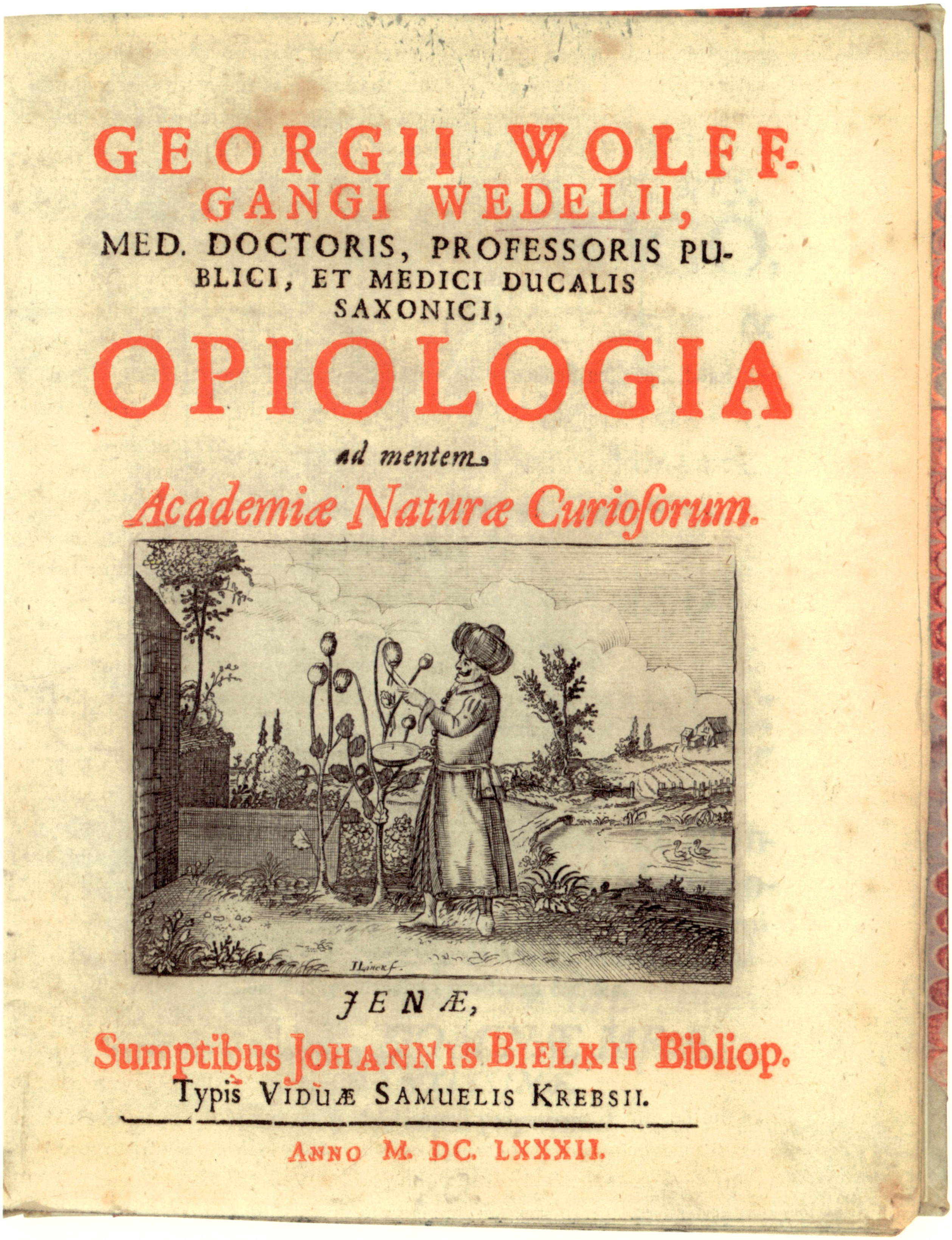

GEORGII WOLFF-
GANGI WEDELII,
MED. DOCTORIS, PROFESSORIS PU-
BLICI, ET MEDICI DUCALIS
SAXONICI,

OPIOLOGIA

ad mentem
Academiæ Naturæ Curiosorum.

JENÆ,
Sumptibus JOHANNIS BIELKII Bibliop.
Typis VIDUÆ SAMUELIS KREBSII.

ANNO M. DC. LXXXII.

GEORG WOLFGANG WEDEL, *Opiologia*, 1682. Title-page (*Papaver somniferum*)

theories and more detailed experiments, interest in simple remedies concocted from medicinal herbs waned and physicians were eager to consult, in addition to herbals with their traditional lists of plants and their uses, works dedicated to single species which contained instructions on how to extract their active compounds.

Although in the introduction to his work Wedel affirms his belief in the efficacy of medicinal herbs, he states that they should not be used indiscriminately and advocates a new approach based on a thorough understanding of the properties of the plants. He begins by observing that opium—the juice contained in the capsule of the poppy ('succus excapitum papaveris thyrisis seu scapibus vulneratis' p. 10)—was an extraordinarily potent 'simple' whose effects had already been described by the classical and Arab authorities, whom he cites extensively. A discussion of the 'temperament' of opium and whether it was 'warm' or 'cold' in accordance with the medical theories of Antiquity follows, but Wedel quickly passes to an examination of the soporific and narcotic properties of the drug, carefully explaining how opium in various forms, including its volatile salts, could be used to treat pain, insomnia, blood fluxes, migraine, and epilepsy. In the second book of the third chapter, he elucidates the risks associated with the use of opium and recommends that great caution be exercised in its administration. What interested Wedel most of all, however, were recipes for the compounds and medicinal remedies, in either liquid or solid form, that could be obtained from the plant and a number of these are described.

The title-page of *Opiologia* presents—in addition to the title, author, and publisher printed in red and black—a charming vignette signed by 'I. Linck' (perhaps a member of the Swiss family of engravers of the same name) that depicts an agrarian landscape with a river and, in the foreground, a walled garden in which a dignified figure in oriental dress can be seen drawing the precious liquid from the pod of a tall poppy plant.

Bound together with this work is another treatise on the same subject, *De Virtute Opii Diaphoretica Dissertatio*, by Michael Ettmüller (1644–1683), a physician from Leipzig who had studied in Wittenberg, spent long periods in Italy and—like his friend Wedel, to whom he dedicated his monograph—taught botany, surgery, and anatomy and wrote treatises on medicine and iatrochemistry.

20. Nicolas de Blégny (1652–1722)

Le bon usage du Thé du Caffé et du Chocolat pour la preservation & pour la guerison des Maladies. Par Mr de Blegny, Conseiller, Medecin Artiste ordinaire du Roy & de Monsieur, & préposé par ordre de Sa Majesté, à la Recherche & Verification des nouvelles découvertes de Medecine. [*Typographical ornaments*]. A Lyon, Chez Thomas Amaulry, ruë Merciere, au Mercure Galant. [*rule*] M.DC.LXXXVII. Avec Privilege du Roy.

12° 15.1 × 7.9 cm. ã[12] A–P[12] *i–xxiv* 1–120, 131–226, 219–357 [5] (273 as 27, 291 as 219, 342 as 42) [384 pp.].

PLATES: Engraved frontispiece of a Chinese drinking tea with tea leaves being harvested in the background, signed in the lower right: 'Bouchet Fec.' and 13 engravings related to tea, coffee, and chocolate, most signed 'I. H. fec.,' and one signed 'I. Hainzelman fe.'

BINDING: Quarter vellum and blue-gray boards with leather label on spine.

REFERENCE: Hunt 376.

NICOLAS DE BLÉGNY was one of the most curious and controversial figures in French medical circles at the end of the seventeenth century. The son of a caretaker at the Faculty of Medicine of the University of Paris, he successfully completed his studies in medicine and surgery, and then began his rapid professional and social ascent by establishing a certain reputation in the capital city for his treatment of hernias. A prolific author, he wrote a great number of books and treatises on a wide variety of subjects, although these were for the most part compendia based on more serious and thoroughly researched studies by other scholars. His long bibliography included works on medical subjects, such as a treatise on the powder known as the 'remède anglois' (quinine), which was extracted from the bark of the cinchona tree and used to treat fever. His name was furthermore closely associated with the first medical journal published in France, *Journal des nouvelles découvertes concernantes les sciences et les arts qui font partie de la médicine*, which appeared for several years beginning in 1673.

De Blégny also engaged in activities aimed at the general public; he organized public lectures on such topics as surgery and pharmacology, as well as workshops with more practical aims, such as wig-making for coiffeurs. He wrote (under the pseudonym of Abraham du Pradel) a guide to useful addresses in Paris (*Livre commode des adresses*), among which could be found the names of musicians and pastry-cooks (1690), a short work on beauty secrets (1688–89), and another on astronomy and the use of the telescope (1681).

On the title-page of *Le Bon Usage du Thé, du Caffé, et du Chocolat pour la Preservation & pour la Guerison des Maladies* the author describes himself as 'Medicin Artiste ordinaire du Roy ... préposé par l'ordre de Sa Majesté, à la Recherche et Verification des nouvelles découvertes de Medicine.' Here De Blégny was referring to the 'Laboratoire des Quatres Nations,' a sort of scientific academy of which he was director, whose aim was to study the medicinal properties of exotic plants. Its members offered free consultations and sold novel remedies that they themselves had prepared.

As ambitious as he was enterprising, De Blégny succeeded in having himself appointed *chirugien ordinaire* to the Queen in 1678, then to the Duc d'Orléans in 1683, and finally to King Louis XIV himself in 1687. He suffered an abrupt fall from grace, however, when he was accused of fraud and imprisoned for eight years at the Chateau de Angers. Once freed, he retired to the relative obscurity of Avignon, where he died in 1722.

By the end of the seventeenth century the drinking of tea, coffee, and chocolate had become immensely popular all over Europe and De Blégny seized the opportunity to write a work on these exotic beverages, which were furthermore believed to be endowed with all manner of health-giving properties. Written for the benefit of 'the doctors in medicine of the provincial and foreign faculties who practice at the Royal Court in Paris' and published in Lyon in 1687, *Le Bon Usage du Thé* is divided into three parts, each dedicated to one of the three drinks. The title-page to De Blégny's work is decorated with a charming engraving signed 'Bouchet' (perhaps A. Bouchet, an engraver active in Lyon in the eighteenth century), which depicts a Chinese drinking 'la liquer du thé'; in the background is a scene showing the harvesting of the precious leaves. As was his usual practice, the author cobbled his text together from other sources, for example citing frequently from a popular work by Jean Baptiste Tavernier, *Les six voyages ... en Turquie, en Perse et aux Indes*, which was published in Paris in 1676.

In his *Avertissement* De Blégny observes that references to these beverages could be found in the accounts of travellers, but these consisted mostly of information haphazardly gleaned from local doctors and herbalists. The subject in his opinion deserved a more complete and in-depth treatment, which he sought to provide with this work. The first part of *Le Bon Usage* is dedicated to tea, a common drink in China since earliest times that was made, as the author had learned from Chinese ambassadors, from the long, dentate leaves of a small tree belonging to the genus *Ternstroemia*, which did not exceed in height that of a rosebush. De Blégny discusses in considerable detail the nature of the plant, its cultivation, and its different names and varieties, affirming that the best tea came from China and Japan, where it was called 'cha.' The leaves, he explained, were heated over a low fire and then gently shaken in small bags of thin cotton cloth until they were completely dried. In this way 'une graine noirastre' was obtained, the form in which tea was exported to European markets. The author not only informs the reader as to the best qualities of tea (those made from the smallest, most tender leaves) but also their price, which might vary from ten to eighty French francs per pound. Tea was prepared by infusion and a pleasing illustration shows teapots of different shapes made of silver, tin, or porcelain. According to De Blégny, not only did the beverage have a fortifying effect, but tinctures and pills for headache, stomach or intestinal pains, catarrh, and fluxes could be made from the leaves. He also describes a 'sirop du thé' of his own invention that he claimed was effective in reducing fever.

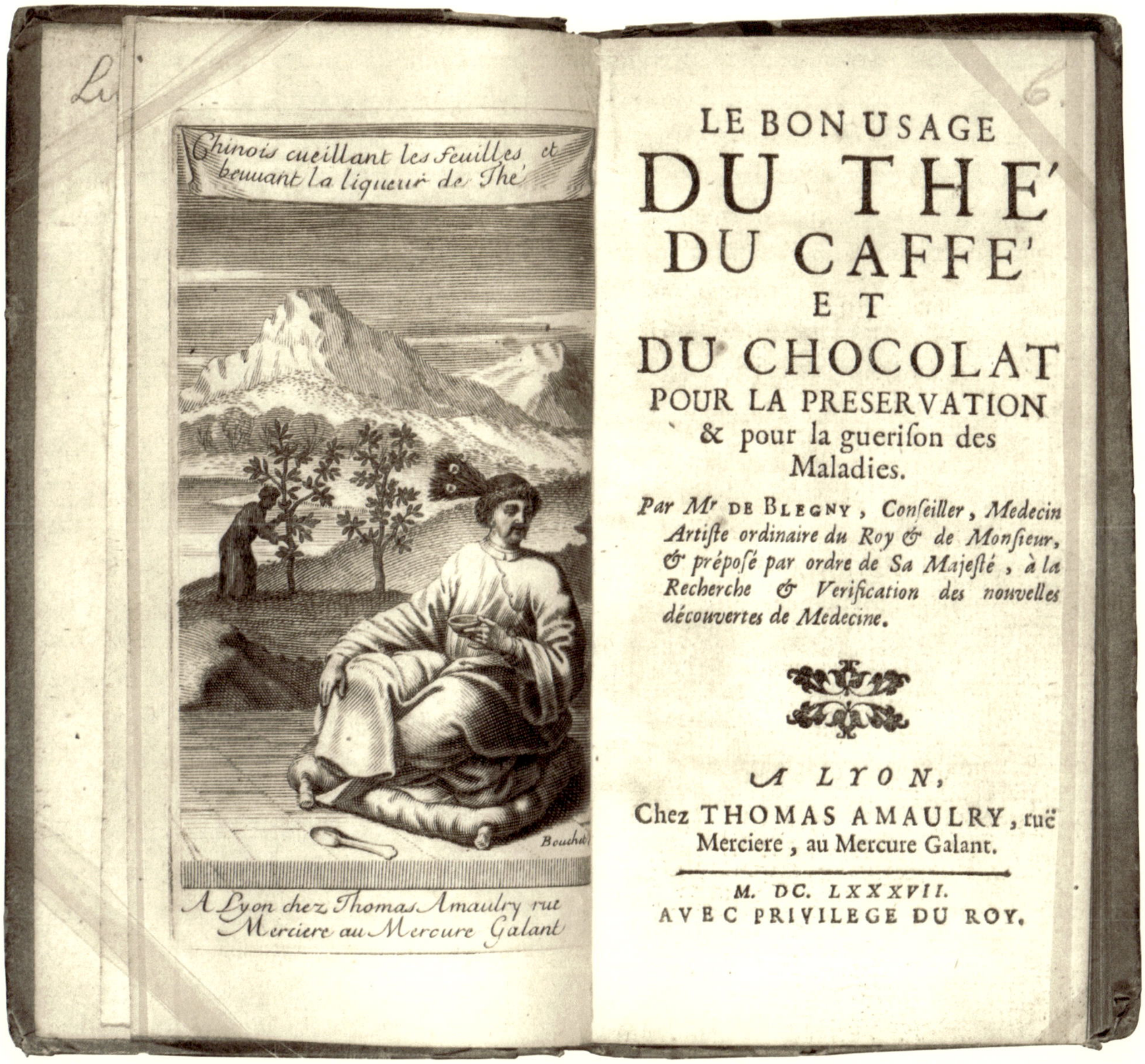

NICOLAS DE BLÉGNY,
Le bon usage du Thé du Caffé et du Chocolat,
1687. Frontispiece
and title-page,
Camellia thea

The second part of De Blégny's work is devoted to coffee, a drink made from the seeds of a tropical evergreen shrub belonging to the genus *Coffea* of the *Rubiaceae* family. De Blégny writes that the coffee plant could be found growing abundantly in the kingdom of Yemen in Arabia and that it has leaves similar to those of the cherry tree. In Yemen the seeds were called in different regions 'Bonchum,' 'Buncho,' or 'Buncha,' in Egypt 'Elkarie,' and in Arabia 'Cachua' or 'Chaue,' from which was derived the name 'café' by which the plant came to be known all over Europe.

An engraving signed 'I.H. fe'—which signature is expanded in one engraving to 'I. Hainzelman,' i.e., Johann Hainzelmann (Augsburg 1691–Berlin 1693)—depicts a branch of the plant with its characteristic fruit, which according to the author were hard and tasteless. The seeds, however, could be roasted, ground, and dissolved in a small quantity of water to produce the aromatic beverage. De Blégny writes at length about the remarkable popularity of this drink and the fashionable 'coffee-houses' devoted to its consumption, which had sprung up in every European city. In London, for example, 'il ya plus de trois mille maisons destinées à boire du Caffe, dans lequelles il ya des grandes salles, où l'on voit tout le jour & une bonne partie de la nuit, un grand nombre de beuveurs, & et l'on sait qu'à Paris il s'en fait une prodijeuse consommation, non seulement chez le Marchants de liqueurs, mais encore dans les maisons particulières & dans les communautés' (... there are more than three thousand houses destined for the drinking of Coffee, in which there are immense rooms where one may see all day & for a good part of the night a large number of customers, & it is also well known that in Paris it is consumed in prodigious quantities, not only in establishments where alcohol is sold, but even more in *maison particulières* & in the community).

According to the author, coffee was a nutritious and fortifying beverage that also had significant medicinal properties. Salts, unguents, and syrups made from the bean could be used to treat hysteria, hypochondriacal illnesses, intermittent fever, rheumatism, and scurvy, although as with all medicines it should be administered with care. He notes that in London the beverage was sometimes prepared using copper utensils, a practice that was to be avoided as the metal could alter its taste; he also expressed himself contrary to the addition of other substances or flavorings such as cinnamon or cardamom. The average price of ground coffee, he noted, was four pennies per pound. When brewing coffee it was important to conserve the bean's volatile elements, and therefore the powder should not be added to the pot until the water has come to a boil, after which it should be allowed to cook for only five or six minutes.

An engraving, once again signed by 'I.H.,' shows 'une caffettière montée sur son fourneau pour faire le caffé avec esprit de vin' (a coffeepot resting on its own little spirit burner, for making coffee). Another depicts a portable coffeepot, while a third illustrates a 'four-

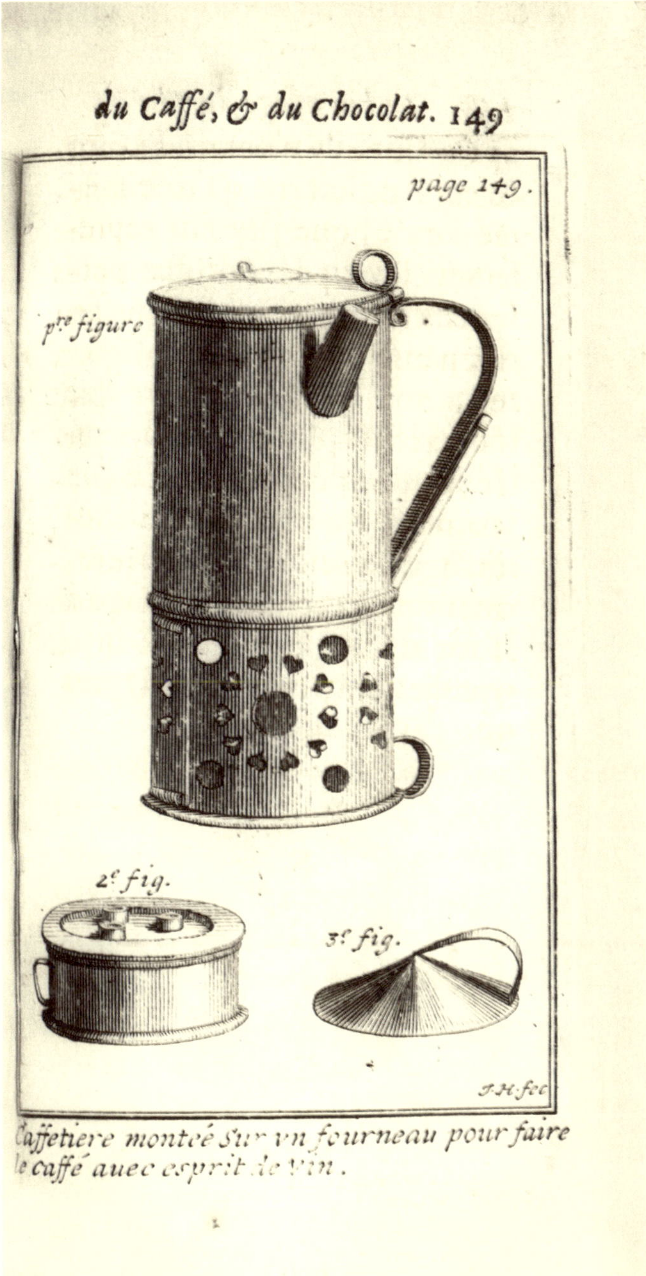

NICOLAS DE BLÉGNY, *Le bon usage du Thé du Caffé et du Chocolat,* 1687. Coffeepot with spirit burner, page 149

neau à lampe pour entretenir la chaleur des caffetières' (heater with a flame to keep the coffeepot hot), and the fourth a set of 'cabarets à caffé' or coffee cups.

The third section of De Blégny's treatise is devoted to the 'Nature, proprietés & bon usage du chocolat.' Cacao (*Theobroma cacao*) originally came from the West Indies, in particular Guatamala according to the author, where the natives called it 'cacavaquahuit.' They added cacao to their bread, together with sugar, vanilla, musk, and amber. One of the engravings that accompany the text in this section shows the 'Cacauifère ou arbre du cacao' (cacao tree) with fruit hanging from its branches ready to be gathered and dried, while another illustrates a detail of the fruit; both are signed 'I.H.'

De Blégny describes the various steps involved in the production of ordinary cacao. First one must roast the fruit until it is reduced to a paste; an engraving shows a man spreading this paste with a rolling-pin. The oil is then separated out by means of a press, which is illustrated in another engraving. The end product could be pressed into various shapes for sale and consumption—tablets, rolls, or 'lumps.' The most delicious way to consume cacao was in the form of a hot drink, which should be foamy in consistency; various mills designed to make this froth are shown in another engraving. In his capacity as physician, De Blégny noted that cacao had a significant anti-venereal action.

In the last part of his book De Blégny presents a list of the 'Marchandises qui sont actuellement dispensées par les artistes du Laboratiore Royal des Quatre Nations' (Products which are at present dispensed by the artisans of the Royal Laboratory of the Four Nations), the institute of which he was director. He does not hesitate to provide a catalogue of his other books, together with their prices, including: *Recueil des journaux de Medicine, contenant toutes les nouvelles découvertes des medicins & Artistes de ce siècle; Le remède Anglois; La Doctrine des Rapports de Chirurgie; Les observations qui ont été faites sur les astres depuis l'invention des lunettes d'approche; Dissertation sur la maladie venerienne; L'histoire anatomique d'un enfant qui a été 25 ans dans le ventre de sa mère*; and *L'Art de guerir les Hernies ou Descentes* (Collection of Medical Journals, containing all of the latest discoveries of physicians & artists in this century; The English Remedy; The Doctrine of Surgical Relationships; Observations that have been made of the stars since the invention of binoculars; Dissertation on Venereal Disease; The anatomical case history of an infant that remained for 25 years in the womb of his mother; and The Art of Healing Hernias). At the end of this list, De Blégny adds that readers might also obtain from him 'quelques autres aussi curieuses' (other equally surprising) works.

21. Thomas Malie (*c.* 1700–1789)

43 × 30.5 cm. Eight drawings in pen and India ink finished with gray wash.

The American Plum – 1726
Anona Maxima – circa 1730
The Bichy Tree – circa 1730
Cacao – circa 1730
The Citron – 1728
Mammaia Sapota – circa 1730
Mammone – 1730
Seaside or Mangrove grape tree [and] *Cedar Tree* – 1741

THOMAS MALIE, the artist responsible for this series of eight drawings in pen-and-ink of tropical plants, was an English physician who spent many years in the Americas as an army surgeon. Malie was born around 1700 and after completing his medical studies in 1723 was appointed to serve in the King's Dragoon Guard. He departed with his company for the West Indies in 1725. This expedition was followed by at least one further voyage to Central and South America. In addition to practicing medicine, Malie studied the flora and fauna of these tropical regions, documenting his research in the form of drawings, each labelled with a name and many accompanied by miscellaneous observations. Most of his work is in a bound album consisting of thirty-nine sheets conserved at the Natural History Museum in London; additional drawings were also sold at an auction by Sotheby's in June 2000.

The illustrations depict birds, insects, plants, and marine life. From the annotations we can retrace the places visited by Malie. In 1726 he was in Panama as well as Colombia, where he travelled to the cities of Cartagena, Baranquilla, and Santa Marta, and lastly Maraquita Island off the coast of Venezuela. Malie travelled extensively the following year visiting Marcaibo, and Caracas, Venezuela, Santa Fe (Argentina), and Guatemala. In 1728 he visited Caracas once again; in 1729 he made his third trip to Caracas and also visited Buenos Aires; and in 1730 he travelled to Mexico and Cuba, before finally returning to London.

When England engaged in war with Spain, Malie, due to his previous experience in the West Indies, was appointed Surgeon General to Lord Cathcart, Commander-in-Chief of the British forces. During the first months of this campaign Malie was consumed with his medical duties and during this period the physician executed a single drawing of a fish captured in St. Rupert's Bay. The English soldiers not only suffered greatly from the tropical climate, but also lost their commander Lord Cathcart in December 1740. The troops set up camp in Terra Bomba near Bocca Chica (Colombia), after failing to take Fort St. Lazar, they retreated with their fleet to Jamaica. Here Malie was able to resume his naturalistic studies, producing drawings of plants and animals in Cartagena, Jamaica, and Dona Marina

THOMAS MALIE,
Eight Plant Studies,
1728–1741.
'The American Plum'
(*Spondias purpurea*)

THOMAS MALIE,
Eight Plant Studies,
1728–1741.
Anona Maxima' (*Annona muricata*)

THOMAS MALIE,
Eight Plant Studies,
1728–1741.
'The Bichy Tree,' Kola
nut (*Cola acuminata*)

THOMAS MALIE,
Eight Plant Studies,
1728–1741.
'Cacao' (*Theobroma cacao*)

Bay (Cuba). His last works date from 1742 and illustrate flora and fauna from the West Indies island of Dominica. It was presumably after this date that the physician returned to England, where he settled after his long period of service abroad. His death in 1789 was marked by an obituary in the *Gentleman's Magazine*.

The eight drawings in the Oak Spring Garden Library supplement the large body of works preserved in the Natural History Museum, London. Taken together, his work represents a genuine contribution to the botanical sciences of this period, as many of the plants that he studied were still unknown in Europe. The drawings testify to Malie's scientific interests and his modest but conscientious talent as a botanical artist. As an amateur botanist, Malie must have been acquainted with the literature on the subject and it is interesting to note, while some of his plants were drawn from life, others are actually copies of illustrations from *Metamorphosis Insectorum Surinamensium*, the remarkable work on the flora of Surinam (Dutch Guiana) by the botanical artist Maria Sibylla Merian, that was published in Amsterdam in 1705 (see *An Oak Spring Flora*, No. 101, pp. 380–383). For example, Malie's drawing of the purple mombin (*Spondias purpurea*) was based on the illustration of the Yellow Mombin Plum Tree and Blue Moth in *Metamorphosis* (engraving no. 13). Dated 1726, it is accompanied by an extended note at the foot of the sheet:

> The American plum—some call it in Jamaica the bubby-plum, it grows as tall as a walnut tree & all together of the size the leaves and flowers resemble our Elder flowers, but have no smell the fruit hang in clusters. This fruit is binding but helps perspiration. The stone makes almost half of the fruit, the flesh of which is full of fiber ... 1726.

Merian's original painting—a vibrant work in opaque watercolors enlivened by the addition of butterflies and other insects, which do not appear in Malie's version—is conserved in the archives of the Academy of Sciences of Saint Petersburg (Inv. no. P IX, 8, 31). The purple mombin, a shrub-like tree native to southern Mexico, Peru, and Brazil, is also cultivated today for its edible, plum-like fruit in southern Florida.

The drawing of the 'Citron,' dated 1728, bears the annotation:

> The Citron—the trees are equally as high as our apple trees & grow spontaneously in the woods, the leaves & flowers resemble those of the lemmon, excepting being larger & thicker, & in proportion to the size of the fruit the pulp is very small. The rind is a fine preserve. A strange sort of an insect very poisonous for if it only touches the skin it swells it & inflames it, it feeds on the tree ... A surprising od sort of a scaraberry [beetle], red and yellow & beautifully spotted ... This is an exact replica of engraving No. 28 in Merian's work, which depicts a harlequin beetle (*Cerambyx longimanus*) and *Citrus medica*, a tree that can be found growing in Mesoamerica and Central America, although it is not indigenous to these regions.

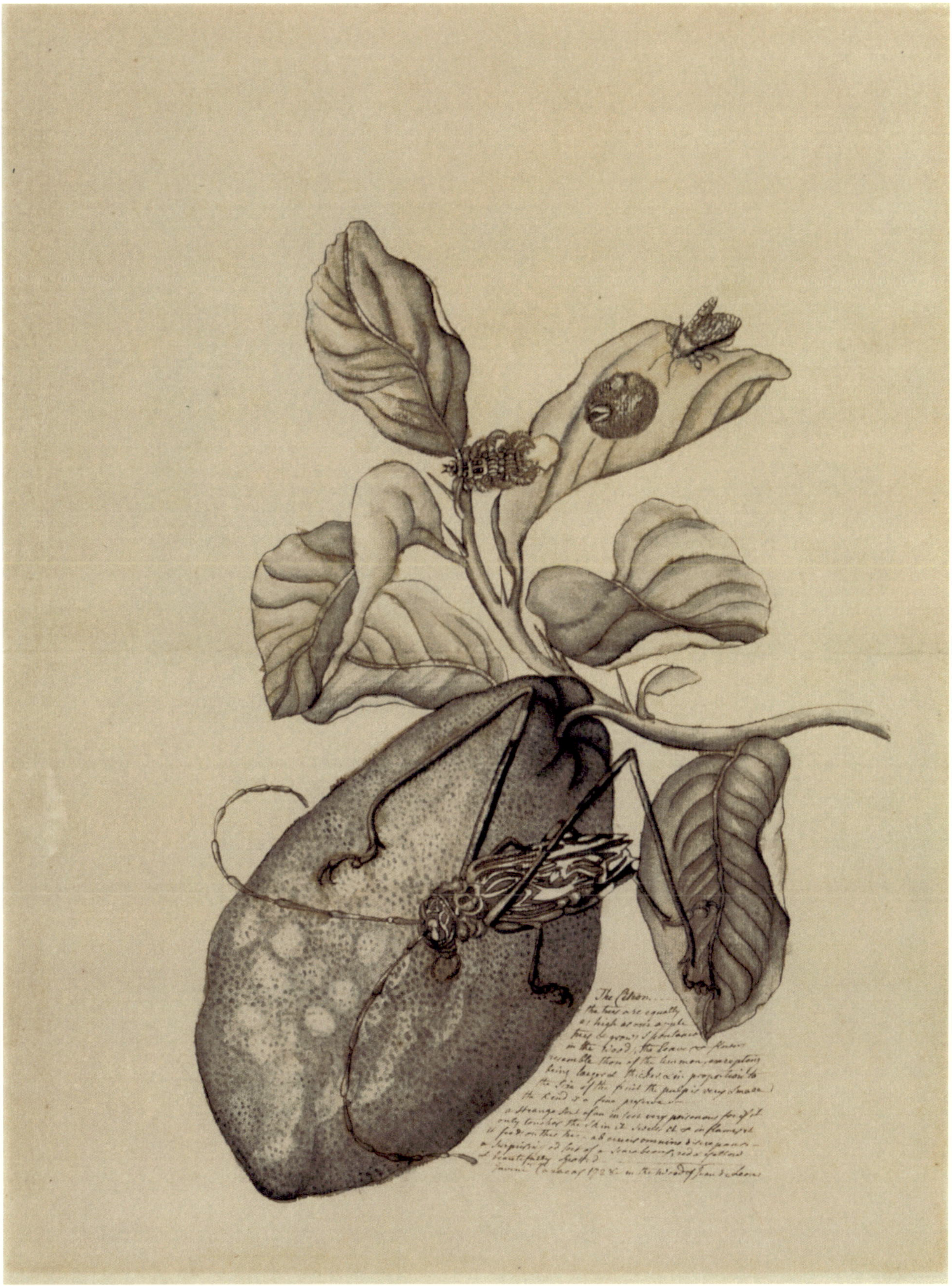

THOMAS MALIE,
Eight Plant Studies,
1728–1741.
'The Citron' (*Citrus medica*) with insects

THOMAS MALIE,
Eight Plant Studies,
1728–1741.
'Mammaia Sapota,' Sapote (*Pouteria sapota*)

THOMAS MALIE,
Eight Plant Studies,
1728–1741.
'Mammone,' Cherimoya
(*Annona cherimolia*)

THOMAS MALIE,
Eight Plant Studies,
1728–1741.
'Seaside or Mangrove grape tree,' Seagrape (*Cocoloba uvifera*) and 'Cedar tree' (*Cedrela*)

A third drawing is recognizable as a copy (without the insects) of engraving no. 14 (Soursop with Owlet Moth) from *Metamorphosis*. Dated 1730, it accompanied by a long legend in Dutch. The soursop (*Annona muricata* of the family Annonaceae), also known by its Spanish name 'guanában,' is very common in tropical countries. The Spanish official and courtier Gonzalo Fernández de Oviedo y Valdés (1478–1557), an observant student of the natural history of the New World, described this plant in *Historia general y natural de las Indias* (Toledo, 1626), noting that it grew everywhere in the West Indies. Its fragrant, pleasantly tart fruit has always been appreciated by the natives and is now sold in markets all over the world.

The other plants in the collection were probably drawn from life by Malie himself. Less refined in detail and execution, they are nevertheless pleasingly composed and in many cases portrayed with sufficient accuracy to allow their identification. For example, the *Mammaia sapota* depicts an elegant spray of leaves, scaly fruit, and large seed, and may be identified as *Pouteria sapota*. This plant, whose name seems to be based on the Aztec word 'tzapotl,' is widespread in Mexico and northern Nicaragua, where it can be found growing to quite tall heights. It is also cultivated in Florida and the Philippines for its sweet fruit, which may be round, oval, or elliptical in form.

It is not possible to identify the plant in the drawing labelled 'Bichy Tree. Jamaica and Santa Maria,' even though Malie also shows the leaf, fruit, and seed. Due to the fact that there are over 100 species in the Anona family, the illustration '*Anona maxima*,' portrayed with its spiny, heart-shaped fruit is difficult to identify. Another drawing shows a branch of the *Theobroma cacao* with its leaves, fruit, seed, inner pod, and flowers.

In the last drawing, 'Seaside or Mangrove grape and Cedar Tree,' dated 1741, the leafy parts of three different plants are illustrated with explanatory notes: :'3..4..5 Sea Side or Mangrove grape tree: of Carthay: 1741 Cuba.....' Inscribed vertically next to stem of second drawing: 'Fi. 1... Pruno forte attinis arbor, folio.... alato, flore herbaceo pentapetalo.... racemoso.... Carthagena April 4...1741...' Inscribed below third drawing: 'Cedar Tree 2 ..... 220.....'

22. CHINESE SCHOOL, CANTON & MACAO (*c.* 1770–1774)

Drawings of Chinese plants and fishes by a native artist (Mauk-Sow-U) under the direction of John Bradby Blake (1745–1773).

40 × 36 cm. 205 numbered leaves of drawings in four volumes.

PLATES: 173 of plants, 18 of fishes, two of plant boxes, one of a turtle, all watercolor and gouache.

BINDING: Four folio volumes in multi-colored original silk brocade covered boards, plus seven volumes of manuscript text: (1) manuscript listing the drawings in volume one, old marbled wrappers, 32.7 × 20.7 cm; (2) manuscript listing the drawings in volume two, old marbled wrappers, 32.9 × 20.7 cm; (3) manuscript listing the drawings in volume three, old marbled wrappers, 32.5 × 20.6 cm; (4) description of another Chinese herbal in five volumes, old marbled wrappers, 33 × 20.8 cm; (5) illustrated manuscript of Chinese/English vocabulary, silk wrappers, 22.3 × 19.8 cm; (6) lists of seeds and plants sent to England from China, old marbled wrappers, 32 × 20.4 cm; and (7) various descriptions of plants with letters, old marbled wrappers, 36.9 × 24.5 cm.

PROVENANCE: From the collection of the late Sir Frederick William Beresford Cripps (b. 1873) of Ampney Crucis, Gloucester. Manuscript label on front free end papers in volumes 1–4: 'G A Nutt.'

REFERENCE: Cox, p. 43.

THIS UNIQUE COLLECTION of material on the natural history of China consists of a total of eleven volumes. Four folio volumes are paintings of mostly flowers, the fourth of which also contains fishes and a turtle. A separate manuscript describes and indexes each of the first three folios containing illustrations of plants, but, interesting to note, the fourth folio does not have a manuscript dedicated to an indexed description and appears unfinished with several blank pages. Included in the seven ancillary volumes of this wonderful compilation is an illustrated dictionary in Chinese and English, a list of plants and seeds sent to England from China, various descriptions of plants with letters, and a description of a Chinese herbal, all written and compiled by a remarkable English plantsman, John Bradby Blake (1745–1773).

Blake was born in 1745 and educated at Westminster School. In 1766, at the age of twenty, he embarked as a supercargo or the representative of the owner of an East India Company merchant ship bound for China. His primary ambition was to study the plants used by the Chinese in medicine, cuisine, and manufacturing and to collect specimens of these economic plants for distribution in Europe and America. Blake remained in China until his death at the age of twenty-nine. It is tragic that he died so young, for he would surely have become a great collector. As his notes demonstrate, through assiduous study and reading, his knowledge of natural history was both profound and up-to-date. He managed to accomplish an astonishing amount in a brief period of time despite the highly unsettled local conditions.

During his sojourn in Canton, Blake initiated another worthwhile project. He arranged for a talented artist, Mauk-Sow-U, to come to his house daily and make drawings of the plants that he was studying and to translate his botanical notes into Chinese. He must have supervised the work of the artist very closely, for these drawings show a remarkable degree of botanical accuracy.

Although the details of his activity are not yet clear because there needs to be further sifting of the immense amount of material that he left behind, it does appear Blake was quite successful in shipping seeds and plants to his colleagues in Europe. No doubt an important factor in this achievement was the care with which he packed his precious cargo to protect it from damage during the long voyage by sea. Numerous passages from his letters and notes testify to the ingenuity he applied to this task; in a note written in 1772, for example, we read of boxes of seeds packed in burnt paddy (rice) chaff, wax, or dry ashes.

Many botanists in England benefited from the young man's efforts, not only the directors of His Majesty's Garden at Kew, the authorities of the Apothecaries' Company, and the Physic Garden of Chelsea, but also the owners of various private gardens in England and the colonies. One of his most important contributions was the introduction of Cochin-China rice and the tallow tree to Jamaica and the Carolinas, where they were success-

CHINESE SCHOOL, Drawings of Chinese Plants, circa 1770–1774. Binding of silk brocade boards

fully acclimatized. Among Blake's illustrious correspondents was the naturalist and voyager Sir Joseph Banks (1743–1820), and it is interesting to note that the Banks collection at the library of the Natural History Museum in London includes a volume of tempera paintings of Chinese flowers that are perhaps attributed to Mauk-Sow-U (Banksian MS 12).

From the assorted papers belonging to the young botanist we may deduce he was planning to compile a comprehensive herbal of Chinese plants. It is unfortunate he died before this project could be completed, for it would have constituted a genuine monument to the flora of Asia and to the horticultural links between the Orient and Europe. In this period the only works available on the subject were *Flora Sinensis* published in Vienna in 1656 by the Jesuit priest Michel Boym (1612–1659), which described for the first time the fruit of *Litchi chinensis*, and an herbal of Japanese plants compiled by the physician Engelbert Kaempfer (1651–1716), published in Lemgo in 1712, which included illustrations of the camellia, the bitter orange, and *Ginkgo biloba*.

Blake himself referred frequently to Kaempfer's book in his manuscripts and letters. In a letter sent to his father, John Bradby Blake, in 1772 he announced:

> By the ship Osterly, under the care of cap. Fortesque, I have sent you ten plants of Tsubakki, described in Kaempfer, Amoenitatum Exoticarum, page 850-852, which you desired me to procure for you, on the request of Mr. Aiton, Botanic gardener at Kew, as it would be a desirable article to add to the valuable collection in His Majesty's garden at that place,

CHINESE SCHOOL,
Drawings of Chinese Plants, circa 1770–1774. A variety of 'Teene' (*Leguminosae*), volume 1, folio 40

CHINESE SCHOOL, Drawings of Chinese Plants, circa 1770–1774. 'Gardenia,' possibly a Common Gardenia (*Gardenia jasminoides*), volume II, folio 3

under his care. I have also sent you six plants of the Tallow Tree and two of the Gardenia jasminoides that have been in the same potts two years in my possession in a box of Earth. Under the particular care of Mr. Rogers, Chief Mate of the Osterly, I have sent you two boxes of valuable seeds collected from different Provinces of the Empire.

The historical value of the notes left by Blake is greatly enhanced by the iconographic material that accompanies it, containing nearly two hundred large paintings, mostly of plants. These works testify to Mauk-Sow-U's innate talent and sensitivity as a botanical artist. He made a noteworthy effort, under Blake's guidance, to record such important morphological details as the flower, fruit, and seed, often sectioned to expose their pulpy and woody parts. To convey an idea of the actual size of a leaf or another part of a plant, the artist employed a most unusual and fascinating expedient. Next to or behind the plant in the drawing he portrays the leaf or flower in its natural dimensions, divided into two parts, if necessary, to fit the page, using innumerable tiny dots of pale black ink stippled with the tip of a fine brush to create an ethereal image that contributed greatly to the elegant *mise en page* of his works. These works are outstanding from both a scientific and an artistic point of view. Mauk-Sow-U clearly possessed noteworthy gifts that can be appreciated in the layout of the plants, the painstaking rendition of details, and the skillful application of brilliant colors, often set off with subtle washes and touches of lucent varnish. Particularly unique was his ability to represent various tones of white against the natural white paper.

Among the species depicted are many varieties of flowering plants (the camellia, hibiscus, chrysanthemum, magnolia, mallow, and gardenia) and exotic fruits and vegetables ranging from citruses, peaches, and the litchi (*Litchi chinensis*) to pepper plants with anomalously shaped leaves. Next to each plant we find its name written in Chinese characters and its transcription into the Roman alphabet.

Blake devoted great pains to ensuring the scientific correctness of these drawings and in the indices noted not only the date on which a plant was gathered but also the date on which it was portrayed. For example, in the case of the gardenia 'Whangchee,' he wrote:

> This is the Gardenia, as described by Mr. Ellis FRS [see No. 25] and Dr. Solander in a letter to P.C. Webb Esq. I had been anxious to procure the same in flower for which purpose ... The gardenia or Whangchee N.- was drawn June 1rst 1771. I had two plants brought to me about two days journey from a very hilly situation, they are drawn in every respect of full size.' (11, fol.3)

For the tea plant or 'E Chau' Bradby Blake writes:

> July, the eighteenth, 1771. Transplanted 10 tea Plants of about 4 or 5 inches in height (which were sowed about the beginning of February last) into two small Garden Pots five

CHINESE SCHOOL, Drawings of Chinese Plants, circa 1770–1774. Two leaves from the Chinese dictionary of various plants, insects, and fowl, 42 verso and 43 recto

in Each—as they seem to thrive best in Clumps or Parcels ... [and] are so described to be sewn in Japan by Keampfer .... Dec. eighteenth: Painted the Tea Plant of one years Growth from some planted myself from the seed.' (II, fol.17)

One of the most delightful items in this collection is an illustrated dictionary that contains tiny sketches of human figures, clothing, shoes, household implements, weapons, musical instruments, animals, and plants—a veritable micro-universe replete with details concerning the daily life and customs of China. Next to each word is its Chinese name followed by a Latin transcription and the corresponding word in English. Undoubtedly this is a one-of-a-kind remarkable assortment of collected works that reflect the inspirations that a foreign land can have on an avid naturalist.

23. GIOVANNI DOMENICO CIVININI (*fl.* 1730s)

Della Storia e Natura del Caffè Discorso Accademico di Gio: Domenico Civinini All' Altezza Reale di Gio: Gastone I. Granduca di Toscana. [*Printer's device*] In Firenze. MDCCXXXI. Nella Stampería di Bernardo Paperini, all' Insegna de Pallade, e d'Ercole. [*rule*] Con Licenza De' Superiori.

4° 24.8 × 18.2 cm. A–F⁴ *1–6* 7–46 [*2*] [48 pp.].

PLATES: One engraved folding plate of a coffee plant, signed bottom right: 'Hya: Tofani Incidit.'

BINDING: Contemporary off-white stiff wrappers.

REFERENCE: Haller II.234.

GIOVANNI DOMENICO CIVININI was born in the Tuscan town of Pistoia into a dynasty of medical doctors; his father Lodovico was a distinguished surgeon, and his brother Marcantonio also became a physician. Giovanni studied under Tommaso Puccini (1669–1724), an anatomist and physician at the court of the Medici.

Very little is known about the life of Giovanni Domenico Civinini, apart from the fact that he was a member of the Società Botanica Fiorentina, which was founded in 1716 by a group of scientists that included the physician Sebastiano Franchi of Lucca, the Florentine abbot Gaetano Moniglia, the physician and malacologist Niccolò Gualtieri of Florence (1688–1744), and the botanist Pietro Antonio Micheli (1679–1737), who was the author of a history and catalogue of the Botanical Garden of Florence. The Società Botanica Fiorentina was the first botanical academy to be established in Europe. After conducting its activities for some years in various private gardens, beginning in 1718 by the order of Grand Duke Cosimo III de' Medici, the Società was charged with the management of the Botanical Garden of Florence, which had been constructed in the sixteenth century near the Church of San Marco.

Civinini dedicated his two most noted works to the members of this prestigious society whose aim was the furtherance of the sciences in Tuscany: *Della storia e natura del caffè*, published in Florence in 1721 and republished ten years later, and *Della storia degli Agrumi*, published in 1734. Both of these works were described by the author as 'discorsi accademici' and therefore, it may be assumed, constituted the texts of talks delivered before the Società Botanica. Another, shorter academic lecture on the subject of the reaction of fear in humans and animals was published in Florence in 1735 in *Raccolta di opuscoli scientifici e filologici* (vol. XI, pp. 481–491).

Civinini's published works therefore consisted of 'academic discourses' that were intended to be read aloud and to hold the attention of an audience. While brief, they were furnished with many well-chosen citations from classical and contemporary authors that testify to the author's considerable learning, not only in the sciences but also in the humanities. The language in which they were couched, as well, reflects his full mastery of the techniques of academic rhetoric current during the baroque period.

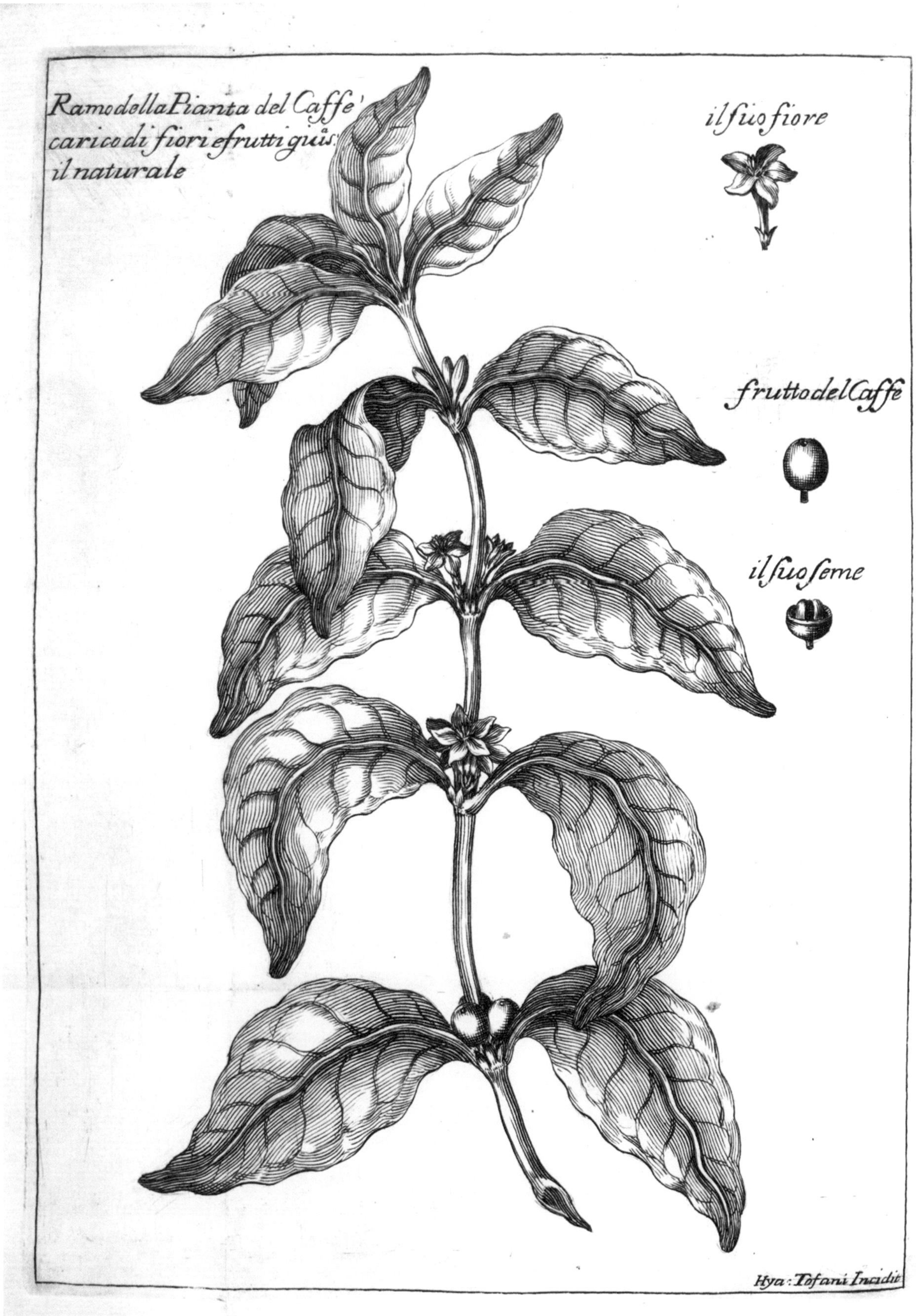

GIOVANNI DOMENICO CIVININI, *Della Storia e Natura del Caffè*, 1731. Coffee plant, possibly *Coffea arabica*, between pages 6 and 7

Della Storia e Natura del Caffè was dedicated to Gian Gastone, the last grand duke of the Medici family, whose death in 1737 marked the end of the dynasty. The work opens with an imposing sentence in Latin designed to serve as an epigraph and very probably written by the author himself: 'Natura pinxit remedia, in floribus, in seminibus, visuque ipso animos invitavit' (Nature has painted her remedies in flower and seed, by the very sight of which she has attracted the minds of men). A folded engraving signed by an unknown artist 'Hya: Tofani' portrays a 'ramo della pianta del caffè carico di fiori e frutti a guisa il naturale' (branch of the coffee plant loaded with flowers and fruit [portrayed] in a natural manner), together with details of its flower, fruit, and seed.

As Civinini observes at the beginning of his lecture, one of the most singular plants produced by a bounteous nature was the evergreen *Coffea arabica* of the family *Rubiacae*, whose history and nature he intended to briefly review before his learned colleagues. He was quite aware that many works had already been written about this exotic species and cites a number of them, in particular the Latin poem, *De saluberrrima potione cahé seu café*, written by Anton Fausto Naironi and published in Rome in 1671. Based on what others had written, Civinini affirms that the plant is native to a part of 'Arabia Felice' called Yemen. He provides an accurate botanical description: the flowers, which are white in color and similar in appearance to the Spanish jasmine, grow at the axils between the leaves and the stem. The fruit are produced three times a year and indeed one may often see both flowers and fruit on a plant at the same time; the mature berries resemble small cherries, being dark red in color with a granular pulp.

According to Civinini, the beverage made from this fruit was called 'couhe' by the Orientals and the Turks and had been used since ancient times for its medicinal properties; he notes that it may have even been cited by Avicenna. The coffee plant first arrived in Italy via the port of Venice, from which it spread across the peninsula during the second half of the sixteenth century. However, it was actually discovered and brought to Europe by the Dutch consul Nicolaes Vitsen (Witsen) (1640–1717), director of the East India Company, who entrusted it for cultivation to the botanical garden of Amsterdam. There, gardeners acclimatized and propagated the plant, sending a specimen to Louis XIV of France. In 1715 the grand duke Cosimo III de' Medici managed to acquire a specimen, which he donated to the Botanical Garden of Pisa. At that time the director of the garden was the professor of botany Michelangelo Tilli (1655–1740), who described the plant in detail in his *Catalogus Plantarum Horti Pisani* (1723), calling it 'Jasminum arabicum.' The catalogue includes a beautiful illustration of the coffee plant showing its leaves, flowers, and fruit (see No. 32).

Civinini attests that 'il sapore di questo frutto ha dell'amarognolo, e dell'austero' (the taste of this fruit is somewhat bitter and sour), but nonetheless quite palatable. His opinion is in marked contrast to that of the Florentine naturalist and author Francesco Redi

(1626–1697), who in a famous poem entitled *Bacco in Toscana* sings the praises of wine and declaims: 'Beverei prima il veleno? Che un bicchier, che fosse pieno? Dell'amaro, e reo Caffe' (Shall I first drink poison | In a glass, that may be full | Of bitter and evil-tasting coffee?).

The physician notes that one may often find a certain amount of confusion in the texts by various authors regarding the coffee plant, and that it has sometimes been erroneously identified as the 'nepentes' mentioned in Homer's *Odyssey*. Among the botanists who have written about it, Civinini cites Prospero Alpini and his book on Egypt (see No. 15); the author of another work on Egyptian plants, the eminent German professor at the University of Padua, Johannes Vesling (1598–1649); and Antoine Galland (1646–1715), author of *Origine et du progrez du café*, which was published in Caen in 1699. Galland drew upon the information contained in two manuscripts, one written by an Arab and the other by a Turk, which the author had discovered in the Bibliothèque Royale of Paris. Civinini also reports that a group of members of the 'Società di Londra' (the Royal Society of London) conducted a series of experiments on the fruit and 'per così dire, notomizzato' (one might say, veritably dissected it) (p. 24). On the basis of these experiments it was concluded that the fruit was composed of a 'fixed sweet-tasting salt' and an 'oily, sulphurous part.'

In the final part of his talk, Civinini reviews the medicinal properties of the beverage, claiming that it had a salubrious effect on the stomach, the liver, and the digestion, and could also relieve cough and headache. He declared that 'il latte caffettato è di grandissimo uso appo gli Oltramontani' (milk flavoured with coffee was widely consumed by people of Northern Europe), thus anticipating the universal popularity of cappuccino, and recommended that the beverage be drunk hot with a small amount of sugar, accompanied if desired by a biscuit. Finally, Civinini sensibly observes that, like all foods, it should be consumed in moderate quantities since its abuse could be harmful to the constitution.

24. Niccolò Gavelli (*fl.* 1749–1818)

Storia Distinta, e Curiosa del Tabacco Concernente la sua scoperta, la introduzione in Europa, e la maniera di coltivarlo, conservarlo, e prepararlo, ec. per servirsene, con altre ottime, ed utili osservazioni sopra il medesimo; In questa nova impressione accresciuta di diversi esperimenti fatti circa la sua virtù e con la Figura della Planta del Sudetto Tabacco. [*Vignette*] In Ferrara MDCCLVIII. [*rule of dashes*] Per il Giglio. Con Lic. de' Sup. A spese di Francesco Altieri.

8° 19 × 12.1 cm. A–E[8] F[4] *1–4* 5–88 (61 as 16) [88 pp.].

PLATES: One woodcut on leaf A2, entitled 'Nicotiana Tabacco.'

BINDING: Original white wrappers and marbled paper spine.

PROVENANCE: Inscription on title-page 'Ad usu[m] F. Philippi Agelli Min: Cotis[?].'

NICCOLÒ GAVELLI directed a printing shop in the town of Pesaro in the region of the Marches during the second half of the eighteenth and the first years of the nineteenth centuries. Like many other publishers during the Enlightenment, he produced works on a wide range of subjects from antiquities and history to the natural sciences, quite conscious of the intellectual and civic importance of his activity. He also edited a news-sheet of cultural and current events modelled on the celebrated French Jesuit publication *Mémoires de Trevoux* (1701–1767). Gavelli's *Memorie per la storia delle scienze, e buone arti* appeared at more or less regular intervals from 1743 to 1756, its purpose being to furnish the public with up-to-date information on 'tout qui paraît de curiuex ... dans le monde en quelques genre de sciences.'

It may be hypothesized that *Storia Distinta e Curiosa del Tabacco*, a short anonymous treatise on the plant *Nicotiana tabacum*, which has played such an important role in Western society and economics ever since its introduction to Europe, was compiled by Gavelli, drawing upon one or more of the many—in this case probably French—works written on the subject. In Italy there had been great interest, both scientific and commercial, in the the tobacco plant ever since the end of the seventeenth century, and this was further reinforced by a series of initiatives undertaken by Pope Benedict XIV (Prospero Lambertini, 1675–1758), creating a situation that as a publisher Gavelli was quick to exploit. The pope was a great advocate of the pleasures of smoking and sponsored the construction of the first papal tobacco manufactory on the Gianicolo, one of the hills of Rome. He abolished the tax on tobacco in 1757 by eliminating the state monopoly. Henceforth it would be possible for anyone to plant, harvest, and sell the raw or processed product in Rome and in the Papal States. He also eliminated the customs' tariff on imported tobacco.

Storia Distinta e Curiosa del Tabacco appeared in Pesaro in 1758. Its full title may be rendered in English as *The Distinguished and Curious History of Tobacco concerning its Discovery, the Introduction in Europe, and the Manner of its Cultivation, Conservation, and Preparation, &c. with other Excellent and Useful Observations on the Same; in this New Edition supplemented with various Experiments conducted on its Property and with the Illustration of the Plant of the aforesaid Tobacco.* It was immediately reprinted by Francesco Altieri in Ferrara, a city belonging—like Pesaro—to the Papal States. The Oak Spring Garden Library has a fine copy of this second edition in its possession. The volume opens with the modest woodcut illustration of a tobacco plant and a note explaining that this work, which was being published simultaneously with the papal suppression of the tobacco monopoly, had been written to provide the 'benigno lettore' (indulgent reader) with a useful source of information on this exotic species. The anonymous author begins with a brief history of its introduction and cultivation in Europe. He informs the reader of the names by which it was known in different countries; when the plant first arrived in France around 1560 it was called 'petum'

NICCOLÒ GAVELLI,
Storia Distinta e Curiosa del Tabacco, 1758.
'Nicotiana Tabacco'
(*Nicotiana tabacum*),
A2

or 'yoli,' while the Spaniards, who had discovered it in the Yucatán Peninsula, were already calling it 'tabacco.' In France other names sprang up over time, one of these being 'la plante de Tornabuoni' because in 1579 the cardinal Niccolò Tornabuoni sent some seeds of the tobacco plant from Paris to his nephew—Alfonso, the bishop of San Sepolcro—in Florence. Two other names, 'l'herbe de l'ambassadeur' and 'l'erbe de la reine,' were particularly significant for they linked the plant to Jean Nicot, the French ambassador to Portugal, who acclimatized the tobacco plant in his garden and presented a specimen to Catherine de' Medici (1573–1642), the Queen Consort of Henri II and regent of France; the genus *Nicotiana* was named after him (see No. 16). Nicot believed the leaf had great curative powers and recommended it in particular for—ironically—respiratory ailments. Tobacco continued to be associated with the Papal State, for in Portugal it was called the 'plant of the Holy Cross' in commemoration of the arrival of the first apostolic nuncio.

We know that in Europe between the end of the sixteenth and the beginning of the seventeenth centuries tobacco consumption spread rapidly and became a mass phenomenon. It was chewed, snuffed, or smoked in a variety of forms: shredded, wrapped in leaves, or fermented and pulverized with flavorings. Such practices led to heated debate and the appearance of vociferous opponents, one of the most determined of these being King James I of England (1566–1625). He was the first to establish a decree prohibiting its use; he also imposed taxes on its cultivation, a fiscal measure that was quickly adopted by governments all over Europe. Pope Urban VIII (1568–1644) on many public occasions expressed his disapproval of the vice of smoking.

Let us return to our treatise, however, which was the product of the initial wave of enthusiasm for this novel pastime. *Storia Distinta* continues with a description of the characteristics of the plant, which the author states 'belonged to the second class' in Tournefort's botanical system. In America four different varieties were cultivated that could be distinguished by the form of their leaves. A large portion of the treatise is devoted to a comparison of the cultivation, drying, and curing processes used in the New World and Europe. The author noted that in America the leaves were usually gathered into bundles, twisted like cord and left to dry, after which they could be processed. In the colonies of Virginia and Maryland the seeds were planted in the month of February and the seedlings, when two inches high, were transplanted to greenhouses by Negro slaves, 'attentive and faithful persons, who serve the colonists on every plantation, and particularly those of tobacco' (p. 23). The crop was ready for harvesting when the leaf began to change color and curl so that its tip came to a sharp point. The leaves were carried into 'tobacco houses' or 'drying houses,' wooden barns with ample ventilation and earthen floors where they were spread to dry. They were then gathered into bundles and packed in barrels for transport.

The author goes on to describe the methods used in the Papal States. Here the seeds were sown in February or the beginning of March, transferred to pots or seedling boxes, and then replanted in orderly rows in fertile soil. When mature, the leaves were harvested, carried in baskets to a barn whose floor was covered with straw, and left to dry in bundles that were tied to poles with string. The dried leaves were reduced to a powder using a mill, grindstone, or mortar. The processed tobacco could then be sold by posting a notice with the grades being offered and their prices (the larger leaves produced a lower quality tobacco) in the shops of the Papal States (first quality, Canadian, Milanese, Florentine, flaked, red, and so on). The author furthermore observed that most of the tobacco for sale in Amsterdam came from Virginia, Cuba, and Brazil, and could be purchased in the form of leaves, cord, or powder.

The chapter dedicated to 'Osservazioni fisico-mediche intorno alla pianta del tabacco,' or the properties and pharmacological virtues of the plant, is particularly interesting. According to many authors it offered protection against the plague. Unguents containing tobacco were applied to heal ulcers, inflammation, tumors, skin diseases, and gout, while in powdered form it was effective in killing fleas and lice. The leaves, when chewed or smoked, could reduce phlegm—the humor responsible, according Hippocratic medicine, for weakness and lassitude. 'Questo è il motivo, per cui, usate nelle dette duemaniere, se ne fa un gran consumo; quantunque per altro i più fumino il tabacco piuttosto per passatempo, che per bisogno' (This is the reason why, used in the above described manner, it is consumed in large quantities, although most [people] smoke tobacco as a pastime rather than out of need) (p. 68).

In closing the reader is invited to take advantage of what doctors had learned about the beneficial properties of this plant, but nevertheless to exercise moderation 'poichè sopra l'uso, o l'abuso del tabacco non si possono stabilire regole generali' (since on the use, or abuse of tobacco, it is not possible to establish general guidelines, p. 79).

25. John Ellis (*c.* 1705–1776)

'Dionæa Muscipula or Venus's Flytrap' 1769

50.5 × 38.8 cm. Laid paper with fleur-de-lis watermark folded to 38.7 × 25.5 cm. Pen-and-ink drawing with manuscript text. Signed and dated on verso: 'London, Sept[r] 23[d] 1769 John Ellis.'

[*and*]

'Directions for bringing over Seeds and Plants…' 1769

38.8 × 25.2 cm. Pen-and-ink drawings of plant collecting boxes with manuscript text, accompanied by another folded leaf with additional manuscript text. Signed lower right: 'by John Ellis, F.R.S.'

An Historical Account of Coffee 1774

An Historical Account of Coffee. With An Engraving, and Botanical Description of the Tree. To which are added Sundry Papers relative to its Culture and Use, as an Article of Diet and Commerce. Published by John Ellis, F. R. S. Agent for the Island of Dominica. London: Printed for Edward and Charles Dilly. MDCCLXXIV.

4° 27.3 × 21.1 cm. A² B–K⁴ *i–ii* iii–iv *1–69 70* 71–*72* [76 pp.].

PLATES: Two folded engravings of a coffee plant, one uncolored and one hand-colored, signed 'J. Miller Sc.' and the hand-colored plate with the added engraved signature 'Simon Taylor pinxt.'

BINDING: Half calf and marbled boards; re-backed in modern calf.

PROVENANCE: Inscribed on the leaf preceding the title-page: 'From the Author.'

REFERENCES: Cleveland 514; Desmond, p. 231; Dunthorne 110; Henrey II.283–288 and Nos. 644, 681–682; Hunt 638; Nissen 588; Pritzel 2,667; Stafleu & Cowan 1,662.

JOHN ELLIS was an Irish-born linen merchant and an enthusiastic amateur naturalist. Ellis published in 1755, for which he is most known, *An Essay Towards a Natural History of the Corallines.* He made a significant contribution to the understanding of marine biology of coastal Great Britain and Ireland, and would later make several contributions to the study of botany. Ellis embarked upon the buying and selling of plants and natural history specimens with great success, eventually becoming a Fellow of the Royal Societies of London and Uppsala and a member of the Society for the Encouragement of Arts, Manufactures, and Commerce. In his capacity as a Royal Agent for West Florida in 1764, and for the island of Dominica in 1770, he ardently gathered tropical plants and insects for importation to Europe. Since his livelihood depended upon it, he developed a profound interest in techniques for the conservation and conveyance of plants and seeds. This represented a genuine challenge in a period when the only means of transport was a slow and hazardous journey by sea; not to mention it threatened delicate cargo with problems of dampness, vermin, and salt spray. He even raised this topic in his correspondence with the *princeps botanicorum,* Carlus Linnaeus, who described Ellis in the most flattering terms as 'a bright star of natural history' (see Desmond, p. 231).

Naturalists began to take an interest in the problem of how to transport plants and seeds in the sixteenth century, as a community of scientific men arose, eager to exchange information and specimens with colleagues all over Europe. It became a regular practice to enclose with their correspondence not only books and scientific drawings, but also small boxes containing botanical specimens (see No. 22). Many of these delicate specimens, however, failed to reach their destination intact as their recipients were forced to report. The problem became even more imperative as naturalists began to collect and study species from distant continents. Samuel Pullein, who published in 1759 *An essay towards a method of preserving the seeds of plants in a state fit for vegetation, during long voyages: For the improvement of the British colonies in America,* was one of the earliest authors to publish on the subject, and his account is in part a polemic against a certain 'Mr. E—s,' whom he claimed appropriated his researches.

Yet Ellis in a letter to Linnaeus dated 24 October 1758, states: 'I am at this time endeavoring to find out a method to bring exotic seeds from China, and other distant parts of the world, in a vegetative state' (see Henrey, II.284). Linnaeus and many other botanical scholars promoted this type of inventive design. This is evident as Linnaeus responded to Ellis in his undated letter, 'I am glad that you take such pains about preserving seeds. If your attempt should succeed, the whole world will be obliged to you, for our gardens will, through your means, be enriched ...' It seems appropriate that Ellis would publish a pamphlet on the subject along with a description of the Venus Fly Trap, a perfect example of why it was so important to master the transportation of plants: *Directions for bringing over seeds and plants, from the East-Indies and other distant countries, in a state of vegetation: Together with A Catalogue of such Foreign Plants as are worthy of being encouraged in our American Colonies, for the Purposes of Medicine, Agriculture, and Commerce. To which is added, the Figure and Botanical Description of a new Sensitive Plant, called Dionaea Muscipula, or, Venus's Fly-Trap.* This pamphlet was printed in London in 1770 by L. Davis of the Royal Society, and reprinted a year later in a new edition, in which the section on the Venus's Fly-Trap was replaced by *The Method of Catching and Preserving Insects for Collections* (Henrey, No. 644).

A unique, three-folio manuscript pertaining directly to this printed work mentioned above demonstrates the contribution of an avid collector. The manuscript in the Oak Spring Garden Library is part of a most interesting story based on a single plant, told thoroughly in Nelson and McKinley's book, *Aphrodite's Mousetrap: A Biography of Venus's Flytrap.* The story weaves together a cast of prominent characters. Arthur Dobbs (1689–1765), the provincial Governor of North Carolina, first described the plant to Peter Collinson (1694–1768), an English Quaker and naturalist, 'we have a kind of a Catch Fly sensitive which closes upon any thing that touches it.' Collinson enlisted the aid of his brother American Quaker, the nurseryman John Bartram (1701–1777), in order to obtain a sample. In his correspondence with Collinson, Bartram applied the vulgar slang 'Tipitiwichet' to the plant, which likened the leaves to the female genitalia, and Collinson was even more eager for a live sample so as to 'forme Some Idea of the Wagish Plant as Wagishly Described.' A dried specimen of the leaves finally reached Collinson after many miscarriages and they were examined and described by the Swedish botanist Daniel Carl Solander (1733–1782), who was about to embark on a voyage with Joseph Banks and James Cook to Tahiti in order to observe a rare transit of Venus. Solander, covering Bartram's patois with a respectable, scientific Latin cloak, named it *Dionaea crinita.* Collinson also sent a leaf to Linnaeus—'only to him would I spare such a jewel!'

In the meantime, William Young successfully transported live Flytraps to London, one of which blossomed. Ellis was now in a position to sketch the plant and engage James Roberts to make a polished engraving. Roberts was a fairly well-known artist who showed his work regularly at the Royal Academy of Art between 1773 and 1799, and was awarded

JOHN ELLIS,
Venus's-flytrap
(*Dionaea muscipula*)

the annual prize of the Society of Arts in 1766. John Ellis published a letter in a London paper in September 1768 describing the unknown wonder and officially naming it—keeping with his colleagues' *jeu d'esprit—Dionanae muscipula.* Ellis evangelized the discovery, sending the engraving and a description in letters to various personages, including Linnaeus, Henri Louis Duhamel du Monceau, and the Dutchess of Norfolk.

Half of the first sheet of the manuscript, conserved at the Oak Spring Garden Library, is taken up with four drawings of boxes designed for the transporting of plants. Since the manuscript does not correspond exactly to the printed text, we may hypothesize that it constituted a first draft or summary that the author subsequently elaborated upon for publication. Meticulously executed in pen, ink, and gray wash, these drawings reproduce the illustration that appears on the verso of the title-page to *Directions for bringing over seeds and plants*. Each drawing signed 'by John Ellis FRS' is accompanied by an explanatory text that is repeated, with only slight variations, in the printed work. Beneath the drawings is the text in the author's hand that begins with the title *'Directions for bringing over Seeds and Plants from the East Indies and other distant Countries in a State of Vegetation'* and continues on the back of first folio and then for the two sides of the second folio.

Ellis not only provides concrete illustrations of boxes, but he also cites numerous examples of how he successfully shipped various plants from distant continents. Both the manuscript and pamphlet mention the 'The Alpine Strawberry sent to England from Turin to Henry Baker Esq. FRS by pressing the pulp with Seeds thin on paper, and letting it dry before they were inclosed' and how 'The paper mulberry from China was brought much in the same manner.' Ellis recommends that 'The inside of the Boxes should be washed with water that has been impregnated with Sulphur,' and notes that one could also line plant boxes with tobacco leaves since 'Tobacco is universally known to destroy Insects.' The pamphlet concludes with an invitation (that does not appear in the manuscript) to 'curious gentlemen, as well as intelligent seedsmen and gardeners' to take into consideration and apply its recommendations, that were the fruit of the author's personal experience.

The third folio bears a drawing in pen, ink, and gray wash of the 'Venus's Fly trap,' that reappears in the engraving preceding the title-page to the second section of the pamphlet, *A botanical description of the Dionaea Muscipola, or Venus's Fly-Trap. A newly discovered sensitive plant: In a Letter to Sir Charles Linnaeus, Knight of the Polar Star, Physician to the King of Sweden, and Member of most of the Learned Societies of Europe*. The manuscript, entitled '*Dionaea Muscipula or Venus's Fly trap, a newly discovered Sensitive Plant from the Swamps of N. Carolina North America,*' is an abridged version of the published text, although, unlike the latter, it does not take the form of a letter. It presents a description of this curious plant that constituted a novelty to be studied alongside already known sensitive plants such as the *Mimosa pudica*. The text continues on the back of the folio and is signed and dated 'London

JOHN ELLIS,
Travelling containers
for seeds and plants

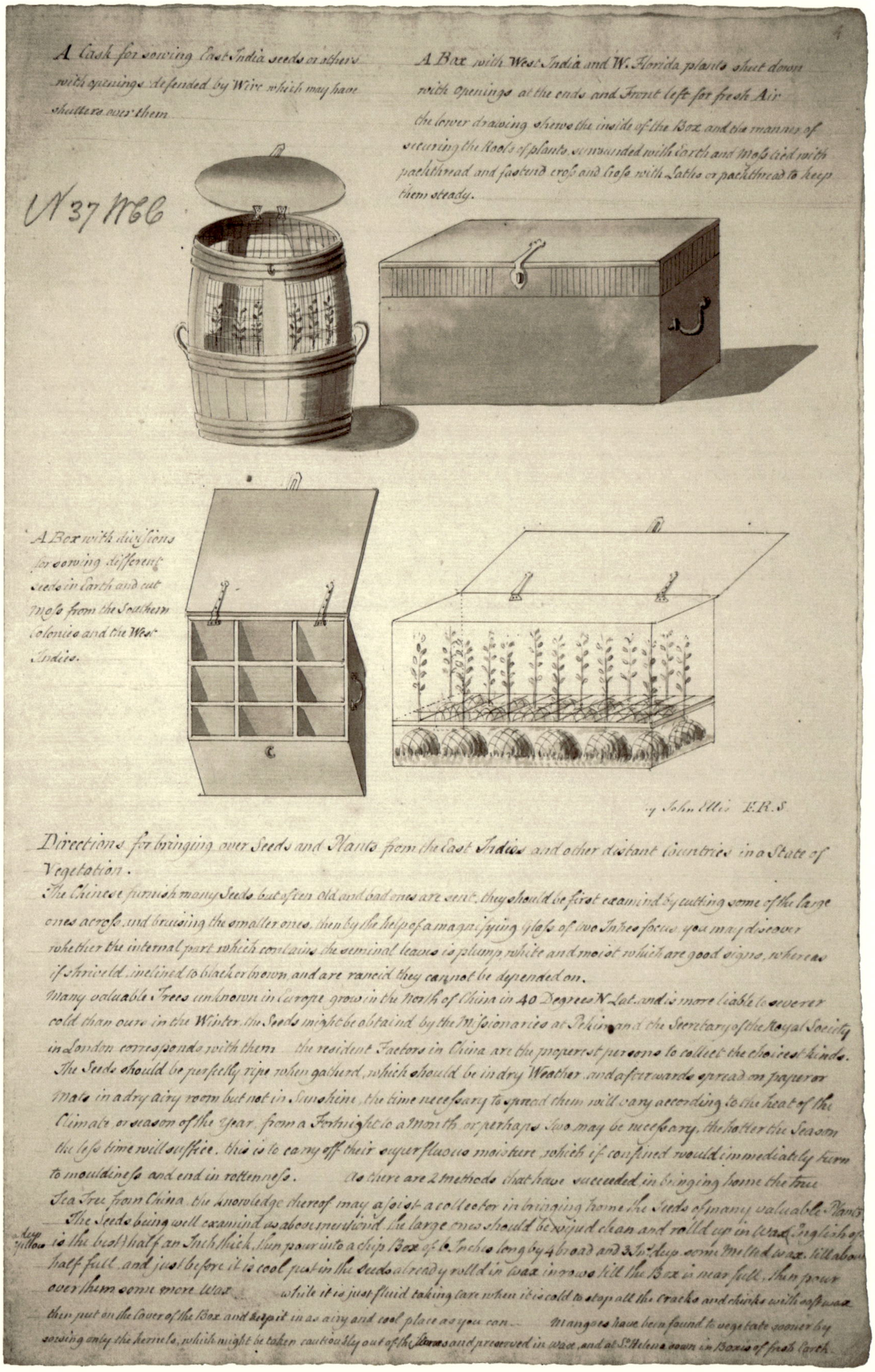

Sept. 23th, 1769 John Ellis' with the notation 'Copy of what was wrote to Sr. Cha. Linnaeus.' Like the plant boxes, this botanical illustration is beautifully and accurately drawn, testifying to Ellis' considerable talent as an artist.

As an agent for the Island of Dominca, Ellis became interested in certain economic plants that could then be introduced into the British Colonies in the West Indies. For this purpose Ellis published, *A Description of the Mangostan and the Breadfruit* in 1775, and in the preceding year *An Historical Account of Coffee*. In this latter work, conserved at the Oak Spring Garden Library, Ellis presents the botanical characteristics of the genus *Coffea*, a detailed description of the plant, and a short history of the beverage from earliest times to its introduction in Europe. According to the author, the coffee plant was first brought to Amsterdam in 1690 by Robert Witsen, the Director of the East India Company; this assertion is repeated by Civinini (see No. 23). The author also states that the official authorities of Amsterdam presented King Louis XIV with the gift of a coffee plant. This created a sensation at the Académie des Sciences, and *Coffea* was planted in the University of Paris' botanical garden at Marly by Antoine de Jussieu (1686–1758), professor of botany. This exposition is followed by insightful traveller tidbits, including *The Civil and Natural History of Jamaica* by Patrick Brown (1720–1790), a letter received by Ellis from the physician and naturalist John Fothergill (1712–1780), plus another letter written to Fothergill by Georg Scott (1720–1770), the ex-governor of Dominica. The work concludes with a list of items sent from England to the islands of the British West Indies. Considering the letters mentioned, we may assume that Fothergill contributed extensively to this small book and parts of it along with the plates are published in *The Works of John Fothergill*, 1783.

An Historical Account of Coffee was written with the intent to encourage the growth of coffee in the West Indies, and although the book contributes nothing new in terms of scientific knowledge, it opens with a beautiful illustration, an engraving of a branch of the coffee plant bearing fruit and flowers, with details of both in the lower part of the page. The illustration is signed by the engraver, John Miller (*c.* 1715–*c.* 1790). This plate is followed by the same engraving, this time hand colored and carefully refinished by the botanical artist Simon Taylor (1742–1796), as inscribed by the engraved signature 'Simon Taylor Pinxt.' It may be noted that in this period the artist Johann Müller of Germany, who changed his name to John Miller, and his English colleague Simon Taylor were the two most important botanical illustrators in England. They had the opportunity to work with the great artist Georg Dionysius Ehret when he was engaged to paint the botanical collection of Dr. Fothergill's thirty-acre garden in Upton near Essex, on the eastern edge of London (see Calmann, p. 64 and *An Oak Spring Flora*, p. 206).

Ellis contributed significantly to the study of botany and natural history, writing the descriptions of several other species, and later publishing in 1786, *The Natural History of*

AN
HISTORICAL ACCOUNT
OF
C O F F E E.
WITH
An Engraving, and Botanical Defcription of the Tree.
TO WHICH ARE ADDED
Sundry Papers relative to its Culture and Ufe, as an Article of Diet and of Commerce.
Published by JOHN ELLIS, F. R. S.
AGENT FOR THE ISLAND OF DOMINICA.
LONDON:
PRINTED FOR EDWARD AND CHARLES DILLY.
MDCCLXXIV.

JOHN ELLIS, *An Historical Account of Coffee,* 1774. Title-page and frontispiece of coffee plant (*Coffea Arabica*)

Many Curious and Uncommon Zoophytes. His own collection of plants, sent to him by colleagues, is conserved in the Smith herbarium. After his death it was his daughter that gave his papers to Sir James Edward Smith (1759–1828), which contained summary sketches of the Venus Flytrap now preserved at the Linnaean Society, along with a dried herbal containing plants sent by Ellis to Linnaeus (Nelson, figs. 6–8). Linnaeus was said to have called the plant a 'miraculum naturae,' as Darwin would later declare this carnivore of the vegetable kingdom 'the most wonderful plant in the world.'

26. Baldassarre Cattrani (*fl.* 1770–1810)

Exoticarum atque indigenarum plantarum sthegmata ad naturam forma ac coloribus expressa collegit Baldasar Cattrni [*sic*].

41.3 × 17.3 cm. Manuscript title-page leaf and 63 pencil and watercolor drawings on light grey paper.

REFERENCES: *An Oak Spring Flora,* No. 59; Tongiorgi Tomasi, 2002.

1. ALOE AFRICANA.
2. AMARANTHUS OCIMOIDES.
3. APOCINUM VENETUM.
4. BERBERIS VULGARIS.
5. BORAGO OFFICINALIS.
6. BORRAGO ORIENTALIS.

7. BRASICA ERUCA.
8. CAMPANULA BONONIENSIS.
9. CAMPANULA GLOMERATA.
10. CAMPANULA GRANDIFLORA.
11. CARDAMINE PRATENSIS.
12. CENTAUREA.
13. CENTAUREA NIGRA.
14. CHELIDONIUM MAYUS.
15. CONUALLARIA POLYGONATUM.
16. CONUOLVULUM ARUENSI.
17. CONUOLUULUS SEPIUM.
18. CONUOLVULUS TRICOLOR.
19. CYNANCHUM ERECTUM.
20. ERIGERON.
21. GLECOMA HEDERACEA.
22. HEDERA QUINQUFOLIA.
23. HYACINTHUS DE VOLOVUR POURPRÉ.
24. HYACINTHUS FLORA.
25. HYACINTHUS OPHIR.
26. HYACINTHUS PASTOR FIDO.
27. HYOSCYAMUS ALBUS.
28. HYPERICUM ANDROSÆMUM.
29. LEUCOJUM ASTIRUM.
30. LYCHINIS PURPUREA.
31. LYCHINIS VISCOSA.
32. MATRICARIA PARTHENIUM.
33. MELISSA ROMANA.
34. MIRABILIS JALAPPA FLO. LUTEO.
35. MIRABILIS JALAPPA FLO. RUBRO.
36. MOLDAUICA FLO. CAERULA.
37. NERIUM FLO. ALBO.
38. NERIUM FLO. CARNEO.
39. NERIUM FLO. LUTEO.
40. NICOTIANA TABACUM.
41. ORNITHOGALUM VMBELLATUM.
42. OXALIS ACETOSELLA.
43. PANCRAZIO MARITIMO.
44. PAPAUER SOMNIFERUM.
45. PHYSALIS PUBENSCENS.
46. PHYTOLACCA DECANDRA.
47. POLYGONATUM ORIENTALA.
48. PRUNELLA.
49. RANUNCULUS BULBOSUS.
50. RANUNCULUS SYLUESTRIS.
51. RAPUNTIUM COCINEO.
52. RESEDA VNDATA.
53. RUTA GRAUEOLENS.
54. SALVIA PRATENSIS.
55. SATUREY HORTENSIS.
56. SCARLATEA CARDINALIS.
57. SILENE ARMERIEA.
58. SOLANUM DULCAMARA.
59. SONCHUS MARITIMUS.
60. VERONICA TEUCRIUM.
61. VINCA MAJOR.
62. VIOLA ALBA.
63. ZINIA FLAUESCENTE.

The following drawings are illustrated in *L'Orto Rappresentato*, edited by Luciana Sitran Rea: *Aponicum venetum*, [*Aquilegia canadensis*], *Campanula glomerata*, *Convolvulus sepium*, [*Veratrum nigrum*], pp. 173–183.

THE ARTIST Baldassare Cattrani was considered for the first time in the volume of *An Oak Spring Flora* (No. 59, pp. 217–220) where a manuscript, *Stirpium Imagines*, containing a collection of seventy-five botanical paintings by this gifted artist was discussed. These works were executed by Cattrani at the Botanical Garden of Padua in 1776, when it was under the direction of the *prefetto* Giovanni Marsili. The present collection of drawings can be linked to this manuscript, both in terms of style and chronology. They portray exotic and indigenous species drawn from life, as one may read in the title, where the name of the author is written incorrectly as 'Cattrni' instead of 'Cattrani.' In both of these collections the specimen is portrayed with a cut stem and hence without its roots, but often with enlarged details showing significant parts of the plant, such as its seeds and fruit.

Disappointingly, very little is known about the artist who is simply described as being 'from Rome' in the title to *Stirpium Imagines*, although it may be hypothesized that he was a relative of the more famous Tomaso Cattrani (1723–1824), an architect originally from Città di Castello in Umbria who spent most of his career in Rome. Recently on the grounds of stylistic affinities it has been possible to attribute to Baldassarre Cattrani's hand another voluminous corpus of more than 200 drawings on paper contained in the miscellaneous collection *Icones Plantarum Selectarum* at the library of the Botanical Garden of Padua—specifically drawings 132 to 360 (with the exception of Nos. 166, 177, 309, 344, 350, 358, and 360). Although the execution is more mature and polished, the manner in which the species have been documented is quite similar to that of the two manuscripts in the Oak Spring Garden Library; the plants are shown without their roots, and often details—sectioned and enlarged—are presented in the lower part of the page. The name of the artist is recorded in a document to be found in the Paduan library, where a list of the drawings waiting to be bound in 1837 includes the entry 'Catrani Icones Plantarum.'

It seems to us reasonable to hypothesize that the two manuscripts in the Oak Spring Garden Library—*Stirpium Imagines* and the very similar *Exoticarum atque Indigenarum*—were executed by the artist at an earlier stage in his career than the drawings in *Icones Plantarum Selectarum*. An article by the botanist Pier Andrea Saccardo, who possessed a profound knowledge of the history of the botanical sciences in Italy and was the author of *La Botanica in Italia* (Venice 1895), testifies to how greatly Baldassarre Cattrani was esteemed as an artist at the beginning of the nineteenth century. In a discussion of the botanical drawings of the Abbot Angiolo Franciosi, who was a contemporary of Cattrani, Saccardo cites the appraisal of Angela Nardo Cibele of Venice, an expert in botanical painting with whom he corresponded regularly: 'Bellissime sono le 42 tavole di Cattrani che vorrei che Ella vedesse, e che possono stare al pari di quelle del Franciosi' (The forty-two drawings by Cattrani that I would like you to see are most beautiful and quite the equal of those of Franciosi—Saccardo, 1902). This declaration also allows us to infer that another collection of forty-two botanical drawings by Cattrani were found in Venice at the time this letter was written.

There is evidence to show that Cattrani was still working for the Botanical Garden of Padua at the beginning of the nineteenth century, this time for the *prefetto* Giuseppe Antonio Bonato. In 1806 he completed a series of botanical drawings that were eventually acquired by a private collector in Italy, but which were then dispersed following their sale

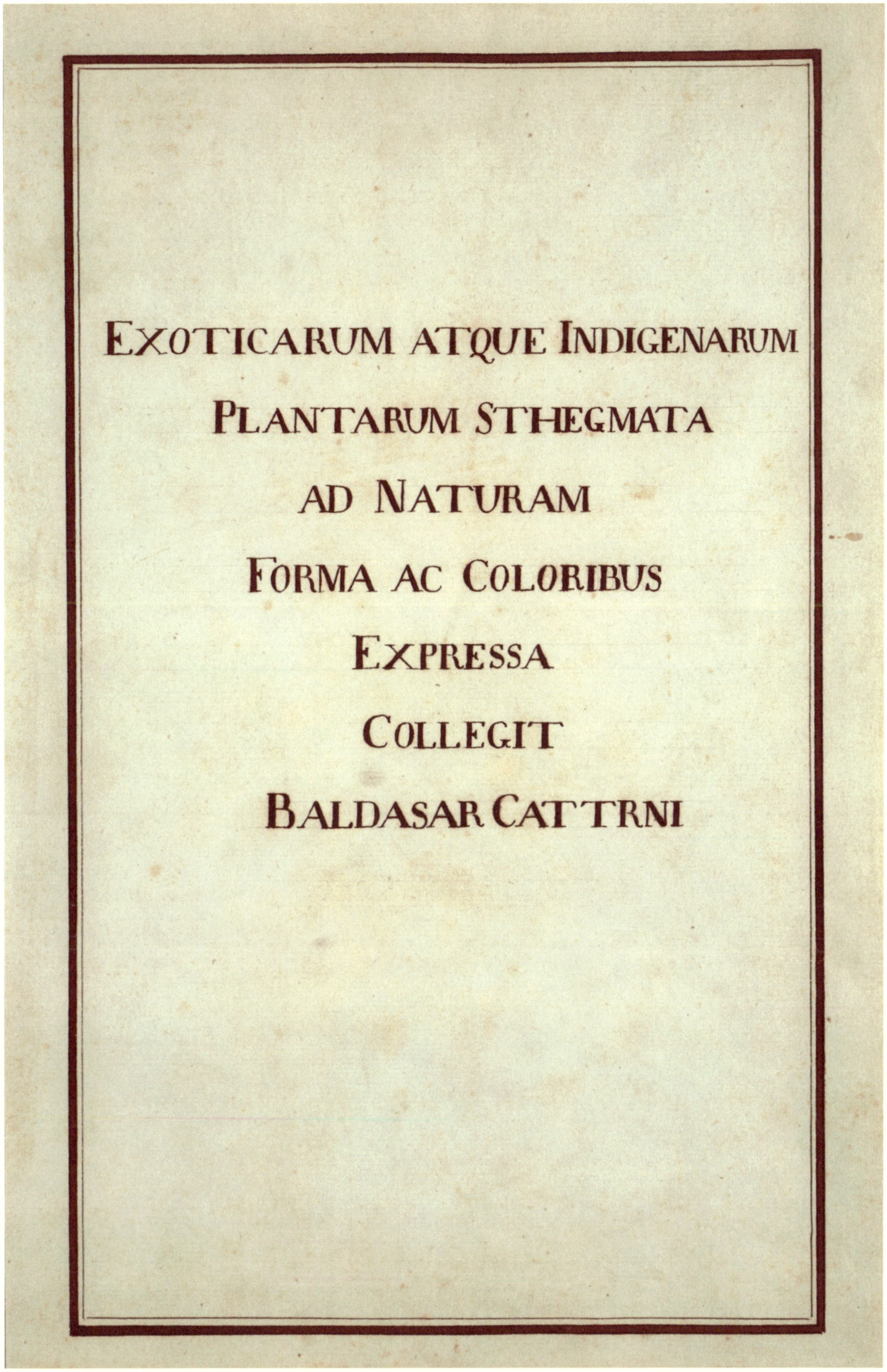

EXOTICARUM ATQUE INDIGENARUM
PLANTARUM STHEGMATA
AD NATURAM
FORMA AC COLORIBUS
EXPRESSA
COLLEGIT
BALDASAR CATTRNI

BALDASSARE CATTRANI, *Exoticarum,* 1776–1800. Manuscript title leaf

BALDASSARE CATTRANI, *Exoticarum,* 1776–1800. 'Convallaria Polygonatum,' Angular Solomon's seal (*Polygonatum odoratum*)

BALDASSARE CATTRANI, *Exoticarum,* 1776–1800. 'Hypericum Androsæmum,' St.-John's Wort or Tutsan (*Hypericum androsaemum*)

at an auction by Sotheby's in Florence in 1977. It has been possible to retrace an alphabetical manuscript index of the plants in this series, together with the original title-page, in which a charming floral garland composed of roses, tulips, anemones, lilies, pansies, and irises surrounds the title *Stirpium Imagines quas ex Horto Patavino delineavit ex vivis coloribus ad naturam fidelissime expressit Balthassar Cattrani Romanus Curante et dirigente Josefo Antonio Bonato ... Anno Millesimo Octingentesimo Sexto.*

From this body of evidence it may be concluded that Cattrani worked for the Botanical Garden of Padua from at least 1776, the date which appears on the first Oak Spring Garden manuscript, until 1806, although perhaps not always without interruption and not always full-time. For example, another substantial body of drawings unconnected with the work of documenting the plant collections at the Botanical Garden can be attributed to this prolific and tireless artist. He in fact devoted himself to the execution of an imposing and refined series of paintings, mostly on vellum, a large number of which today are conserved in five folio volumes in the collection of rare books at Dumbarton Oaks (Washington, D.C.); a further seven drawings from the same corpus can be seen at the Oak Spring Garden Library.

This collection, which occupies an important place in the panorama of early nineteenth century botanical illustration, originally came from the magnificent library of the Viceroy of Italy and Duke of Leuchtenberg, Eugéne de Beauharnais, the son of Josephine Bonaparte by her first husband, Alexandre Beauharnais. In May 1935 a sale was held in Zurich of the manuscripts and printed books belonging to the prince, and on that occasion included in Lot No. 71 was:

> *Cattrani, Baldassarre, Collection de Botanique. Recueil de 1640 acquerelles originales, peintes sur vélin et représentant des fleurs et des plantes. Grandeur des feuilles: 53.5 × 38 cm, 24 vols. Peint en 1806. Dos orné en or et peinté froid. Magnifique oeuvre de botanique. Chaque feuil est encadré d'un double filet noir et porte en bas le nom de la plante, dans quelques volumes ils sont omis. Le premier volume contient plusieurs aquerelles peint sur papier et en général celles de ce volume ont un peu suffert par l'humidité. Les aquerelles sont exécutées avec beaucoup de gout et de soin et l'artiste a su fixer avec son pinceau léclat des couleurs et les nouances que nous admirons dans les fleurs. Comme les Liliacées de Redouté, cette oeuvre a été peinté Malmaison sur les ordres de l'imperatrice Josephine.* (*Bibliothèque Eugéne Beauharnais ... provenant du Chateau de Seeon en Baviére. Vente aux enchéres 23 et 24 mai 1935 Zûrich, Milan, Hoepli, 1935, p. 21*).

On the basis of the detailed description of the sheets and their dimensions provided, it is possible to link both the three hundred drawings in Dumbarton Oaks and the seven at the Oak Spring Garden Library to the dispersion of this extraordinary collection of works painted at the Empress Josephine's magnificent estate at Malmaison. In each of these drawings a flowering stem of the plant is portrayed, without its roots but with certain parts shown in enlarged detail, as was the artist's habitual practice.

In order to reconstruct the voluminous and diverse production of this artist, it is therefore of great interest to analyze this new manuscript of drawings of exotic and indigenous species, *Exoticarum atque Indigenarum* which, as we have seen, is quite similar to the drawings that were produced by Cattrani in *Icones Plantarum Selectarum* for the Botanical Garden of Padua. The artist shows his innate talent for the composition of large and airy pictures. The plants, whether exotic examples such as *Aloe Africana, Mirabilis Jalappa Flore Luteo*, and *Mirabilis Jalapa Flore Rubro*, or familiar Italian species such as *Borago Officinalis* and *Convallaria polygonatum*, are portrayed in a frontal view and enhanced by the addition of such significant elements as the seeds or flowers in enlarged detail. The clarity and sobriety of the *mise en page* is in keeping with the iconographic tradition of the herbal and its scientific functions. The drawing of the tobacco plant was clearly intended to document the observations by a botanist of this curious plant from the New World; the artist has portrayed a large leaf, with its ornate network of veins meticulously reproduced and the detail of a small branch with two pink flowers elegantly arranged to the right.

Baldassare Cattrani was an accomplished draftsman with a striking sense of color; after setting down a quick sketch that captured the flexible contour of the plant, he chose his colors and applied them with great assurance. His masterly technique can be seen in the portrayals of *Vinca Major* and *Convolvulum Aruensis* [sic].

27. Ambroise-Marie-François-Joseph Palisot de Beauvois (1752–1820)

[*Volume I*]: Flore d'Oware et de Benin, en Afrique, Par A.M.F.J. Palisot-Beauvois, Correspondant (ancien associé) de l'Institut national, membre de la Société des Sciences et Arts du Cap-Français de St.-Domingue, membre de la Société Philosophique de Philadelphie, des Sociétés Galvanique et des Observateurs de l'Homme; correspondant du Muséum d'Histoire Naturelle de Paris, de la Société Philomatique, des Sociétés des Sciences et Arts de Bordeaux, de Lille, et professeur à l'Athénée des Étrangers. [*wavy rule*] A Paris, De L'Imprimerie de Fain Jeune Et Compagnie. An XII. – 1804.

[*Volume II*]: Flore d'Oware et de Benin, en Afrique. Par A.M.F.J. Palisot-De-Beauvois, Membre de l'Institut de France; des Sociétés des Sciences et Arts du Cap-Français, de Saint-Dominigue, Philosophique de Philadelphie, Galvanique, des Observateurs de l'Homme, et de l'Athénée des Arts; Correspondant de Muséum d'Histoire Naturelle de Paris; des Sociétés Philomathique, des Sciences et Arts de Bordeaux, de Lille, Douai; et Professeur d'Histoire Naturelle à l'Athénée de Étrangers. Tome Second. [*wavy rule*] A Paris, De L'Imprimerie De Fain Et Compagnie. M. DCCC. VII.

47 × 30 cm. Volume I: π^2 χ^2 (+ $2\chi 1$) 1–25^2 [7] x–xii *1* 2–100 [110 pp.] 120 plates. Volume II: 1–24^2 *1–5* 6–92 *93* 94–95 *96* [96 pp.] 121 plates.

PLATES: 241 stipple engraved, engraved, and etched, in two states, printed in color and finished by hand, after

Mirbel, Prêtre and others. In volume two there are three double-page plates of No. 78, *Napoleona imperialis*, two uncolored and one colored. Loosely inserted are three drawings for plates 14 (*Manisuris polystachia*), 16 (*Xilopia undulata*), and 18 (*Hedisarum lasiocarum*).

BINDING: Original quarter sheep and marbled paper boards.

PROVENANCE: 'Bibliotheque de Drapiez' stamped in volume one on verso of front free endpaper; tipped in volume one between front free endpaper and title-page is a letter from Palisot de Beauvois to Pierre Auguste Joseph Drapiez (1778–1856); pencil note on half title by Drapiez stating there are three drawings by Mirbel that were given to him by the author.

REFERENCES: Cleveland 683; Lack 51; Nissen 1,481; Stafleu & Cowan 7,212.

THE FRENCH naturalist and traveller Ambrose-Marie-François-Joseph Palisot de Beauvois was born in Arras to a family of the *petite noblesse*. He studied jurisprudence and after earning his degree was appointed an advocate to the parliament of Paris in 1772, and later receiver-general. While at university, however, Palisot had also taken up the study of botany under the guidance of François Joseph Lestiboudois (1759–1815), a professor of natural history at Lille, and he continued afterwards to pursue his scientific interests. Indeed, although well placed for a successful career in the French civil service, he longed to visit foreign climes. After failing more than once to gain an assignment as a naturalist to the expeditions of Casten Niebuhr (1733–1815) and Jean François de La Pérouse (1741–1788), Palisot finally managed, in 1786, to embark on a French ship that was being sent to the Gulf of Guinea on a mission to establish diplomatic relations with the king of Oware. He would spend nearly two years in West Africa, mostly in Oware and Benin, studying the natural history of the region.

Then in 1788 Palisot set off for the Caribbean, where he settled in the French colony of Saint-Dominigue (now Haiti) until 1791; a privileged resident, he was admitted to the colonial assembly and the Conseil Superieur. Given his background, it is not surprising that he was vehemently opposed to the abolition of slavery and was incensed at the activities of English abolitionists, whom he accused in a pamphlet of fomenting discontent among the black inhabitants. Fearing that the island was on the brink of upheaval, he travelled to the United States to petition for support in case of an eventual native rebellion. When he returned, Saint-Dominigue was indeed caught up in a violent cataclysm; he witnessed the massacre of white colonists, was himself imprisoned, and saw his valuable scientific collections and manuscripts destroyed.

When he was finally allowed to leave the island, Palisot returned to Pennsylvania with the intention of making his way back to France, but once again the tide of history had turned against him. Paris was in the grip of the Reign of Terror and he was denounced *in absentia* as an *émigré*. His property was confiscated and, deprived of his private income, Palisot had to make a living for a time as a musician in a circus orchestra in Philadelphia. He continued with his study of natural history undeterred, however, and resumed collecting plants, insects, reptiles, birds, and minerals.

Palisot's fortunes turned once again when the diplomat and chemist Pierre Adet (1763–1832) obtained funds for a scientific expedition to the interior of the United States and invited him to participate. The French naturalist sojourned extensively in the American wilds, even living for a period with two different Indian tribes, the Creek and the Cherokee. When he returned to Philadelphia, Palisot was elected a member of the American Philosophical Society and before this forum he presented many of his findings, including data on several cryptogams, and the description of several new species of rattlesnakes, and a hitherto unknown plant, *Heteranthera reniformis*. He established close ties with the physician and man of learning, Caspar Wistar (1761–1818), professor of anatomy at the University of Pennsylvania Medical School and the author of the first American text on anatomy, who shared his passion for the botanical sciences.

Palisot was involved in many scientific projects during this period, including the reorganization of the natural history cabinet of the painter, naturalist, and soldier Charles Willson Peale (1741–1827, cf. *A scientific and descriptive catalogue of the Peale Museum*, by C.W. Peale and A.H.F.J. Beauvois, Philadelphia, 1796). This collection was immortalized by Peale himself in a celebrated painting, 'The Artist in his Museum' (1822), today conserved in the Pennsylvania Academy of Fine Arts.

While in the midst of preparations for another scientific expedition, news arrived from France that his name had been expunged from the Assembly's list of *émigrés*, and Palisot decided to return to Paris, bringing back with him his rich collection of natural history specimens. Dividing his time henceforth between the capital and his estate in Picardy, he was able to devote himself to the study of botany while remaining in contact with many naturalists, in particular Adrien-Laurent-Henri de Jussieu (1797–1853), whose dried herbal included many specimens that had been sent to him by Palisot.

During the remaining two decades of his life Palisot wrote an impressive number of works, among them the paper 'Description du tour naturel dans la Caroline du Nord,' published in 1811 in *Annales du Museum d'Histoire Naturelle*, and an important work on tropical insects, *Insectes recueillis en Afrique*, which was published in parts between 1805 and 1821. His monumental two-volume *Flore d'Oware et de Benin* was printed between 1804 and 1821, while *La Muscologie ou traité sur les mousses* appeared posthumously in 1822. The naturalist also presented reports on the native tribes that he had encountered during his voyages and on the musical instruments used by them at the prestigious Societé des Observateurs de l'Homme, a scientific academy devoted to the study of primitive peoples. In 1821 a biography of the life and work of Palisot was published by the botanist and voyager Arsenne Thiébault de Bernaud (1770–1850).

La Flore d'Oware et de Benin unquestionably represents Palisot's most important contribution to the science of botany. It opens with an 'Extrait de la Decade Philosophique' written by his friend de Jussieu, and is followed by a short account of Palisot's trip to Guinea and a few biographical notes that testify to the naturalist's indefatigable courage and determination. For example, he recalls that as soon as he set foot on dry land after the long voyage by sea from Lisbon, he began searching for plants, sending specimens of his most interesting finds to de Jussieu in Paris. In Oware he fell ill during a virulent epidemic—'attributable to the humidity' and the many marshy areas in the region—that killed many inhabitants. Having recovered his health, he set off again into the wilderness to continue his explorations, sending a constant stream of botanical and zoological specimens to de Jussieu. After a time, however, exhausted by the insalubrious climate Palisot was forced to leave this fascinating country. Discovering that there was a ship bound for Santo Domingo, he set sail for the New World and fresh adventures.

In his *Avant-Propos* the author pays tribute to the tradition of the great French naturalist-explorers, citing in particular Joseph Pitton de Tournefort's (1656–1708) journey to the Levant, the intrepid expeditions of Joseph de Jussieu (1704–1779) to Peru, and Joseph Dombey (1742-1793) to Chile. He then proudly underlines that 'Les royaumes d'Oware et de Benin, dont je publie aujourd'hui la Flore, n'avaient été vus et visités par aucun naturaliste avant moi,' a circumstance that permitted him to present a number of hitherto unknown genera such as *Omphalocarpum, Myrianthus*, and *Spathodea* (the African tulip tree). The text is accompanied by an impressive set of illustrations, the author explaining that the botanical drawings were the work of Charles François Mirbel (also known as Brisseau, 1776–1854), a naturalist at the Musée d'Histoire Naturelle and professor at the Athénée in Paris. According to Palisot, Mirbel possessed 'le talent d'un parfait dessinateur ... et la touche d'un habile artiste' and had been appointed by Mme Bonaparte as supervisor of her garden at Malmaison (1802–1806, p. XII); the naturalist-artist also compiled a catalogue of the garden for his illustrious patroness. Two other artists contributed to *Flore d'Oware et de Benin*: Mademoiselle Sophie de Luigné, who was also the author of many scientific drawings for the *Annales du Musée*, and Jean Gabriel Prêtre. Their drawings were transformed into stipple engravings by several artists of whom today nothing is known—L'Epine, Lambert, Bouquet, and Canu. Many of these were printed at the famous works of Langlois in Paris, who supervised much of Redouté's output.

In the splendid copy of *Flore d'Oware et de Benin* conserved at the Oak Spring Garden Library all of the illustrations appear in two states—black and white, and then in color, retouched in many cases by hand. The method of color-printing developed in France for stipple engraving was known as *à la poupée*. By this technique, colored ink was applied to areas of the copperplate with a small piece of felt twisted into a cone shape (called a 'dollie'), wiped clean, and then a second or third color was applied to different areas of the plate. Thus, a multi-colored print could be obtained from one pull of the press. Black ink

AMBROISE MARIE FRANÇOIS JOSEPH PALISOT DE BEAUVOIS, *Flore d'Oware et de Benin,* 1804–1807. *Napoleona imperialis*, a cauliflory, showing the blue flowers on the limbs was used for the plate title, number, details of the fruit, seed, etc., and hand-coloring was often used to finish the details. Particularly noteworthy are three double-page illustrations (two in black and white, and one in color) of *Napoleona Imperialis*, an unusual species characterized by a 'cauliflory' inflorescence. This plant still remains something of an anomaly today, being assigned by some botanists to the genus *Lecythidaceae* and by others to *Barringtoniaceae*. The magnificent color plate (vol. 2, pl. LXXVIII) depicts a stem of azure flowers and bears the names of the artist, the engraver and the printer: 'J.G. Prêtre,' 'Lambert,' and 'Langlois.'

IV

HERBALS BY HERBALISTS, PHARMACISTS, AND PHYSICIANS

ANTOINE DU PINET,
Historia Plantarum, 1561.
Title-page

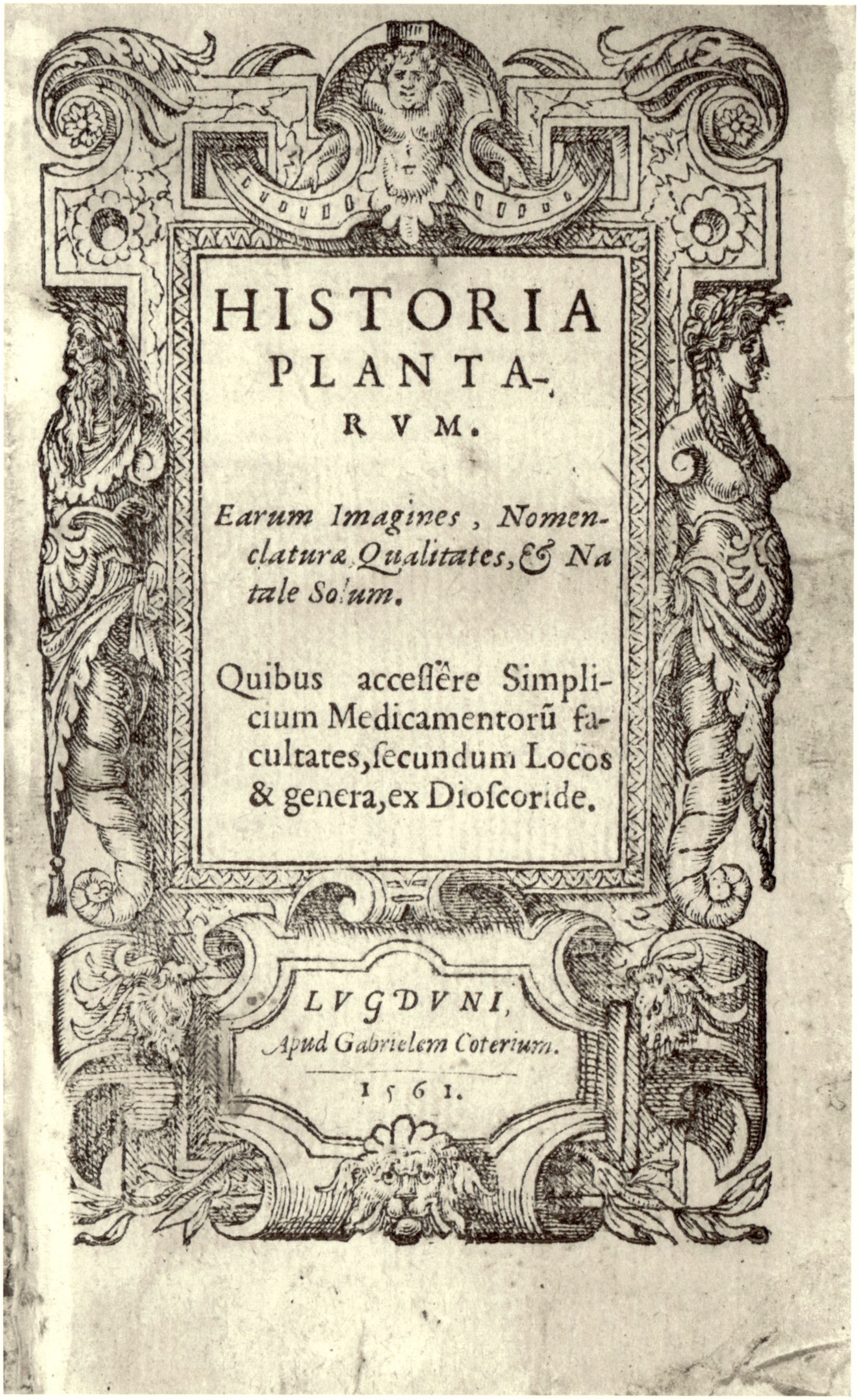
HISTORIA
PLANTA-
RVM.

Earum Imagines, Nomenclaturæ Qualitates, & Natale Solum.

Quibus acceſſere Simplicium Medicamentorũ facultates, ſecundum Locos & genera, ex Dioſcoride.

LVGDVNI,
Apud Gabrielem Coterium.
1561.

28. Antoine Du Pinet de Noroy (1515?–1584?)

[*Within architectonic woodcut border*]: Historia Plantarum. Earum Imagines, Nomenclaturæ Qualitates, & Natale Solum. Quibus accessêre Simplicium Medicamentoru[m] facultates, secundum Locos & genera, ex Dioscoride. [*within cartouche compartment*]: Lugduni Apud Gabrielem Coterium. [*rule*] 1561.

12° 10.7 × 7.6 cm. a–z⁸ A–R⁸ 2a–2q⁸ *1–2* 3–640 (35 as 36, 93 as 91, 401 as 371, 566 as 556, 597 as 592) 1–229 (66 as 82, 67 as 83, 70 as 86, 71 as 87, 74 as 90, 75 as 91, 78 as 94, 79 as 95, 82 as 66, 83 as 67, 86 as 70, 87 as 71, 90 as 74, 91 as 75, 94 as 78, 95 as 79) [*27*] [896 pp.].

PLATES: 636 woodcuts of plants in text.

BINDING: Quarter calf and brown paste paper over boards.

REFERENCES: Cleveland 87; Hunt 85; Nissen 565; Plesch, p. 213; Pritzel 2,539; Scrase, pp. 14–15.

Antoine du pinet de noroy (or du Pinay), who was born in Besançon and died in Paris, was not only a polymath but also a competent draftsman. Judging from an album of forty-eight floral paintings in watercolor preserved at the Fitzwilliam Museum in Cambridge, it is evident that he was a talented artist (Scrase, pp. 14–15). The album, made to celebrate the arrival and marriage of Louise of Lorraine, Queen of France (1553–1601), begins with a letter addressed to her and signed with the initials 'A.P.' Du Pinet's knowledge was wide-ranging and included, for example, an interest in town planning, for he compiled a work entitled *Plantz, pourtraitz et descriptions de plusieurs villes et forteresses, tant de l'Europe, Asie, et Afrique, que des Indes, et terres neuves...*, published in Lyon in 1564. Du Pinet must have been well respected, for authors such as Georg Braun and Franz Hogenberg attributed to him the drawings of the Mexican city of Cusco, illustrated in the first volume of the celebrated atlas, *Civitates Orbis Terrarum* (Cologne, 1572). His inquiring mind led him in other directions, for he prepared the French edition of an Italian treatise on navigation, *Traité des travaux et privilèges des galères de Don Antoine de Guevara* (Lyon, 1560), a subject of consuming interest in this period of sea exploration. He even authored the French translation of a Latin tract that criticized the Catholic Church and its practice of selling indulgences, *Taxe des parties casuelles de la boutique du pape* (Lyon, 1564).

Although little is known of Du Pinet, except for his varied published works and translations, he must have been interested in the use of medicinal plants. Du Pinet dedicated his work, *Historia Plantarum*, whose title-page contains an elegant architectonic frame, to Nicolas Henry, Seigneur de Cremyeu and a French magistrate in Vienna. It appears that Seigneur Henry presented the author with a copy of Pietro Andrea Mattioli's *Commentarii* (see No. 7) and encouraged him to translate this fundamental botanical text into French. Du Pinet chose to prepare a condensed compendium in Latin, in what might be regarded as an early 'pocket edition' intended for use by botanists in the field, for as he wrote: 'etiamnum manualem ipsum facere' (and now I have made [of it] a veritable manual) (p. 3).

Oleaſter, ſiue Olea Sylueſtris: Grecis, ἀγριελαία, ἢ κότινος: Italis, Oliuo ſaluatico: Germanis, Vuilder oelbaum: Hiſpanis, Azebuche: Gallis, Oliuier ſauuage.

Pli. li.16. cap.18. Gale.li.6. ſimp.me.

Locus.

Naſcitur in montibus cāpeſtribus, & inſulis.

Qualitates.

Oliuæ rami quantum habent adſtrictionis, tanta & frigiditatis participes ſunt. Fructus verò, ſi quidem perfectè maturus eſt, moderatè calfacit: ſin immaturus, magis tum adſtringit, tum refrigerat.

Olea

ANTOINE DU PINET,
Historia Plantarum, 1561.
Olive (*Olea europaea*),
page 81

Printed in Lyon by Gabriel Coterius in the smaller and more manageable duodecimo format, *Historia Plantarum* presents a series of plants with their names in Latin, Italian, French, and sometimes also Greek and German, a brief botanical description, information on their habitat, and finally an extended discussion of their qualities. In addition to Mattioli, Du Pinet acknowledges other sources for which he considered himself grateful, which included Leonhart Fuchs (see No. 6) and, more rarely, Dioscorides and Galen.

What makes this diminutive herbal so attractive is the large number of woodcuts in the text. These, too, were copied, albeit in considerably simplified form, from the illustrations in Mattioli's *Commentarii,* perhaps Du Pinet referred to the 1554 edition published in Venice by Valgrisi. Following the same sequence used by Mattioli, Du Pinet begins with the iris and concludes with the grape vine. It is probable that the preparatory drawings for these woodcuts were made by the author himself, although it is difficult to establish any stylistic affinities with the album of botanical paintings in Cambridge, because of the very different techniques used in the two works.

At the end of the herbal the author has included five short appendices that are not illustrated. The first two contain several recipes copied in abbreviated form, in actuality little more than lists of ingredients, from Dioscorides (*Simplicium medicamentorum facultates secundum locos ex Dioscoride* and ... *secundum genera ex Dioscoride*). These are intended for cosmetic rather than medicinal purposes: treatments for hair loss, facial blemishes, moles, lice, etc. The third appendix, entitled *Simplicium medicamentorum quæ corpori decorem præstant, ex Dioscoride,* is dedicated to the treatment of imbalances in the humors of the body, while the fifth, *Mensurarum et ponderum typi, ad Dioscoridis mentem, ex Galeno,* lists the units of apothecary weight used by Dioscorides and Galen (the drachm, uncial, obolus, etc.).

Historia Plantarum, with its novel format, was such a success that two French translations appeared in rapid succession, the first published in Paris in 1620 and the second in Lyon in 1627.

29. Castore Durante (1529–1590)

[*Within architectonic woodcut border*]: Herbario Nuovo di Castore Durante Medico, & Cittadino Romano. Con Figure, che rappresentano le viue Piante, che nascono in tutta Europa, & nell'Indie Orientali, & Occidentali. Con Versi Latini, che comprendono le facoltà de i semplici medicamenti. Con Discorsi, che dimostrano i Nomi, le Spetie, la Forma, il Loco, il Tempo, le Qualità, & le virtù mirabili dell'Herbe insieme col peso, & ordine da usarle, scoprendosi rari Secreti, & singolari Rimedij da sanar le più difficili Infirmità del corpo humano. Con due Tavole copiosissime, l'una dell'Herbe, & l'altra dell'Infirmità, & di tutto quello che nell'opera si contiene. Con Privilegio del sommo Pontefice & d'altri Principi, & con Licentia delli Superiori. [*rule*] In Roma, Appresso Bartholomeo Bonfadino, & Tito Diani, M D LXXXV.

[*Colophon, V6 recto, printer's device*]: In Roma, Nella

Stamperia di Bartholomeo Bonfadino, & Tito Diani. M D LXXXV.

2° 35 × 23.1 cm. †–2†⁶ A–2R⁶ 2S–2T⁸ 2V⁶ *i–xxiv* 1–108, 119–286, 289–293, 295–296, 296–492 [*44*] (69 as 68, 181 as 186, 199 as 198) [548 pp.].

PLATES: 859 woodcuts in text, printed on blue-grey paper, except pages 147–150, 291–292, 297–298, which are white and were inserted later; woodcut portrait of Castore Durante on 2†6; woodcut printer's device above colophon.

BINDING: Modern panelled calf; gilt and blind tooling on cover and spine.

REFERENCES: Anderson, pp. 187–192; Blunt & Raphael, pp. 140–141; Hunt 151; Nissen 569; Pritzel 2,552.

CASTORE DURANTE was born in 1529 in Gualdo Tadino (Perugia); his father was a famous jurist, but Castore chose to study medicine (devoting time to literary studies as well) when he enrolled at the university in Perugia. He graduated sometime before 1567, and we know that in 1565 he was already practicing as a physician and herbalist in the city of Viterbo, conducting research on what at the time was referred to as *materia medica* or botanical medicine.

In 1565 Durante published *De bonitate et vitio alimentorum centuria*, which described nearly one hundred different plants and foodstuffs along with the positive and negative effects which they could have on one's health. One year later he published a verse translation of the sixth book of Virgil's *Aeneid*. These two texts established his reputation as a highly cultivated man of letters as well as an extremely knowledgeable physician, and in 1568 he was nominated *protomedico generale* (head physician) of the Papal States. In 1586 his authority and expertise were so great that he obtained the prestigious appointment of personal physician to Pope Sixtus V, a position that he held until his death in 1590. At the same time he was nominated to succeed the great botanist Andrea Bacci as *lettore dei semplici* at the Università la Sapienza in Rome. As *lettore* it was his duty to lecture on medicinal plants *dal vivo* and he conducted his classes in the university's botanical garden, reading what had been written about the plants by the great authorities of Antiquity, such as Galen, Pliny, and Dioscorides, and furnishing his own commentary. Durante only taught for one year, however, and then returned to his practice in Viterbo, where he died in 1590. The *Herbario Nuovo* contains a woodcut portraying the wise, gentle face of this learned physician at the venerable age of fifty-six.

Durante's most famous work, *Herbario Nuovo*, was the fruit of many years' labor. It was dedicated to the Cardinal Girolamo Rusticucci, a friend of the family who had already sponsored his translation of Virgil. The first edition was printed in Rome in 1585 by Bartolomeo Bonfandini and Tito Diani, and was immediately followed by a second edition printed by Bonfandini and Diani for Giacomo Bericchia and Giacomo Tornieri in Rome in the same year.

Another work published by Durante in Rome in 1586 and intended for broader dissemination, *Il Tesoro della sanità*, enjoyed a similar success. This was a handbook describing

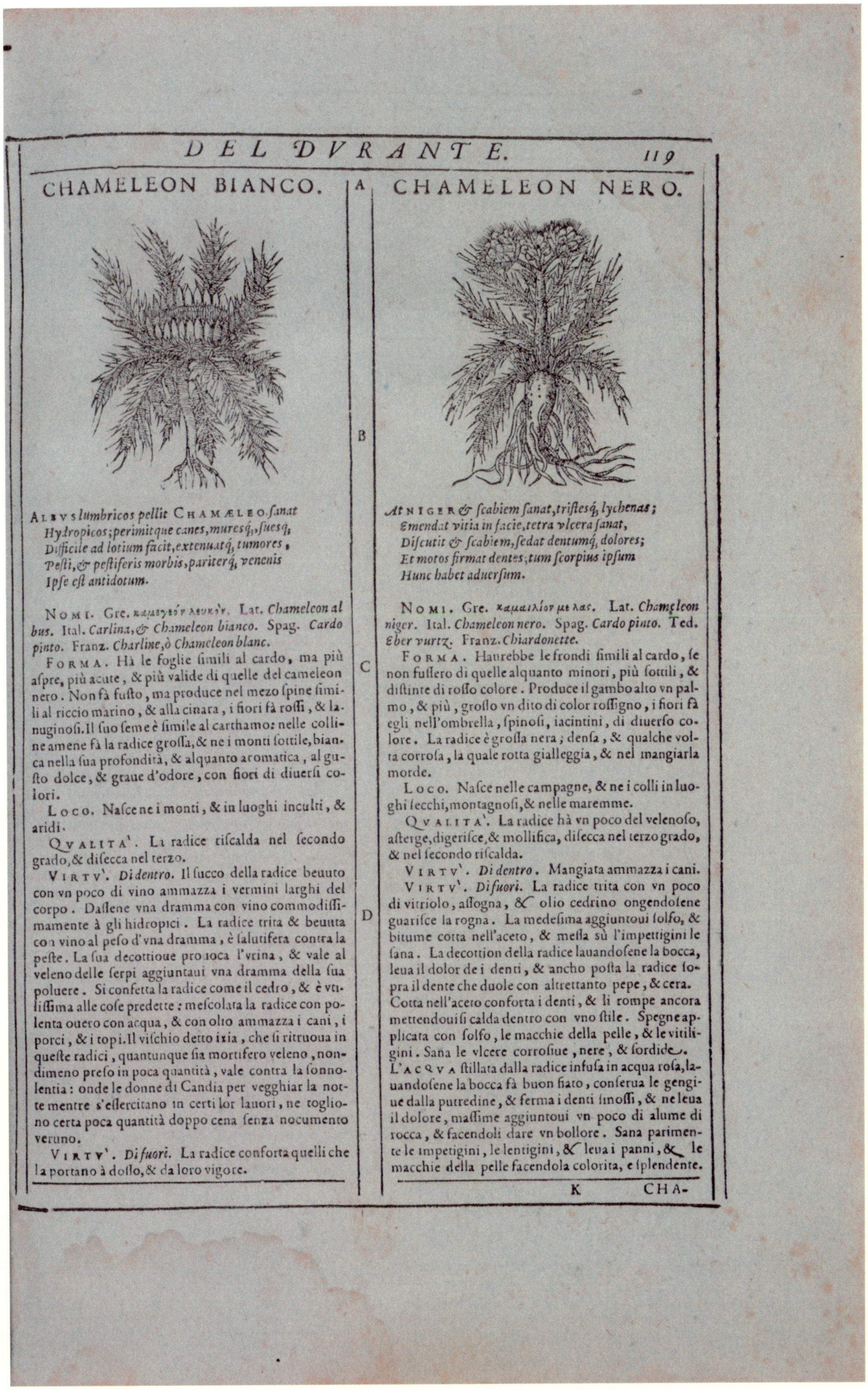

CHAMELEON BIANCO.

Albvs lumbricos pellit Chamæleo, *ſanat*
Hydropicos; perimitque canes, mureſq́, ſueſq́,
Difficile ad lotium facit, extenuatq́, tumores,
Peſti, & peſtiferis morbis, pariterq́, venenis
Ipſe eſt antidotum.

Nomi. Gre. χαμαιλέων λευκὸν. Lat. *Chameleon albus.* Ital. *Carlina, & Chameleon bianco.* Spag. *Cardo pinto.* Franz. *Charline, ò Chameleon blanc.*

Forma. Hà le foglie ſimili al cardo, ma più aſpre, più acute, & più valide di quelle del cameleon nero. Non fà fuſto, ma produce nel mezo ſpine ſimili al riccio marino, & alla cinara, i fiori fà roſſi, & lanuginoſi. Il ſuo ſeme è ſimile al carthamo: nelle colline amene fà la radice groſſa, & ne i monti ſottile, bianca nella ſua profondità, & alquanto aromatica, al guſto dolce, & graue d'odore, con fiori di diuerſi colori.

Loco. Naſce ne i monti, & in luoghi inculti, & aridi.

Qvalita'. La radice riſcalda nel ſecondo grado, & diſecca nel terzo.

Virtv'. *Di dentro.* Il ſucco della radice beuuto con vn poco di vino ammazza i vermini larghi del corpo. Daſſene vna dramma con vino commodiſſimamente à gli hidropici. La radice trita & beuuta con vino al peſo d'vna dramma, è ſalutifera contra la peſte. La ſua decottioue prouoca l'vrina, & vale al veleno delle ſerpi aggiuntaui vna dramma della ſua poluere. Si confetta la radice come il cedro, & è vtiliſſima alle coſe predette: meſcolata la radice con polenta ouero con acqua, & con olio ammazza i cani, i porci, & i topi. Il viſchio detto ixia, che ſi ritruoua in queſte radici, quantunque ſia mortifero veleno, nondimeno preſo in poca quantità, vale contra la ſonnolentia: onde le donne di Candia per vegghiar la notte mentre s'eſſercitano in certi lor lauori, ne togliono certa poca quantità doppo cena ſenza nocumento veruno.

Virtv'. *Di fuori.* La radice conforta quelli che la portano à doſſo, & da loro vigore.

CHAMELEON NERO.

At Niger *& ſcabiem ſanat, triſteſq́, lychenas;*
Emendat vitia in facie, tetra vlcera ſanat,
Diſcutit & ſcabiem, ſedat dentumq́, dolores;
Et motos firmat dentes; tum ſcorpius ipſum
Hunc habet aduerſum.

Nomi. Gre. χαμαιλέων μέλας. Lat. *Chameleon niger.* Ital. *Chameleon nero.* Spag. *Cardo pinto.* Ted. *Eber vurtz.* Franz. *Chiardonette.*

Forma. Haurebbe le frondi ſimili al cardo, ſe non fuſſero di quelle alquanto minori, più ſottili, & diſtinte di roſſo colore. Produce il gambo alto vn palmo, & più, groſſo vn dito di color roſſigno, i fiori fà egli nell'ombrella, ſpinoſi, iacintini, di diuerſo colore. La radice è groſſa nera; denſa, & qualche volta corroſa, la quale rotta gialleggia, & nel mangiarla morde.

Loco. Naſce nelle campagne, & ne i colli in luoghi ſecchi, montagnoſi, & nelle maremme.

Qvalita'. La radice hà vn poco del velenoſo, aſterge, digeriſce, & mollifica, diſecca nel terzo grado, & nel ſecondo riſcalda.

Virtv'. *Di dentro.* Mangiata ammazza i cani.

Virtv'. *Di fuori.* La radice trita con vn poco di vitriolo, aſſogna, & olio cedrino ongendoſene guariſce la rogna. La medeſima aggiuntoui ſolfo, & bitume cotta nell'aceto, & meſſa sù l'impettigini le ſana. La decottion della radice lauandoſene la bocca, leua il dolor de i denti, & ancho poſta la radice ſopra il dente che duole con altrettanto pepe, & cera. Cotta nell'aceto conforta i denti, & li rompe ancora mettendouiſi calda dentro con vno ſtile. Spegne applicata con ſolfo, le macchie della pelle, & le vitiligini. Sana le vlcere corroſiue, nere, & ſordide. L'Acqva ſtillata dalla radice infuſa in acqua roſa, lauandoſene la bocca fà buon fiato, conſerua le gengiue dalla putredine, & ferma i denti ſmoſſi, & ne leua il dolore, maſſime aggiuntoui vn poco di alume di rocca, & facendoli dare vn bollore. Sana parimente le impetigini, le lentigini, & leua i panni, & le macchie della pelle facendola colorita, e ſplendente.

K CHA-

CASTORE DURANTE, *Herbario Nuovo*, 1585. 'Chameleon Bianco,' Silver thistle (*Carlina acaulis*), and 'Chameleon Nero,' (*Carlina* cf. *gummifera*), page 119 on blue paper

the 'names, quality, choice, benefits, harmful qualities, remedies' of different edible plants and how they could be used to preserve one's physical and mental well-being.

In the preface to *Herbario Nuovo* the author states that his aim was to provide a more thorough treatment of the material presented in Pietro Andrea Mattioli's *I Discorsi*. Although Durante's work does contain a series of references and correspondences that reflect a meticulous study of Mattioli's botanical text (and also of *De historia stirpium* by Leonhart Fuchs), it was entirely different in scope and purpose. *Herbario Nuovo* was conceived as a manual that could be easily and quickly consulted, quite different from the detailed systematic approach that characterizes *I Discorsi*.

In Durante's work, which constitutes one of the most interesting herbals to be produced in Italy at the end of the sixteenth century, quite disparate branches of knowledge are intermingled—from botany, medicine, and pharmacology to literature and history. The text is also interspersed with many personal observations, the result of two decades of experience as a practicing physician. Durante in fact possessed an uncommon knowledge of botany, for he clearly and succinctly describes the morphology of hundreds of plants, many of which he himself had studied and used. On various occasions he cites, for example, the place where he personally found a certain species during a botanizing expedition. A total of 859 plants are presented in alphabetical order by their vernacular Italian names. Next to each description is a small but realistic woodcut illustration and a few lines of verse composed by the author in the style of Macer Floridus, a medieval author of the Latin verse herbal *De viribus herbarum*; the pseudonym 'Macer' very likely links the medieval poet to Amelius Macer, a contemporary of Virgil who composed a poem *De herbis*, only a fragment of which survives. Durante also lists the names of the plant in Greek, Latin, Italian, German, French, Spanish, Czech and, in the case of exotic species (particularly those from Arabia, Egypt, and the East and West Indies), the names by which they were known in their country of origin. A description of the plant follows, including its *forma*, *loco* or habitat, the season of its flowering, and its *qualità* and *virtù*. In addition to its uses in medicine, the author furnishes instructions for other applications—veterinary medicine, in the home (cosmetics, insect repellents), and in various trades (such as cloth dying). At the end of the work is a list of the plants discussed and the illnesses that they could be used to treat, as well as a page containing the illustrations of several that are not mentioned in the text. Among the novel species presented in the herbal is tobacco, called 'erba di Santa Croce' in honor of the Cardinal Prospero Santacroce, to whom the author dedicated an epigram *In tabacum*.

Durante's work is also noteworthy for the quality of its illustrations. *Herbario Nuovo* opens with an elaborately composed architectonic title-page consisting of scenes depicting gardens and orchards framed by two hermes. The many woodcuts that follow were

no nell'aceto, & ſparſo per la camera ammazza le pulci.

CORIDALA.

Corydala ex vino ſi quando ſicca bibatur
Siue recens ſit ſumpta cibis, auferre dolores
Eſt coli pollens.

Nomi. Gre. κόρυδαλις. Lat. *Corydalis*. Ital. *Coridala*.

Forma. E' congenere con la fumaria maggiore: Creſce queſta pianta con foglie come di coriandro, ma più picciole, & più ſottili, fà i gamboncelli alti vn ſommeſſo ſottili ramoſi, & ben carichi di foglie con fiori ſimili ad augeletti conformi alle lodole. Hà molte, & copioſe radici, lunghe, bianchiccie, & ſottili.

Loco. Naſce ne i colli, & nelle campagne.

Qvalita'. & Virtv'. La poluere beuuta in vino, ouero magnata l'herba gioua à i dolori colici.

CORNIOLO.

Aſtringunt, ſiccant, & ſiſtunt Corna *valenter*
Vulneris ora ſolent eadem coniungere magni,
Et ſanant morbum cui ceſsit ab impete nomen;
Ebrietasq; illis tardatur; commoda dantur
Hęc dyſentericis, prauo tamen improba ſucco
Sunt; & apes horum flore occiduntur amaro.

Nomi. Gre. κάρνια. Lat. *Cornus*. Ital. *Corgniolo, & i frutti cornole, cornie, & corgnali*. Ted. *Coentlbaua*. Spag. *I frutti cornizolos*. Franz. *Cornoille*.

Spetie. Ritruouaſene di due ſpetie, cioè il maſchio, & la femina.

Forma. Il maſchio ha le foglie ſimili à quelle del ſanguino: hà la ſcorza neruoſa, & ſottile, la materia del ſuo legno è durißima & ſalda ſimile al corno, ond'egli hà tratto il nome. Creſce all'altezza di dodici braccia. La femina come che faccia poco tronco, fà pur'aſ-

A ſai vergelle ſimili al vitice, & ſono aſſai più vencide, & men ſalde di quelle del maſchio. In Ida mõte di Troia il maſchio non fà frutto: ma in Italia producono i frutti l'vno, & l'altro. Il legno del maſchio è ſenza midollo, & imperò migliore, & più forte di quello della femina. Il maſchio produce, & matura il ſuo frutto la ſtate, & la femina nel fine dell'autunno, manco buono aſſai di quel del maſchio. E' d'auuertire, che non ſi piãtino ne ſi laſcino appreſſo à i lochi delle Api, percioche mangiando elle i ſuoi fiori muoiono, quantunque facciono il contrario ne gli huomini. E' queſt'arbore di mediocre grandezza, & il più delle volte vergelloſo, con breue tronco, onde naſcono le verghe ferme robuſte & nodoſe, che ſeruono per gameti nell'arte della lana: & il legno perche è duriſſimo s'adopera com-

B modamente per far i denti delle ruote de i molini. Produce come s'è detto, le frondi quaſi come il ſanguino venoſe, liſce, & mediocremente carnoſe: Fà i fiori piccolini, muſcoſi, & gialli, da i quali naſcono i frutti roſſi, & lunghetti come oliue con duriſſimo nocciuolo, nel principio verdi, & poi roſſi.

Loco. Naſce nelle ſelue, ne i colli, & nelle fratte.

Qvalita'. Hà facultà di diſeccare valentiſſimamente, & d'aſtringere.

Virtv'. *Di dentro*. I frutti ſono efficace rimedio à tutti i fluſſi del ventre, imperòche coſtringono al paro delle neſpole, ò de pruni ſaluatichi. Si conciano come l'oliue, & faſſi della lor polpa conſerua come de i cotogni, con zucchero ò mele, che vale alla diſenteria, & à corroborar lo ſtomaco.

C Virtv'. *Di fuori*. L'humore, che riſuda dalle frondi, ò da i virgulti che ſi bruſciano gioua vnto alle impetigini. L'olio che ſi caua del legno vale à i dolori artetici, & franciosi. Le frondi, & i germini ſuoi conſolidano le ferite grandi, ne i corpi duri.

CORONA REALE.
Pianta maſsima.

CORONOPO DOMESTICA.

D

Aluum conſtringit percocta Coronopvs *eſa*,

Mul-

CASTORE DURANTE, *Herbario Nuovo*, 1585. 'Corniolo,' Cornelian Cherry (*Cornus mas*) and Coronopo Domestica,' Buck's-horn Plantain (*Plantago coronopus*), page 149 on white paper carved by Leonardo Parasole (also called Norsino) based on drawings prepared by his wife, Isabella Cattani. Parasole was an engraver and print dealer from Sant'Angelo in Visso who settled in Rome, while Cattani was a particularly interesting figure in the cultural history of sixteenth century Italy—a woman artist and engraver who also published a book of embroidery designs, *Teatro delle Nobili e Virtuose Donne* (Rome, 1616). In *Le vite dei pittori* (Rome, 1642) the painter and writer Giovanni Baglione noted that she had collaborated with such illustrious intellectuals as Federico Cesi, founder of the Accademia dei Lincei; Baglione described with admiration her 'intagli nel Libro dell'Herbe del Principe Cesi d'Acquasparta, letteratissimo signore' (engravings in the Book of plants of that most learned gentleman, Prince Cesi of Acquasparta).

While most of the illustrations in *Herbario Nuovo* were closely derived from the woodcuts in various editions of Mattioli's *I Discorsi* and Fuchs' *De historia stirpium*, Parasole and Cattani nonetheless gave ample proof of their artistic skill in meeting the specific demands of Durante for this botanical text. In a departure from the traditional format of the illustrated herbal, the woodcuts are quite small and therefore not much detail could be conveyed, but each plant is portrayed with clarity and elegance. Indeed, within the limits of the modest format conceded them, the two artists used their graphic talents and imagination to create compositions of great charm and variety. Some of the plants are shown growing in pots reminiscent of the elegant flower vases of the late Mannerist period, a compositional device that had already been adopted by Mattioli in his illustration of the rose in the 1560 edition of the *Commentarii*. The woodcuts of the aloes, on pages 17 and 18, were clearly inspired by Durante's comment that it may be 'piantasi in tutti i giardini, & tiensi ne i vasi su le loggie, e finestre' (planted in any garden, and kept in pots on terraces, and windowsills). Other plants are shown trained against wooden frames or climbing the stems of other plants, such as the 'Clematis terza' (*Clematis alba*) which 'nasce per tutto nelle siepi, & nelle selve aggirandosi a gli arbori' (grows everywhere in hedges, and in forests winding itself around trees).

Particularly appealing are the landscapes against which many of the plants appear, and which in some cases allude to the natural habitat of the species. The 'Thipha,' that the artists copied from Fuchs, is portrayed next to a watercourse to illustrate Durante's observation: 'poche sono le acque delle paludi, de i laghi, & de gli stagni, che non produchino infinite piante di Tifa' (few are the waters of marshes, lakes, and swamps that do not produce an infinite number of cattail plants). The 'Lentisco del Perù' (Peruvian Pepper Tree, *Schinus molle*) appears against a teeming landscape consisting of a river plied by boats and, in the background, hills dotted with buildings. In some cases a human figure has been inserted, such as the wayfarer accompanied by his dog in the illustrations of the 'Agarico' and the 'Larice.' There are also diminutive genre scenes reminiscent of the [*H*]*ortus Sanitatus* (see

C

D

VVA PASSA

Sunt *Vvæ Passæ stomacho, iecoriq́; salubres,*
Subducunt aluum, pulmones inde iuuantur;
Damascena tamen præmolliet aptius aluum;
Astringunt autem ventrem si nucleus insit;
Pectori & angusto prosunt, tussimq́; repellunt,
Renum & vesicæ vitijs si nucleus absit
Conueniunt, sedantq́; etiam inflammata pudenda.

Nomi. Gre. Σταφις. Lat. *Vua passa.* Ital. *Vua passa.* Arab. *Zibib.* Ted. *Vuein beerlen, mertreubel, & coseincas.* Spag. *Passas.* Fran. *Roisius eu captz.*

Rr 2 FOR-

CASTORE DURANTE,
Herbario Nuovo, 1585.
Detail: Grapes (*Vitis vinifera*) drying on a table, page 483 on blue paper

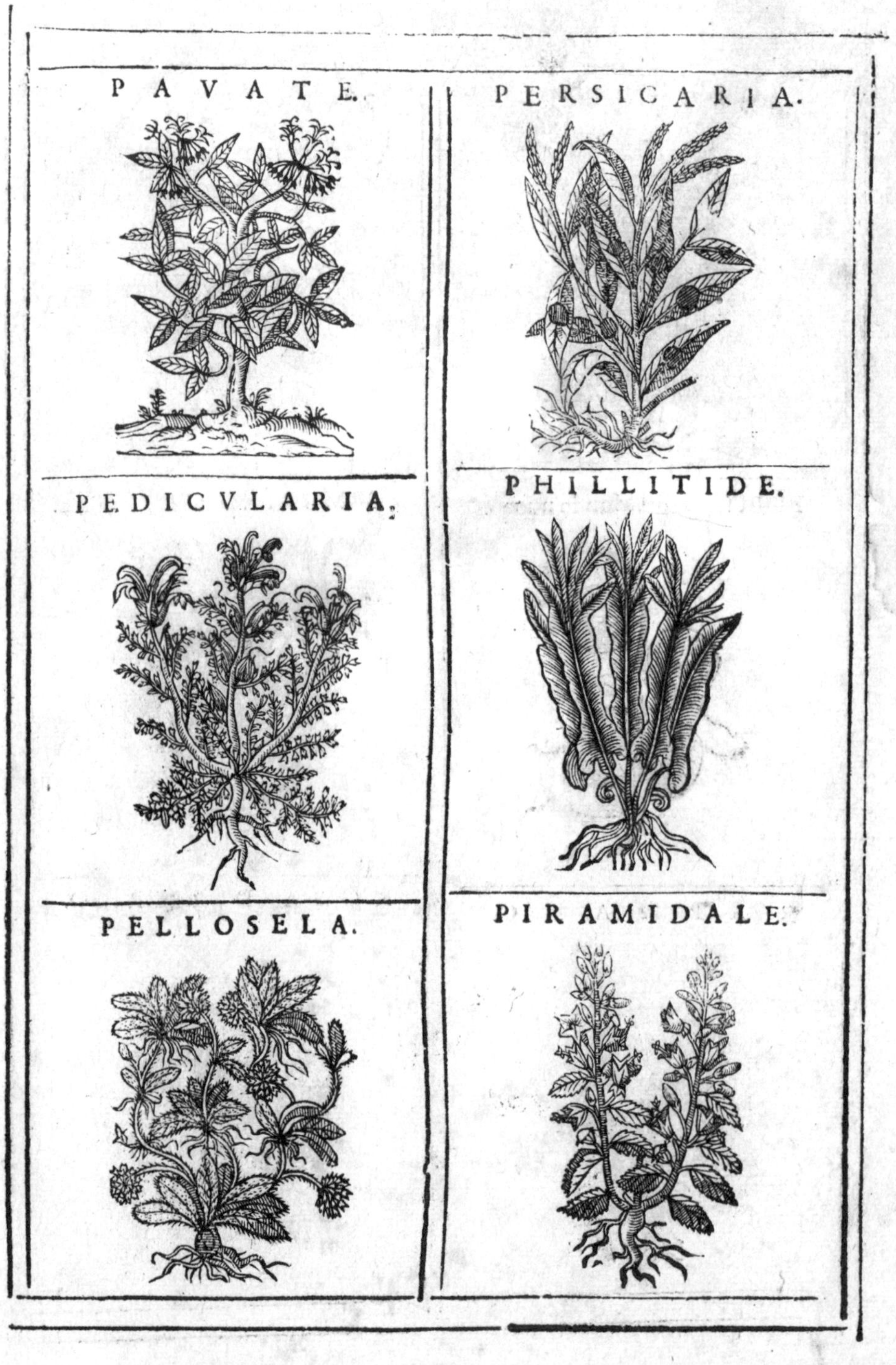

CASTORE DURANTE, *Herbario Nuovo*, 1585. Six medicinal plants: 'Pavate,' not identified (*Morinda citrifolia?*); 'Persicaria' (*Persicaria maculosa*), Red Shank; 'Pedicularia' (*Pedicularis* sp.), Lousewort; 'Phillitide' (*Phyllitis scolopendrium*), a variety of Hart's-tongue fern; 'Pellosela' (*Hieracium pilosella*), Mouse-ear Hawkweed; 'Piramidale' (*Ajuga pyramidalis*), Pyramidal Bugle; Vv4

No. 3)—a cheese stall to illustrate the 'Ricotta,' clusters of grapes left to dry in the sun in the chapter on the 'Uva Passa,' and a beehive in 'Mele liquore' (liquid honey).

Sometimes the author passes on age-old superstitions that were still accepted as scientific fact in his day, such as the bizarre legend of the 'Arbor dell'anitre,' according to which a certain species of goose (Barnacle Goose, *Branta leucopsis*) is born from a shell produced by this tree. The legend was associated with an equally fanciful image that continued to turn up in scientific texts during the sixteenth and seventeenth centuries, including *The Herball* by John Gerard (see No. 12). In reality this shell was a crustacean that sometimes emerged from the sea attached to floating tree trunks. Of particular interest because of its literary associations is the illustration of the 'Salvia,' which is shown growing beside a toad. Here Durante is referring to a tale recounted by Boccaccio in his *Decameron* (Fourth day, Seventh story):

> ... la salvia deve essere accompagnata con ruta negli horti, altrimenti s'infetta dalle serpi, e dai rospi, e l'infettano col loro alito velenoso, e con la loro saliva pestifera, il che con bellissima e memorabile historia insegna Giovan Boccaccio, raccontando che due Amanti, stropicciandosi i denti con la salvia, miseramente morirono e che cavandosi poi quella salvia vi fu trovato un rospo.'
>
> (Sage must be accompanied by rue in the garden, otherwise it will be infected by serpents and toads, which infect it with their poisonous breath and their pestiferous saliva, as Giovan Boccaccio teaches us in a most beautiful and memorable story, recounting that two Lovers who rubbed their teeth with sage died most miserably, and later where that sage plant was pulled up a toad was found).

In his *Nova Plantarum Americanarum Genera* (1703) Charles Plumier (1646–1704) named two plant entities after Castore Durante: *Castorea repens spinosa* and *Castorea racemosa, flore caeruleo*. The genus *Duranta* was named in honor of him and there are approximately thirty species of this type of tree and shrub that grow in Florida, the West Indies, South America, and Mexico.

The copy of Durante's *Herbario Nuovo* in the Oak Spring Garden Library is particularly fine because it was printed, perhaps at the request of the author, on a rare and valuable type of blue paper.

30. Paul de Reneaulme (1560–1624)

Pauli Renealmi Blæsensis Doctoris Medici Specimen Historiæ Plantarum. Plantæ typis æneis expressæ. [*Printer's device*]. Parisiis, Apud Hadrianum Beys, sub signo Stellæ aureæ, via Jacobæa. [*rule*] M. DC. XI.

4° 20.9 × 16.3 cm. ã⁴ A–R⁴ S² T–V⁴ *i–viii* 1–139 *140–150* *151–152* [*4*] (37 as 33, 51 as 52) [164 pp.].

PLATES: 25 engravings of plants.

BINDING: Full red morocco by Lobstein-Laurenchet with their stamp on front free endpaper.

REFERENCES: Blunt & Raphael, pp. 173–174; Blunt & Stearn, pp. 103–105; Hunt 192; Nissen 1621; Plesch, p. 380; Pritzel 7,542.

THE AUTHOR of *Specimen Historiæ Plantarum*, Paul de Reneaulme (who Latinized his name to Paulus Renealmus), was a physician and botanist from Blois, France. He had a thorough knowledge of the flora of the Pyrenees and the Alps, Switzerland, and the region around Paris, based on extensive botanizing expeditions. We can also deduce that he was well versed in the classical languages, because his herbal is written in very cultivated Latin and contains numerous citations from the Greek literature.

In the dedication addressed to the cardinal and archbishop Jacques Davy (1556–1618), Reneaulme laments the fact that his colleagues had an inadequate grounding in botanical nomenclature, often based on erroneous interpretations of the *veteres auctores*. The physician was far ahead of his time in raising this issue, the importance of which his contemporaries failed to appreciate. They accepted unquestioningly the traditional principles of herbal medicine and saw no need to develop a more analytical system for the naming of species and genera. Unfortunately, Reneaulme only succeeded in compounding the confusion that already reigned in the area of botanical nomenclature by proposing a mononomial system of his own that did not differentiate between the levels of genus and species and was based on Greek names of his own invention, rather than on the ancient names of the plants.

Apart from this well-intentioned attempt to resolve the problem of plant classification, *Specimen Historiæ Plantarum* follows the traditional format of the herbal, describing the morphology, uses, temperament, and virtues of different medicinal plants. Reneaulme's scientific approach can be appreciated in the clarity and precision of the descriptions that appear under the heading 'Forma' for each plant. The description includes succinctly accurate notes regarding their flowering season and their country of origin. De Reneaulme's herbal is embellished with a series of twenty-five illustrations that constitute one of the earliest examples of the new technique of copperplate engraving to illustrate a botanical text. While not always technically perfect, because the printer sometimes applies too much or too little ink to the plates, the illustrations are striking for their incisiveness and legibility. The anonymous artist has captured the elegant outline of each plant. The detail of texture and form suggested by means of skilful shadowing insures each species is unmistakably

PAUL DE RENEAULME, *Specimen Historiæ Plantarum*, 1611. Common mullein (*Verbascum thapsus*), page 102

recognizable. Wilfrid Blunt expressed an admiration for the vividness and immediacy of these botanical portrayals (see Blunt and Raphael, p. 173).

Of special interest are the sections devoted to the sunflower (pp. 83–85), which he called χρυσῆϊς, and the tobacco plant (pp. 18–19) (βλεννοχοίς) in which he cites some of Monardes' observations on the species (see No. 16). The illustration of the sunflower, with its masterly layout and fine detail, is particularly admirable. Since its unusual size did not allow the artist to depict the entire plant on a single page, it has been divided into two sections, showing the apex of the stem with its majestic corolla and next to it the base of the plant with its dense root system. There are also three very fine engravings of the gentian, other plates illustrating bulbous plants, and one engraving of *Verbascum album*, once again divided into two parts due to its large size. In the section on the tobacco plant (p. 33), de Reneaulme notes that a decoction made from its root was a sovereign remedy for intermittent fever and snakebite.

Although the engravings in this slender volume bear certain affinities with the emerging genre of the florilegium, the attention devoted by the author to problems of nomenclature and the medicinal properties of the plants bring it closer to the typology of the herbal.

31. Jacques Contant (?–1588) *and* Paul Contant (*c.* 1552–1629)

[*Printed in red and black within a double rule border*] Les Oeuvres de Iacques et Paul Contant Pere et Fils Maistres Apoticaires de la Ville de Poictiers. Divisées en cinq Traictez. 1. Les Commentaires sur Dioscoride. 2. Le Second Eden. 3. Exagoge Mirabilium naturæ è Gazophylacio. 4. Synopsis Plantarum cum Ethymologiis. 5. Le Jardin & Cabinet Poëtique. Avec les figures des Plantes en taille douce. [*Engraved Coat of Arms*]. A Poictiers, Par Iulian Thoreau, & la Vesue d'Antoine Mesnier Imprimeurs ordinaires du Roy, & de l'Université. [*rule*] M.DC.XXVIII. Avec Privilege.

2° 34.2 × 22.3 cm. ★^{8} A–X^6 §^{4} (-§4) A–F^6 G^4 ¶^{2} A^4 B–H^6 ã^6 ẽ^2 A–E^6 *i–xvi* 1–250 (120 as 119) *i–viii* 1–79 [5] 1–90 *i–xviii* 1–59 *60* [526 pp.].

PLATES: Engraved vignette of the coat of arms of Prince Henri Bourbon-Condé on title-page signed 'P Demoges fecit' (with 'P D' as monogram). 15 engraved plates, one signed P Demonges fecit (with 'P D' as mongram) and one signed 'Pinson pinxit.'

BINDING: Contemporary calf.

PROVENANCE: Arpad Plesch.

REFERENCES: Cleveland 177; Contant; Plesch, p. 182; Pritzel 1,850; Schnapper, pp. 223–4.

JACQUES CONTANT and his son PAUL were apothecaries in the town of Poitiers, *amateurs* of natural history, and avid collectors. Their *cabinet* was one of the most important and renowned in seventeenth-century France. It included scientific specimens and unusual

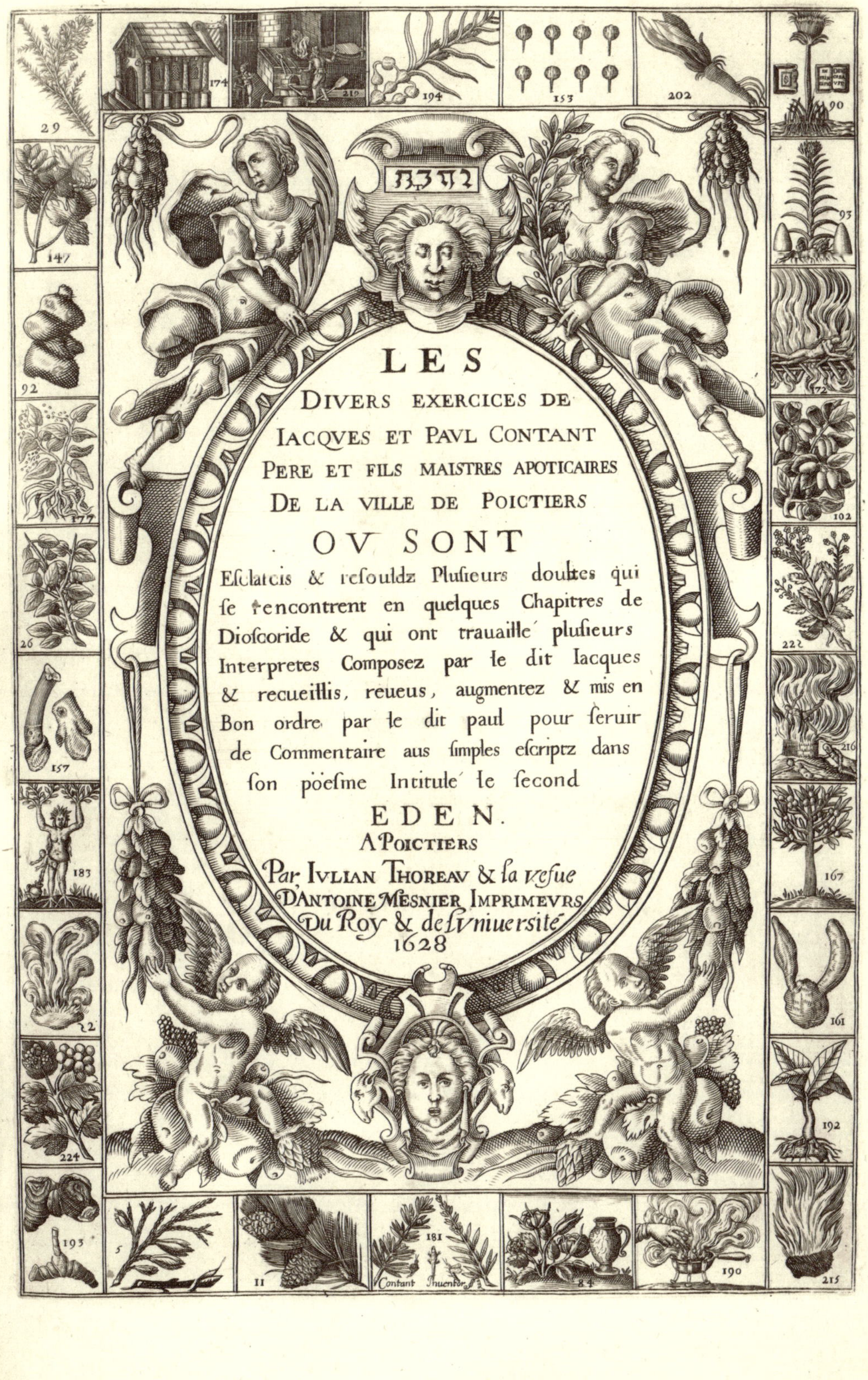

JACQUES AND PAUL CONTANT, *Les Oeuvres*, 1628. Title-page, *7

objects found by them or donated by friends, fellow townspeople, and colleagues associated with the university at Montpellier, which was famous for its faculty of medicine. In *Le Jardin et Cabinet Poetique*, a fanciful plant catalogue in verse, Paul Contant describes his collection, most significantly, as a 'racourcy du magazin du monde' (a brief summary of the Emporium of the World). Indeed, in this period the *Wunderkammern* assembled by emperors and men of learning were conceived of as nothing less than models in miniature of the macrocosm (see Schnapper, pp. 223–5).

There were many pharmacies in the Poiteau-Chareten region of France in the early years of the seventeenth century, including seventeen in Poitiers, and Paul Contant was one of the most eminent apothecaries of the region. His father Jacques was a pharmacist and amateur botanist, who wrote and sent to Joseph Justus Scaliger, a French humanist and scholar of Italian origin (1540–1609), a commentary on Dioscorides. Paul Contant was born around the year 1552 and at the age of twenty embarked on a cultural and scientific tour around Italy. It is known that he visited various places, including Venice where he acquired a seahorse (*Hippocampus*, then considered a great curiosity), and Padua where the director of the university's Botanical Garden, Luigi Anguillara (1512–1570), made him a gift of several rare plants that he preserved carefully and brought back with him to France. In 1586 the town of Poitiers appointed Paul Contant 'maitre apothicaire' and as a result he became quite prosperous, although in his books he repeatedly affirms that he was anything but wealthy. In addition, the successful pharmacist fancied himself a versifier and wrote, among other things, a *Henriade* in honor of Prince Henri de Bourbon, which has since been lost. His surviving efforts make for somewhat tedious reading, as they are lacking in poetic inspiration and overburdened with an eclectic and rather muddled scholarship. Upon his death in 1629, his vast *cabinet* was inherited by his grandson and subsequently dispersed.

The writings of Paul Contant provide us with useful information on the medical and scientific circles of the period; he mentions the names of several physicians and pharmacists from the towns of Poitiers, Saints, and La Rochelle (a port for ships bringing not only commodities but also interesting natural-history specimens from the New World), all colleagues who contributed to the formation of his collection. Paul was also in contact with the famous 'herboriste' Jean Robin of Paris, to whom he dedicated a short poem (see No. 18), and the botanist *sieur* Pascal Le Coq of Forges (1567–1632), founder of the Jardin des Plantes of Poitiers. Le Coq had travelled extensively around Europe and met the illustrious naturalist Carolus Clusius (see Nos. 9 and 14), for whom Contant himself acknowledged a profound admiration.

All of Contant's works, including those pertaining to botany, were closely linked to and indeed might be considered a by-product of the collecting activities that constituted

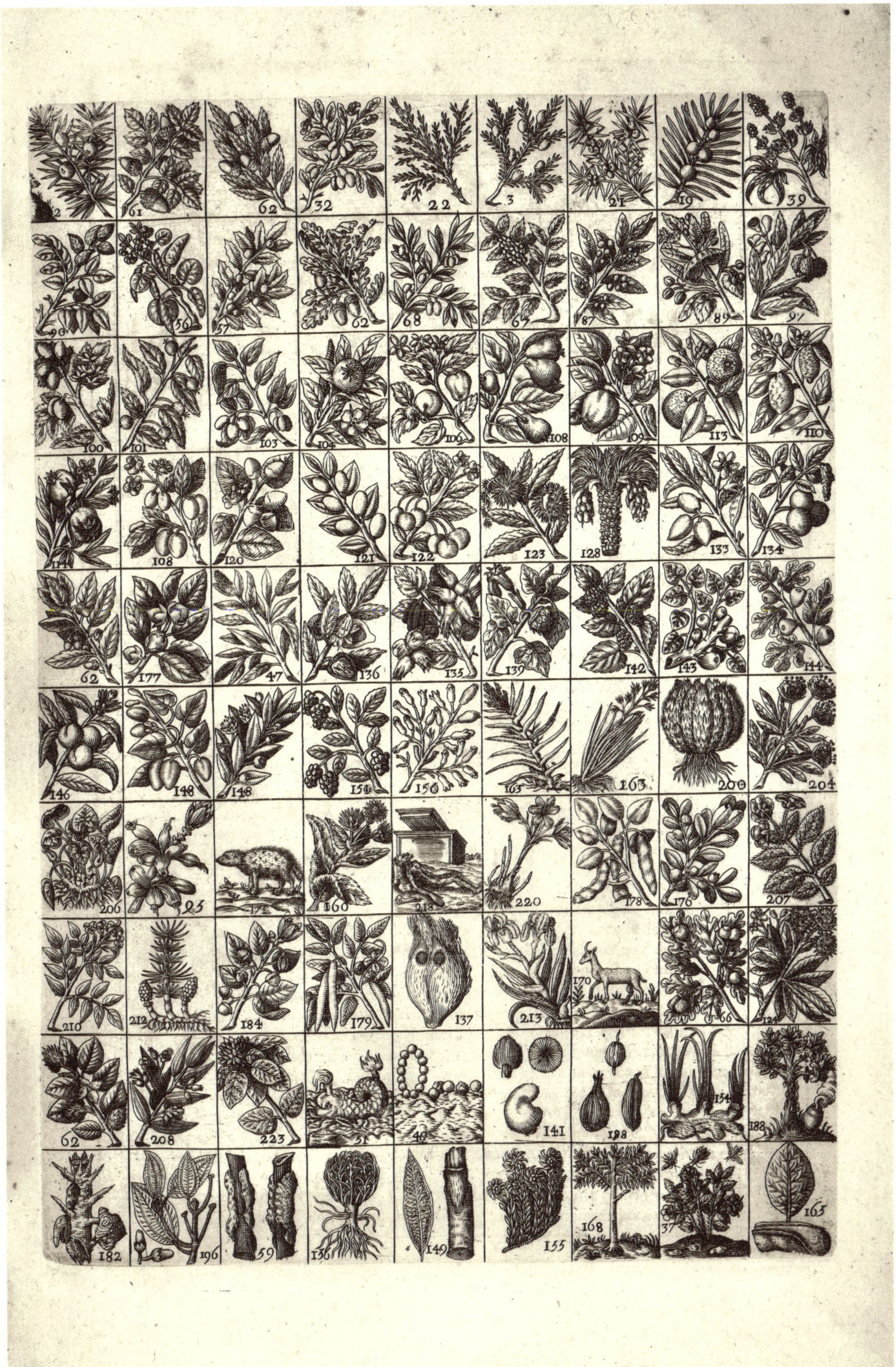

JACQUES AND PAUL CONTANT, *Les Oeuvres*, 1628. 90 plants described in the Contants' text with their reference numbers, *8

the principal interest of himself and his father, both of whom were motivated by the same 'curiosité tumultueuse.' The *cabinet* included numerous plants among its exhibits, and therefore provides us with a fairly exact notion of how the botanical collection of a seventeenth-century pharmacist might have been configured. In addition, the engraving that precedes the *Exagoge* ('Introduction' or 'Preliminary discussion'), executed by an unknown artist who signed his name 'P. Demoges,' shows how some of the exhibits in the museum were displayed. In addition to the *cabinet*, Paul cultivated a botanic garden.

The earliest and most celebrated work by Paul Contant is *Le Jardin et Cabinet Poetique*, which was published in 1609 and dedicated to Maximilien de Béthune, the Duke of Sully and a royal lieutenant of Poitou. A rare, early version of this work entitled *Bouquer Printanier* was published in La Rochelle in 1600. Three decades later, in 1628 the editor Julian Thoreau of Poitiers produced a costly edition of the complete works of Jacques and Paul Contant in a folio volume dedicated to Prince Henri de Bourbon. Bound together in a rather haphazard collage are the *Commentaires sur Dioscoride*; *Le Second Eden*; *Exagoge Mirabilium naturae è Gazophylacio Pauli Contanti Pictauenensis Pharmacopaei*; *Synopsis Plantarum cum Ethymologijs*; and, finally, *Le Jardin, et Cabinet Poetique*. These works consist for the most part of long lists of natural history specimens arranged in no particular order, but in this way they eloquently testify to the problems facing scientists such as Jacques and Paul Contant, who in this period were seeking to classify and arrange their eclectic collections of plants, animals, and minerals.

Les Oeuvres begins with *Commentaires sur Dioscoride*, a scholarly work begun by Jacques and completed by Paul. This is the only work in which the items are presented systematically, the authors having followed the alphabetical order adopted by Dioscorides in *De materia medica*. One full-page engraving is sub-divided into ninety numbered rectangular vignettes that illustrate many of the plants described in the text, each labelled (although not always correctly) with a reference number.

Le Second Eden, is a catalogue of Paul's botanic garden composed in verse. It opens with a fine title-page designed by the author and engraved by a certain Pinson, with scenes showing Adam and Eve engaged in various activities during their pastoral existence before they were banished from the Garden of Paradise. These vignettes are surrounded by different plants, each marked with a number corresponding to the page in the preceding *Commentaires* where their description might be found. This long poem offers a portrayal of the Garden of Eden and the plants that grew there, many of which—the author affirmed—bore the direct sign of God, such as the 'mystique Granadille' or passionflower. There is only one engraving in the text, which illustrates the Cèdre du Liban, described by Contant as a 'conifère sublime,' (p. 38).

The following *Exagoge* lists some of the exhibits that could be found in the pharmacists'

JACQUES AND PAUL CONTANT, *Les Oeuvres*, 1628. Title-page to the second part, containing 24 plants along the outer borders with reference numbers that are described in the text

collection, including seeds, fruits, nuts, plants, roots, plant extracts such as rubber, minerals, marine animals, about one hundred flasks of plant essences and perfumes, and no less than 3,500 bronze statuettes. This catalogue is followed by *Synopsis Plantarum*, a listing of plants in Paul's botanic garden.

The volume closes with *Le Jardin et Cabinet Poetique*, in which Paul Contant describes in verse some of the flowering plants (from rare species such as the prized tulip to simple herbs endowed with 'mille vertus') and other natural history specimens in his collection, indulging in circumlocutory flights of poetry whose meaning is often quite obscure. Significantly, however, the author draws an analogy between his cabinet of preserved specimens and the botanical garden with its 'living' exemplars, describing them as two indissoluble entities that formed a single chapter in the great 'book of the world' and that must be studied together in order to understand the wonders of nature. *Le Jardin* closes with the splendid engraving of a bouquet of flowers, each one of which is described in the poem but, since they blossom at different times of the year, constitute elements in a marvelous 'bouquet impossible' (cf. *An Oak Spring Flora*, p. 22). The volume ends with nine engravings depicting objects from the *cabinet*, including mutated animals (and conjoined human twins); reptiles, fishes, birds, fabulous creatures, and various curios.

32. French School, Eight medicinal plants

Oil on canvas. 105.4 × 62.8 cm. REFERENCES: Uphof; Leung and Forster; Bruni and Nicoletti.

This series of eight large oil paintings (105.4 × 62.8 cm) by an unknown artist formed part of the furnishings of a pharmacy that served a town in the Loire Valley during the seventeenth century, perhaps Saumur. Nothing is known of the building that housed the pharmacy, as it was destroyed during the course of the French Revolution.

The first apothecary shops appeared during the Middle Ages, many of them annexed to convents and monasteries in whose gardens medicinal herbs were cultivated. Medieval hospitals as well were often furnished with a dispensary, at first only for the use of their staff and patients, but which in time began to distribute medicines to the townspeople. Images of pharmacies may be found in many manuscripts; the portrayal of one, greatly foreshortened, appears in the [*H*]*ortus Sanitatus* (see No. 3 [Volume II, 3a1 verso]), while there is a charming illustration of a chemist serving a client in the *Tacuinum Sanitatis* (Biblioteca Casanatense, Rome, MS. 4182). A fresco in the Castle of Issogne in Val da Aosta depicts a fifteenth-century pharmacy with a wealth of fascinating detail—against a background

FRENCH SCHOOL,
Eight medicinal plants.
Mandrake (*Mandragora officinarum*)

of shelves lined with apothecary jars, sponges, and other objects hanging from hooks, an elderly pharmacist studies a recipe while his young assistant serves a client. Much later the Venetian artist Pietro Longhi (1702–1785) depicted a *Bottega dello speziale* (apothecary's shop) in a charming genre painting now conserved in the Museo dell'Accademia in Venice.

During the course of the sixteenth century, in a renewed and more 'modern' approach to the treatment of patients, the first pharmacies not affiliated with a religious establishment or hospital began to appear. These evolved into elaborately equipped shops with a number of functions; medicines were not only laboriously prepared there, but also carefully conserved and stored, sold to the public, and even administered. The pharmacy generally consisted of a room whose walls were lined with shelves on which numerous containers were arranged in orderly rows. These included jugs, flasks, pitchers, and the cylindrical *albarelli* typical of the earliest pharmacies, often made of decorated and glazed majolica or terracotta, in which liquid medicines and ingredients were stored. Sometimes the jars bore cartouches labelled with their contents in a florid script. There were also containers made of glass, such as flasks, pitchers, ampoules, and bottles, and boxes where unguents, pills, and powders were stored. At one counter customers were served, while the remainder of the shop was devoted to the preparation of medicines. Every pharmacist owned several mortars—the most important tool of his trade, which was usually made of some heavy stone such as marble, or else of bronze—in which ingredients could be ground to a paste or powder. With the spread of the science of chemistry, burners, alembics, and retorts for the distillation of active principles also became a common sight in apothecary shops.

We know that pharmacies were sometimes decorated with pictures, usually portrayals of the Virgin Mary, Jesus, or the patron saints of the medical arts (such as Saints Cosma & Damiano), or others venerated as the healers of a particular illness, such as Saint Antonio (for barren women and amputees), Saint Apollonia (for toothache) and Saint Biagio (for those suffering from sore throat). No evidence, however, that botanical paintings may have hung in any of these early pharmacies has come down to us and, therefore, the series of eight paintings of medicinal plants now conserved at the Oak Spring Garden Library—a veritable 'herbal painted in oils'—offers a particularly interesting example of the decorative furnishings of a pharmacy during the first half of the seventeenth century.

It should be recalled that many early pharmacists took an interest in natural history and their shops might contain displays akin to those of the collector's *Wunderkammer*. These included eclectic assemblages not only of botanical and zoological specimens, but also exotic, extravagant, or monstrous objects that could be the handiwork of nature or man, with a stuffed crocodile not infrequently suspended overhead. The Spanish naturalist Garcia da Orta (see No. 17) provides us with a description of the pharmacy of Francesco Calzolari

FRENCH SCHOOL,
Eight medicinal plants.
Silver thistle (*Carlina acaulis*), Carline thistle (subsp. *caulescens*)

FRENCH SCHOOL,
Eight medicinal plants.
Aloe, Common aloe or Medicinal aloe, not identified; leaves possibly of *Agave*, inflorescence of *Yucca*

FRENCH SCHOOL,
Eight medicinal plants.
False Helleborine
(*Veratrum album*)

in Verona 'at the Sign of the Golden Bell' (see No. 33). There were: '... diversi medicameti simplici, e coposti, & eti adio di diversi minerali, mezi minerali, pietre pretiose, animali rarissimi, uccelli visti da pochi, pesci non conosciuti, diverse sorti di terre, & legni, minere d'ogni qualità, & insomma di tutto quello che si può vedere di bello, di raro, & di buono, appresso i più dotti, e più risvegliati ingegni de' nostri tempi ...' (... different herbal medicines and mixtures, and birds just recently discovered, unknown fishes, different kinds of soil, and of wood, minerals of every quality, and in short of all those beautiful, rare and good things that one may see [collected] by the learned, and the most keen intellects of our time... [Garcia da Horta, *Dell'Historia de I Semplici Aromati...*, Venice, Scotto, 1597, p. 381]). Another seventeenth-century pharmacy with a vast natural history collection was that of Paul Contant in Poitiers (see No. 31).

A few of these old pharmacies have survived to the present day, and, although they may have undergone modifications with the passage of time, their original aspect remains relatively intact and can often be reconstructed in considerable detail. One particularly complete example has been re-assembled in the Medical-Pharamaceutical Museum in Amsterdam. As we have said, however, there are no known examples of the artwork that might have decorated these early pharmacies and for this reason the eight paintings from Saumur constitute evidence of noteworthy relevance to the history of the pharmacy as an institution, as well as to the history of botanical painting.

Each painting shows a medicinal plant framed against a pale blue sky; they are portrayed considerably larger than life-size, growing from the earth but with their roots exposed. In accordance with the typology of the floral still life, the artist has added highly naturalistic (although not always identifiable) butterflies and insects to animate each scene. Since we have no information regarding where these paintings were hung, or in what sequence, we shall examine them in random order.

It seems appropriate to begin with the imposing portrait of *Mandragora officinarum*, which since Antiquity has been prized for the magical and therapeutic properties attributed to its root. A poisonous *Solanacea* with an unpleasant odor that has also earned it the name 'Satan's apple,' the mandrake is common all over southern Europe. Although no longer used in medicine because it contains a toxic alkaloid similar to atropine, it was a popular treatment in the past for colic, asthma, and cough. The fleshy tap-root often splits into two or three forks, thus irresistibly suggesting the form of a human being. The mandrake is mentioned in almost every antique herbal, from the *Italian Herbal* (see No. 2) to the [*H*]*ortus Sanitatis* (see No. 3) and the *Grete Herball* (see No. 4), and is generally portrayed as a bearded man or a woman with long hair. In the painting from Saumur the plant is female, shown modestly covering her private parts with her 'root arms.' The remainder of the portrayal is quite naturalistic, the artist depicting the dark green leaves and round yellow fruit of the plant and four hovering butterflies, including on the right a *Papailionidae*, in a simple, yet realistic manner.

The second painting depicts *Carlina acaulis*, with its lance-shaped, dentate leaves, long cylindrical root, and large flower head. A member of the *Compositae* family, it can be found growing in meadows and grasslands in many of the mountainous zones of Europe. Its root, gathered in the autumn, cut into small pieces and dried, has an antipyretic and tonic effect and was widely used to reduce fever. Hovering around the plant are three unidentifiable butterflies.

Another painting depicts a medicinal plant from the family *Liliaceae*: Common aloe or medicinal aloe, whose therapeutic properties have been known since the dawn of civilization; it is mentioned in the Ebers papyrus, an Egyptian medical manuscript dating to 1500 B.C. The leaves of the genus *Aloe* contain two elements, a juice and a gel, which can both be used as moisturizers, emollients, laxatives, and antiseptics; they are particularly effective in the treatment of burns and wounds. The unknown seventeenth-century artist has created an impressive portrait of this tropical plant with its long inflorescence, fleshy leaves, and large root, standing out against the gray-blue background. A dragonfly (*Odonata*) is poised on the tip of a leaf while a menacing black rhinoceros beetle clings to the stem of snow-white flowers. Lower down is a butterfly and, climbing up the root, a caterpillar of an unknown species.

The next painting shows a curious plant native to Europe, *Veratrum album* or False Helleborine, and it is characterized by dense, compact flower heads and a hair-like root system. The artist has portrayed a large butterfly of the family *Papilonidae* (swallowtail) alighting upon a leaf to the left, while on the right are a dragonfly and two small butterflies in dancing flight. Two mushrooms can be seen growing near the base of the plant. The root of False Helleborine is very poisonous and is known to have a paralyzing effect upon the nervous system.

A painting of the evergreen shrub *Ruscus hypoglossum* follows, its bright red, pedunculate berries forming a striking contrast to the dark green leaves. The root contains several curative and vasoprotective principles and is still used today to treat varicose veins. Once again the work is enlivened by the addition of various insects, including a dragonfly on the left, three butterflies on the right, and a snail creeping through the grass.

The next work depicts *Lavandula stoechas*, a fairly common plant of the genus *Labiatae*. Its compact inflorescences, portrayed in careful detail by the artist, contain an essential oil with a heady perfume that is used as an aromatic component in unguents, creams, and lotions. In herbal medicine it is administered as an antispasmodic, a digestive, and a stimulant.

FRENCH SCHOOL,
Eight medicinal plants.
Laurel, Butcher's Broom
(*Ruscus hypoglossum*)

FRENCH SCHOOL,
Eight medicinal plants.
Lavender (*Lavandula* cf. *stoechas*)

FRENCH SCHOOL,
Eight medicinal plants.
Datura metel

FRENCH SCHOOL,
Eight medicinal plants.
Fritillaria pallidiflora

Near its root another plant, perhaps a cyclamen, can be seen while a dragonfly and two butterflies hover in the air.

The artist has also executed a fine painting of the exotic *Datura metel* from Central America, with its beautiful, trumpet-shaped white flowers and spherical seed capsules covered with spines. The leaves and seeds of some species of this plant contain an alkaloid with narcotic and sedative effects that can be administered to treat bronchial asthma, epilepsy, and insomnia. Indian tribes of the southwestern United States used it as a hallucinogen in their religious ceremonies. A dragonfly, a swallowtail, and an unidentifiable butterfly hover around the plant, while at its root sprouts a mushroom with a reddish cap.

The last painting depicts *Fritillaria pallidiflora*, a flowering species from Central Asia that was greatly prized for its magnificent, bell-shaped blossoms. The artist also shows the plant's large bulb, which was known to contain various poisonous alkaloids. The bulb of *Fritillaria imperialis* was cited in the old *pharmacopoeiae* under the name '*Radix coronae imperialis*.' A swallowtail and two dragonflies flutter around the plant, while a caterpillar laboriously climbs up one of the stems.

Some of these paintings could have been based on the woodcuts published in Dodoens' *A Niewe Herball*, 1578 (see No. 10), such as *Lavandula staechas* (French lavender), p. 266, *Veratrum album* (False Helleborine), p. 347, *Mandragora officinarum* (Mandrake), p. 437, *Carlina acaulis* (White Caroline Thistle), p. 529, and *Ruscus hypoglossum* (Horse Tongue), p. 675. The hues in these paintings have become somewhat darkened over time, obscuring the bright, simple range of their original colors, but we can nevertheless appreciate the practiced style of this unknown seventeenth-century botanical artist, who applied his paints in thick, confident brushstrokes. The fact that the roots form a prominent element in most of the works reflects the medicinal applications of these plants.

33. Antonio Donati (1606–1659)

Trattato de Semplici, Pietre, et Pesci Marini, che nascono nel lito di Venetia, La maggior parte Non conosciuti da Teofrasto, Dioscoride, Plinio, Galeno, & altri Scrittori. Diviso in due libri. Nel Primo si contengono le Figure de Semplici, che nascono nel detto Lito, con le sue facoltà. Nel Secondo le Figure di alcune Piante peregrine, Pietre, & Pesci Matini, con le sue virtù. Di Antonio Donati Farmacopeo All'Insegna del S. Liberale in Venetia. Con Licenza, et Privileggio. [*Arabesque vignette*] In Venetia, MDCXXXI. [*rule*] Appresso Pietro Maria Bertano.

4° 21.4 × 15.9 cm. a⁴ A–P⁴ *i–viii* 1–2 *3*–120 [128 pp.].

PLATES: 32 engravings throughout text.

BINDING: Contemporary stiff white wrappers.

REFERENCES: Cleveland 180; Nissen 519; Pritzel 2,368.

FROM the title-page of the *Trattato de Semplici, Pietre, et Pesci Marini che nascono nel lito di Venetia* ... we learn that Antonio Donati was a 'farmacopeo all'insegna del S. Liberale in Venetia'—a pharmacist with a shop in the city of Venice at the 'Sign of Saint Liberale.' During the sixteenth and seventeenth centuries, apothecaries' shops were known by fanciful names that were depicted in picturesque signboards above their entrances, just as pharmacies today in Italy can be identified by the green cross above their doors.

The city-state of Venice was renowned throughout Europe for its pre-eminence in the art of pharmacology; the *Collegio dei farmacisti* was held in great esteem and its members were granted special privileges for the meritorious exercise of their profession and for their contributions to the emerging science of chemico-pharmacology. Indeed, the flourishing business conducted by its numerous pharmacies contributed in no small measure to swell the coffers of the Serenissima, which had been engaged for centuries in trade with the Near and Far East in spices and medicinal plants.

The *Trattato de Semplici* constitutes a noteworthy contribution to the botanical literature by one of these 'addetti al mestiere' (experts in the profession), even though in the dedication addressed to his friend and fellow Venetian, Pietro Caffi, the author makes a somber reference to the 'funebre stagione' then in course (the entire peninsula was in the grip of a plague epidemic) and explains that this disastrous situation prevented him from writing as complete a work as he had envisaged. He trusted, however, that when conditions improved he would be able to amplify and improve upon his treatise, whose aim was to examine the ties between natural history and the science of pharmacology. The contents, as the author underlined, were entirely original and 'tanto necessarie all'uso medicinale, che a punto ricercano altra stagione piu' serena, altro talento di penna non così scarso, & altro spatio di carta non così angusto' (so necessary to the practice of medicine that, in fact, [to do the subject justice] would require another, more tranquil season, a more generous talent with the pen, & more space on a less meager expanse of paper).

The dedication is followed by two short encomiastic poems addressed to the author, and a letter to the readers that draws attention to the wealth of nature in the coastal zones around Venice, particularly its flora—species which, even if some were not mentioned by the classical authors, were replete with 'virtù salutifere' (health-giving virtues). Donati explains that, since it would be impossible to do justice to them all, he has decided to present just some of the most important and unusual. He remained convinced, however, that the plant life of the littoral merited further systematic study and cites the exemplary work conducted in the region of Mount Baldo near Verona by colleagues such as Francesco Calceolari (1521–1600), a celebrated 'farmacopeo' with a shop in Verona 'at the sign of the Golden Bell' (*Il viaggio di Monte Baldo...*, Venice 1566), and Giovanni Pona (1565–1630), another Veronese pharmacist 'at the sign of the Golden Apple,' who was the author of

Plantae seu simplicia ... quae in Baldo Monte ... reperiuntur, published in Verona in 1595. Donati writes that he painstakingly scoured the coast in order to study the simples that grew there, inspired by their example and also following in the footsteps of other Venetian colleagues: Orazio Bresciani, who owned a shop 'at the sign of the Holy Spirit'; Domenico Valle, a pharmacist 'at the sign of the Globe'; and Marchioro Brochini with a shop 'at the Chalice.'

The meaning of the term 'simple' is carefully elucidated: this might be any plant, animal, or mineral endowed with healing virtues, which Donati recommended be used in unadulterated form for the preparation of medicines. The text is preceded by a note explaining that the entities would be presented in alphabetical order by their Italian name, followed by their names in various other languages. The botanical descriptions are admirably clear and precise, and the author always identifies the places where the plant could be found. Sometimes he also furnishes recipes for the preparation of specific remedies.

In addition to plants typical of coastal habitats, from *Eringio marino* and *Pastinaca* to *Salicornia,* Donati describes a series of anomalous marine organisms which he calls 'muschi' (many of these were probably 'sea balls' made of algae) and various *rarità,* such as the 'Lapis nefriticus' (nephritic stone), which he speculated might be the same as the stones to be found in the stomach of a certain fish in India according to Nicolas Monardes (1493–1588; see No. 16). This mineral—which, judging from the illustration, may have been a kind of pumice—ground to a powder and taken in white wine, was an effective treatment for kidney stones according to the author. Here Donati was applying the doctrine of signatures, which credited nature with distributing various secret signs that could guide the attentive physician to the appropriate use of natural substances as medicines (see No. 51).

The *Trattato de Semplici* closes with a *curiosità* that must have fascinated seventeenth century pharmacists, whose shops and cabinets were full of rare and unusual natural-history specimens, as is demonstrated by the printed catalogue and engraving of the 'museum' of the pharmacist Francesco Calceolari (B. Ceruti & A. Chiocco, *Musaeum Francisci Calceolari junioris veronensis,* Verona, 1622). This was yet another marine organism, a jellyfish (*Ryzostoma pulmo*), which Donati refers to as a *satiro marino* (sea satyr); an engraving shows the animal, which was somewhat damaged during its capture, in a ventral view. Donati recounts that Johannes Wesling (1598–1649), a professor at the university in Padua and director of its Botanical Garden, tried to dissect one of these 'monsters,' but the specimen unfortunately dissolved after a little time 'like ice before a fire.'

The *Trattato de Semplici* is accompanied by a fine set of illustrations prepared using the technique of copperplate engraving, which was considerably more precise than the woodcut and began to gain popularity among the authors of scientific publications at the beginning of the seventeenth century.

82 *Trattato de Semplici*

SOLDANELLA.

Braſſica Marina Belg.

Verca Marina Lob.

In

ANTONIO DONATI,
Trattato de Semplici,
1631. Sea Bindweed
(*Calystegia soldanella*),
page 82

A century would pass before the detailed study of the flora of the Venetian coast so strongly recommended by Donati finally appeared to supplant his own excellent treatise. A little later, the naturalist Gian Giacomo Zannichelli (1662–1729) of Modena compiled a *Istoria delle piante che nascono ne' lidi intorno a Venezia* in which about five hundred species endemic to the littoral are described and illustrated; this work was published posthumously in Venice in 1735 by his son.

34. William Coles (1626–1662)

[*Printed in red and black within a double rule border*]: Adam in Eden: Or, Natures Paradise. The History of Plants, Fruits, Herbs and Flowers. With Their several Names, whether Greek, Latin or English; the places where they grow; their Descriptions and Kinds; their times of flourishing and decreasing; as also their several Signatures, Anatomical appropriations, and particular Physical Vertues; Together with necessary Observations on the seasons of Planting, and gathering of our English Simples with Directions how to preserve them iu [*sic*] their Compositions or otherwise. A Work of such a Refined and Useful Method, that the Arts of Physick and Chirurgerie are so clearly laid open, that Apothecaries, Chirurgions, and all other ingenuous Practitioners, may from our own Fields and Gardens, best agreeing with our English Bodies, on emergent and sudden occasions, completely furnish themselves with cheap, easie, and wholsome Cures for any part of the Body that is ill-affected. For the Herbarists greater benefit, there is annexed a Latin and English Table of the several names of Simples; With another more particular Table of the Diseases, and their Cures, treated of in this so necessary a Work. [*rule*] By William Coles, Herbarist. [*rule*] Then the Lord took the Man, and put him into the Garden of Eden, Gen. 2. 25. [*rule*] London, Printed by J. Streater, for Nathaniel Brooke at the Angel in Cornhil, near the Royal Exchange, 1657.

2° 27.3 × 17.6 cm. A^2 a–b^4 B–T^4 $2T^2$ V–$3O^4$ $3P^2$ (-3P2) 4A–$4H^4$ $4I^2$ (-4I2) 5A–$5K^4$ 4L–$4P^2$ $4R^2$ *i–xx* 1–144 (9 as 19, 64 as 65, 84 as 78, 85 as 75) 75–78 145–165 (161 as 111, 162 as 112, 163 as 131) 116–206 (127 as 125, 128 as 126, 133 as 131, 134 as 132, 162 as 160, 163 as 161, 165 as 167, 202 as 220, 204 as 206) 209–264 269–348 (283 as 238, 327 as 237, 336 as 335, 338 as 328) 347–348 353 350–396 (369 as 368) 1–66 (32 as 23) 551–629 (621 as 620, 623 as 627) [*25*] [636 pp.].

BINDING: Contemporary speckled calf.

REFERENCES: Arber, pp. 252–254; Cleveland 222; Foster; Henrey 1.88–90 and No. 30; Hunt 269 and 266 (*Art of Simpling*); Lowndes; Pritzel 1,787; Rohde, pp. 105, 167.

WILLIAM COLES was a seventeenth-century English botanist and simpler who vehemently opposed the astrological theory of natural history supported by Nicholas Culpeper (see No. 41). Coles was convinced the key to pharmacological botany lay in the doctrine of signatures, a philosophy that bridged the astrological-alchemistic tradition and the emerging empirical approach associated with the birth of the modern sciences. The doctrine of signatures was promulgated in Europe by Giovanbattista della Porta of Naples (see No. 51). According to della Porta, all things in nature were impressed with secret 'signs' by their Creator that, if correctly identified, could reveal to the physician or herbalist the specific curative properties inherent in them. This theory was elaborated upon and

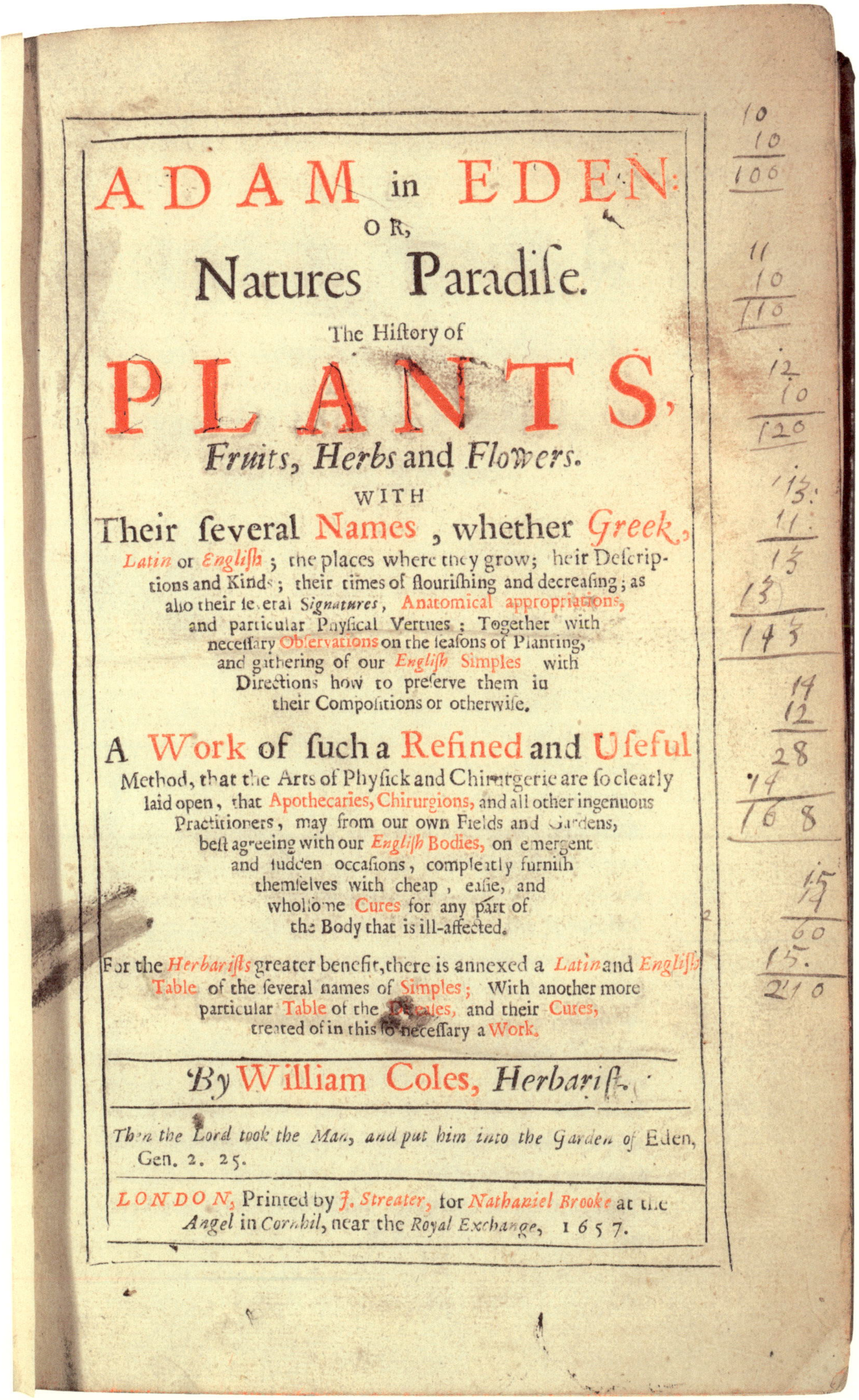

ADAM in EDEN:
OR,
Natures Paradiſe.
The Hiſtory of
PLANTS,
Fruits, Herbs and *Flowers.*
WITH
Their ſeveral Names, whether *Greek*, *Latin* or *Engliſh*; the places where they grow; their Deſcriptions and Kinds; their times of flouriſhing and decreaſing; as alſo their ſeveral *Signatures*, Anatomical appropriations, and particular Phyſical Vertues; Together with neceſſary Obſervations on the ſeaſons of Planting, and gathering of our *Engliſh* Simples with Directions how to preſerve them in their Compoſitions or otherwiſe.

A Work of ſuch a Refined and Uſeful Method, that the Arts of Phyſick and Chirurgerie are ſo clearly laid open, that Apothecaries, Chirurgions, and all other ingenuous Practitioners, may from our own Fields and Gardens, beſt agreeing with our *Engliſh* Bodies, on emergent and ſudden occaſions, compleatly furniſh themſelves with cheap, eaſie, and wholſome Cures for any part of the Body that is ill-affected.

For the *Herbariſts* greater benefit, there is annexed a *Latin* and *Engliſh* Table of the ſeveral names of Simples; With another more particular Table of the Diſeaſes, and their Cures, treated of in this ſo neceſſary a Work.

By William Coles, *Herbariſt.*

Then the Lord took the Man, and put him into the Garden of Eden, Gen. 2. 25.

LONDON, Printed by *J. Streater*, for *Nathaniel Brooke* at the *Angel* in *Cornhil*, near the *Royal Exchange*, 1657.

WILLIAM COLES,
Adam in Eden, 1657.
Title-page

applied systematically to the plant world by the philosopher Jakob Böhme (1575–1624), who published *Signatura rerum or the signature of all things,* London, 1651. Böhme was a master shoemaker from Görlitz near Dresden who, after a religious experience, became a philosophical mystic. His writings exercised a remarkable influence not only on contemporary but also the later intellectual movements.

William Coles was born in Adderbury in the county of Oxfordshire, and in 1642 enrolled at New College, Oxford, where he earned the title of Bachelor of Arts in 1650. He settled in Putney and there dedicated himself entirely to the study of botany, periodically publishing the results of his research. Sponsorship remained an important factor in the success of those engaged in intellectual pursuits and, as Lownes recounts, in one of his early works, Coles sought to gain the goodwill of an influential patron. The first edition of Coles' short treatise, *The Art of Simpling,* did not contain a dedication when it appeared in 1656, but a second edition printed in the same year with various modifications opened with a flattering address to the antiquarian Elias Ashmole (1617–1692), who counted botany among his wide-ranging interests. *The Art of Simpling* is a rare book to find, indicating there were not many printed. It seems the second edition with the dedication to Ashmole did not help with the sale of the book, perhaps this was due to the fact that he was a leading astrologer and did not agree with Coles' views.

Instead of attempting to reprint the *The Art of Simpling,* in 1657 there appears a different herbal, *Adam in Eden,* in which Coles presents a detailed exposition of the doctrine of signatures, which he had only touched upon in his preceding work. In the preface addressed 'To the Reader' the author recalls nostalgically 'the best Hours of my Life being spent in the Fields and in the Physick Gardens, more especially in that Famous One at Oxford' (p. v).

In his preface, Coles launches into a polemic against some of the most prominent English botanists of the preceding century, including Gerard, Johnson (see No. 12), and Parkinson, whom he accuses of having promoted 'Outlandish Plants and Ingredients, which are almost if not altogether impossible to be obtained' to the detriment of 'those more wholesome herbs and plants that [he] hath growing at his own doore, which are more consonant and proper for his body' (p. v). He also criticized the illustrations in their texts, pointing out that they were full of errors and often quite misleading. It is perhaps for this reason that Coles' works have no illustrations. Coles also mentions that the illustrations would make the work too expensive for the common person. At the end of 'To the Reader' Coles interestingly acknowledges the support of Ashmole, perhaps in an effort to be diplomatic, and others, describing them as 'the intelligence of our late times.'

His intention in writing *Adam in Eden* was to present a series of indigenous plants with a discussion not only of their virtues and properties, but also of their growing cycles from germination to flowering and dehiscing. The originality of his work lies in its methodology for, according to Coles, by applying the doctrine of signatures the reader could determine the herbal medicines best adapted to the treatment of various sectors of the human body, comprising the upper, middle, lower, and the limbs, as defined by the anatomists of the period. Coles' book thus follows the parts of the body starting with the head and ending with the feet.

Coles begins his text with a discourse on the walnut tree, 'because the fruit of it doth resemble the Head in Several particulars' (p. 1). After briefly delineating its 'Names, Kinds, Forme, Place and Time, Temperature' Coles embarks on an analysis of the aspect that was of greatest interest to him, the plant's 'Signatures and Vertues.' He states that 'Wall-nuts have the perfect Signature of the Head' going on to describe their points of similarity:

> The outer husk or green Covering represents the Pericranium, or outward skin of the skull ... the Kernel hath the very figure of the Brain, and therefore it is very profitable for the Brain, and refits poisons; For if the Kernel be bruised, and moystned with the quintessence of Wine, and laid upon the crown of the head, it comforts the brain and head mightily (p. 3).

Analogously, in the chapter 'Of Poppie' he writes 'with their Crowns [they] do somewhat represent the Head and the Brain, and therefore the decoctions of them are used with good success in several diseases of the Head' (p. 7). With regard to the 'Pome Citron-tree' or 'Malus Medica,' the author observes that the fruit of the citron, 'being like to the heart in form, is also very soveraing Cordiall for the same, and excellent Antidote against Venome or Poyson, against Plague, or any other infection' (p. 189).

Scrupulously following 'the anatomy of the man's body' in his work, it is appropriate that the last plant described in *Adam in Eden* is the 'Ladies bedstraw' or *Galium*, a decoction of which he reports is useful to relieve the aching feet of travellers. It is noted that Coles failed to identify the signatures of many plants (as these were not always evident), in which case he passed directly to a description of their virtues, in conformity with the more conventional herbals then in general use.

35. John Hill (1714?–1775)

[*Printed in red and black*]: The British Herball: An History of Plants and Trees, Natives of Britain, Cultivated for Use, or Raised for Beauty. By John Hill, M.D. [*Copperplate vignette—Asculapius and Flora gathering from the Lap of Nature, Health and Pleasure. Signed* 'S. Wale del.' 'C. Grignion Sc.']. London: Printed for T. Osborne and J. Shipton, in Gray's Inn; J. Hodges, near London-Bridge; J. Newbery, in St. Paul's Church-Yard; B. Collins; And S. Crowder and H. Woodgate, in Pater-noster-Row. MDCCLVI.

2° 40.6 × 25.7 cm. A^2 B–6U^2 *i–iv 1* 2–30 *31–50 51–54* 55–105 *106–135 136–138 139–140 141–160 161–185 186–*202 *203–216 217–220 221–224 225–231 232–255 256–275 276–*300 *301–*316 *317–338 339–348 349–383 384–394 395–*399 *400–426 427–434 435–445 446–455 456–465 466–*471 472–479 *480–493 494–*507 *508–524 525–529 530–533* [*3*] (493 as 943) [540 pp.].

PLATES: 75 leaves of plants and fungi gathered at the end; plus frontispiece (signed 'S. Wale delin.' and 'H. Roberts sculp'), vignette on title-page (signed 'S. Wale del.' and 'C. Grignion sc.'), coat-of-arms of Earl of Northumberland on dedication, all engraved and etched, hand-colored.

BINDING: Contemporary calf which is re-backed in brown leather.

PROVENANCE: '1865 July 12th H. Holberry. Mattersea' inscribed on front free endpaper, 'Miss H. Holberry' inscribed on top of title-page.

REFERENCES: Blunt & Raphael, pp. 177–181; Blunt & Stearn, pp. 170–171; Cleveland 441; Dunthorne 128; Henrey II.90–109 and No. 798; Hunt 557; Nissen 881; Pritzel 4,063; Stafleu & Cowan 2,769.

The life and work of this English polymath and amateur naturalist, although controversial and often denigrated, is of genuine interest. Born in Petersborough in 1714 or 1716, John Hill first devoted himself to the study of pharmacy, before passing into the service of two members of the aristocracy, Charles Lennox, the second Duke of Richmond, and Robert James Petre, eighth Baron of Writtle, for whom he worked as head gardener and curator of their collections of plant specimens. Eventually, he was forced to abandon this pleasant but not very remunerative activity and open a pharmacy in Covent Garden, at which point he also began to devote himself seriously to the study of botany.

The publication of a translation from the Greek of a short treatise on minerals by Theophrastus earned Hill a degree of fame among scientists and convinced him where his true vocation lay. He became a tireless writer, producing works on subjects ranging from science and horticulture, to theology and naval science. He even turned his hand to the writing of political and economic dissertations, novels, and opera librettos. This activity also provided him with the opportunity to express his artistic talent, which contributed to the success of his works

Hill's fame grew rapidly despite the many envious voices that were raised against him. Typical is the comment of one contemporary, who described his books as 'unscrupulous, impudent and scurrilous.' He won various marks of recognition, and was inducted as a Knight of the Swedish Order of Vasa. He earned an honorary medical degree from the University of St. Andrews and after a knighthood, he would insist on being called 'Sir John.'

Hill continued to take an interest in every aspect of botany. He campaigned for the adoption in England of the system of plant classification introduced by Carolus Linnaeus,

JOHN HILL, *The British Herball*, 1756. A page of *Umbellif-erae:* 'Great Hercules Allheal'; 'Black Libano-tis'; 'Common Dill'; 'Broad leav'd Thapsia,' Penny Cress; 'Narrow leav'd fennell Giant'; 'Broad leav'd fennell Giant'; 'Laserwort'; 'Common Cummin'; 'Masterwort'; 'Lovage'; 'Sermountain'; 'Com-mon Skirret'; 'Com-mon Anise'; 'Common Parsley'; 'Common Bishopweed'; 'Span-ish Toothpick'; 'Candy Daucus'; 'Umbelliferous Pellitory'; 'Black Mas-terwort.' Plate 60

while at the same time criticizing various of its aspects. The methodology adopted by Hill in the compilation of his early, three-volume work *A General Natural History* (1751) is significant, for, in addition to describing, as we learn from the title '... animals, vegetables, and minerals of the different parts of the world: with their vertues, and uses as far as hitherto certainly known, in medicine and mechanics,' he provides an extended treatment of 'the history of materia medica, pictoria, and tinctoria of the present and the earlier ages.' This work was followed in rapid succession by *The Useful Family Herbal, or, An Account of All Those English Plants, Which are Remarkable for Their Virtues ...* (1754); *Eden, or, A Compleat Body of Gardening* (1757), on the cultivation of fruits, vegetables and flowers (see *An Oak Spring Flora*, No. 53, pp. 197–201); *Exotic Botany Illustrated* (1759); *Flora Britanica* [sic] (1760); and *The Vegetable System* (1761–75), his *magnum opus* in twenty-six volumes dedicated to the morphology, life cycle, and classification of plants.

The British Herball was printed in fifty-two installments that were sold as sixpenny (6*d.*) numbers between January 1756 and January 1757 by the publishers T. Osborne and J. Shipton. Beginning in January of 1757, Osborne and Shipton issued a complete edition of the work printed 'on royal paper, with cuts coloured,' but still dated 1756 [issued in weekly number, 1757–1758]. The work is dedicated to Hugh Smithson, Earl of Northumberland, whom Hill referred to as a 'patron of useful knowledge.' Like many of Hill's works, the *The British Herball* is generously illustrated. It opens with a hand-colored frontispiece portraying 'The Genius of Health receiving the tributes of Europe, Asia, Africa, and America and delivering them to the British Reader.' The etching made by Henry Roberts was based on a drawing commissioned from the celebrated and prolific artist Samuel Wale (1721?–1786). The title-page is decorated with a vignette by Wale, this time engraved by Charles Grignion (1716–1810) and featuring 'Asculapius and Flora gathering from the Lap of Nature, Health and Pleasure.'

The herbal presents 1,500 plants from the British flora, classified by genus and species based on the form of their corolla, ovary, stylus, and stigma, but it is not consistently binomial. Hill applies some, but not all of the criteria established by Linnaeus in *Species plantarum*, and indeed does not hesitate to criticize his eminent colleague's system. *The British Herball* is, nonetheless, one of the first publications following the appearance of Linnaeus' *Species plantarum* (1753) to adopt the Linnaean system, if imperfectly. Indeed, Hill's earlier work *Flora Britanica* [*sic*] (1760) was the first flora in Britain to be arranged according to Linnaeus' sexual system. The text is commendable for its detailed and incisive botanical descriptions. Although, belying what is stated in the frontispiece, the author provides very few indications regarding medicinal properties. Following the text are seventy-five leaves of engravings depicting the plants mentioned in the herbal. Arranged in conformity with Hill's usual practice, the plants are displayed in orderly rows, sometimes fifteen or twenty

JOHN HILL, *The British Herbal*, 1756. A page of ferns and allied plants: 'Common Harts-tongue' (*Phyllitis scolopendrium*); 'Polypody' (*Polypodium vulgare*); 'Rough Spleen-wort' (*Blechnum spicant*); 'Smooth Spleen-wort' (*Ceterach officinarum*); 'Dwarf Sea Fern'; 'English Maiden-hair'; 'Forked Maiden-hair' (*Asplenium viride*); 'Common Male Fern' (*Dryopteris filix-mas*); 'White Maiden-hair' (*Asplenium ruta-muraria*); 'The True Maiden-hair' (*Adiantum pedatum*); 'Common Female Fern' (*Athyrium filix-femina*); 'Black Maiden-hair' (*Asplenium adiantumnigrum*); 'Winged Maiden-hair' (*Polystichum* sp.); 'Osmund Royal' (*Osmunda regalis*); 'Adders Tongue' (*Ophioglossum vulgare*); 'Moonwort' (*Botrychium lunaria*); 'Common Duckweed' (*Lemna minor*); 'Large Duckweed' (*Spirodela polyrhiza*); 'Great Water Horsetail' (*Equisetum fluviatile*); 'Wood Horsetail' (*Equisetum sylvaticum*). Plate 74

to a page. The English name of the plant appears written in a graceful script at the foot of each drawing with the illustrations colored by hand.

Portraying such a large number of plants in an herbal was an immense undertaking. To facilitate this task Hill borrowed images from previous published works, an expedient to which he had already had recourse in *Eden, or, A Complete Body of Gardening.* This earned him scathing criticism from the physician Erasmus Darwin (1731–1802), who was the grandfather of the naturalist Charles. Erasmus wrote in a letter sent at the end of the year 1756 to his friend J.A.H. Reimarus (1729–1814) in Germany: 'I believe I forgot to tell how Dr. Hill makes his herbal. He has got some wooden plates from some old herbals, and the man that cleans them cuts out one branch of every one of them, or adds one branch or leaf, to disguise them. This I have from my friend Mr. G—y, watch-maker, to whom this printmender told it,' adding 'I make plants now every day that God never dreamt of' (see Henrey II.94). Darwin found this stratagem deplorable, but it is understandable that the author felt obliged to resort to it, given the enormous number of plants to be illustrated.

The seventy-five pages of illustrations in *The British Herball* are signed with the names of a number of different engravers such as: Boyce, Roberts, Benning, Darley, and Edwards, and some of whom are unknown today. Particularly beautiful are Plates 23 and 50. Of the artists who may have helped to prepare the original botanical drawings, only the names of Darley and Edwards have been recorded. In some of the engravings the letter 'H' appears, presumably the signature of the author himself, which assumption is plausible, since some of the images were drawn from *Useful Family Herbal*, a work published by Hill in 1754.

36. Johann Wilhelm Weinmann (1683–1741)

[*German title-page, Volumes I–IV, printed in red and black*]: Phytanthoza-iconographia, oder eigentliche Vorstellung etlicher Tausend; so wohl Einheimisch-als auslaendischer, aus allen vier Welt-Theilen, in [*Vol. 4:* Sowohl ... Theilen in] Verlauf vieler Jahre, mit unermůdetem [*Vols. 2–4:* unermůdeten] Fleiß von Johann Wilhelm Weinmann, E. Ehrlőbl. Statt-Gerichts-Assessore und åltesten Apotheckern in Regenspurg gesammleter Pflantzen, Båume, Stauden, Kråuter, Blumen, Frůchten und Schwåmme &c. Welche auf das netteste in Kupfer gestochen, und zugleich durch eine långst-verlangte, und neu erfundene Art, nach der Natur mit lebendigen Farben, in anmuthigsten Abbildungen herausgegeben [*Vol. 4:* heraus gegeben] und verlegt von Bartholomå Seuter, Johann Elia Ridinger, und Joh. [*Vol. 4:* Johann] Jacob Haid, Mahlern und Kupferstechern [*Vols. 3 & 4:* Kupfferstechern] In Augspurg, Deren Benennung, Arten, Kennzeichen, Beschreibungen und Gebrauch aus dem [*Vols. 2–4:* den] besten, so alt-als neuen Autoribus, nach Alphabetischer Ordnung, nebst angezeigten bewåhrtestem [*Vols. 2–4:* bewåhrtesten] Nutzen [*Vol. 4:* Nutzen,] in der Artzney-und Apothecker-Kunst, auch in der Chirurgie und Haushaltung, in lateinisch-und deutscher Sprache richtig und deutlich beschrieben worden von D. Johann Georg Nicolao Dieterichs, [*Vols. 3 & 4:* D. Ambrosio Carolo Bieler,] Der Rőm. Kayserl. und Kőningl. Catholischen Majeståt Rath, Ihro Hoch-fůrstl. Durchleucht von Fůrstenberg-Stůlingen Medico ordin. und Reipubl. Ratisbon. Physico. [*Vols. 3 & 4:* Reipublicæ Ratisbonensis Physico Ord.] Erster Band A. B. [*Vol. 2:* Zweter Band C.D.F.E.F.;

Vol. 3: Dritter Band G.H.J.K.L.M.N.O.; *Vol. 4:* Vierdter Band P.Q.R.S.T.U.X.Y.Z.] [*rule*] Auf obbemeldter Verlegere kosten gedruckt In Regenspurg bey Hieronymo Lentzen 1737 [*Vol. 2:* Lentz 1739; *Vol. 3:* Lentz 1742; *Vol. 4:* bey Heinrich Georg Neubauer 1745].

[*Latin title-page, Volumes I-III, printed in red and black*]: Phytanthoza-Iconographia, Sive Conspectus Aliquot millium, tam Indigenarum quam Exoticarum, ex quatuor mundi partibus, longâ annorum serie indefessoque studio, à Joanne Guilielmo Weinmanno, Dicasterii Ratisbonensis Assessore & Pharmacopola Seniore collectarum Plantarum, Arborum, Fruticum, Florum Fructuum, Fungorum. &c. Quæ Nitidissime æri incisæ & simul diu desiderata ac recens inventa arte, vivis coloribus & iconibus, naturæ æmulis, excusæ & repraesentatæ Per Bartholomæum Seuterum, Joannem Eliam Ridingerum et Joannem Jacobum Haidium Pictores & Chalcographos Augustanos. Quorum Denominationes, Characteres, Genera, Species & Descriptiones ex optimis, tam priscis quam neotericis Auctoribus, ordine ac serie Alphabetica, cum probatissimo usu Medico, Pharmaceutico, Chirurgico ac Oeconomico, latino & Germanico idiomate sincere explicantur à D. Joanne Georgio Nicolao Dieterico, [*Vol. 3:* D. Ambrosio Carolo Bielero,] Sacræ Cæsareæ ac Regiæ Catholicæ Majest. Consiliario, Serenissimi [*Vol. 3:* Reipublicæ Ratisbonensis Physico Ord.] Principis de Furstenberg – Stülingen Medico ordinario ac Reipublicæ Ratisbonensis Physico. Vol. I. A. B. [*Vol. 2:* C.D.E.F.] [*Vol. 3:* G.H.I.K.L.M.N.O.] [*double rule; vols. 2–3 single rule*] Apud prænominatos Pict. & Chalcogr. Augustæ venum prostat, quorum sumtibus imprimebatur Ratisbonæ per Hieronymum Lenzium, M DCC XXXVII [*Vol. 2:* M DCC XXXIX.] [*Vol. 3:* MDCCXLII.]

[*Latin title-page, Volume IV, printed in red and black*]: Phytanthoza-Iconographia, Sive Conspectus Aliquot Millium, tam Indigenarum quam Exoticarum, ex quatuor mundi partibus, longa annorum serie indefessoque studio, à Ioanne Guilielmo Weinmanno, Dicasterii Ratisbonensis Assessore Et Pharmacopola Seniore collectarum Plantarum, Arborum, Fruticum, Florum, Fructuum, Fungorum &c. Quæ Nitidissime æri incisæ et simul diu desiderata ac recens inventa arte, vivis coloribus et iconibus, naturæ æmulis, excusæ et repraesentatæ Per Bartolomæum Seuterum, Joannem Eliam Ridingerum et Joannem Jacobum Haidium Pictores et Chalcographos Augustanos, Quorum Denominationes, Characteres, Genera, Species et Descriptiones ex optimis, tam priscis quam neotericis Auctoribus, ordine ac serie Alphabetica, cum probatissimo usu Medico, Pharmaceutico, Chirurgico ac Oeconomico, latino et Germanico idiomate sincere explicantur à D. Ambrosio Carolo Bielero, Reipublicae Ratisbonensis Physico Ord. Vol. IV. P.Q.R.S.T.U.X.Y.Z. [*rule*] Apud prænominatio Pict. Et Chalcogr. Augustæ venum prostat, quorum sumptibus imprimebatur Ratisbonæ Per Henricum Georgium Neubauerum M D CC XLV.

[*Preface and index volume*]: Uber das mit allgemeinem Beyfall aufgenommene, und nunmehro glůcklich zu Ende gebrachte Weinmannische vortrefliche Werck, Phytanthozaiconographia genannt, welches Durch den unermudeten Fleiß, Kunst, und Geschicklichkeit Der Ehrenvesten und Kunstberůhmten Herren Bartholomå Seuters, Johann Eliå Ridingers und Joahann Jacob Haids Wohlangesehenen Bůrger und berůhmten respective Mahlern und Kupferstechern in Augspurg, verfertiget worden, Bezeugte, Aus besonderer Affection und Ergebenheit Regen wohlermeldte Freunde und Gőnner, seine Danck-Begierde und herßliche Freude darůber J. B. M. Ph. [*double decorative rule*] Gedruckt mit Schőnigkischen Schrifften. A. 1745. d. 6 May.

2° 39.8 × 25 cm. Four volumes in five. Volume I: π^2 A–3D² *i–iv* *1–2* 3–200 pp. (26 as 62, 151 as 152) [204 pp.]. Volume II: π^2 A–6O² *i–iv* 1–516 (207 as 205, 231 as 222, 274 as 272, 275 as 273) [520 pp.]. Volume III: π^2 A–6G² *i–iv* 1–488 [492 pp.]. Volume IV: π^2 A–6X² (-6X2) *i–iv* 1–196 (161 as 911) 195–540 (197 as 195, 208 as 268, 455 as 435) [546 pp.]. Index volume: π^2)(²–5)(² (-5)(2) a–f² a–k² *i–lxxxvi* [86 pp.].

PLATES: Mezzotint frontispiece (signed 'Baumgartner del.' and 'I. I. Haid Sculps.') and portrait of Weinmann (signed 'Hirschman pinx.' and 'I. Iac. Haid Sculps. aug. Vind.') in Volume I, and portrait of Ambrosius Karl Bieler (1693–1747) in Volume III (signed 'M. C. Hirschman junior pinxit.' and 'I. Iacob Haid Sculps. aug. Vind.'), all printed in blue; 1,025 engraved, etched, and mezzotint plates, some folding, of plants, fungi, and lichens, printed in color and hand-colored, engraved by Johann Jacob Haid (1704–1767), Johann Elias Ridinger (1698–1767), and J. Seuter, after drawings by Georg D. Ehret (1708–1770) and N. Asamin; 15 are folded, 1–275 in Volume I, 276–525 in Volume II, 526–775 in Volume III, 776–1,025 in Volume IV.

BINDING: Contemporary calf with gilt tooling on spines, Vols. I and II, with blind tooling on Vols. III and IV.

REFERENCES: Blunt & Stearn, pp. 152, 154–155, 159; Calmann, p. 16; Cleveland 388; Dunthorne 327; Hunt 494; Nissen 2126; Pritzel 10,140; Stafleu & Cowan 17,050.

JOHANN WILHELM WEINMANN was a Prussian who moved to the Bavarian city of Regensburg in 1710 to study pharmacy; deciding to remain there, in 1712 he set up a thriving apothecary shop, the 'Elefanten-Apotheke.' He has earned a place in the history of botany as the author of *Phytanthoza-iconographica*, a work described by Wilfrid Blunt as 'a memorable undertaking' and *Thesuarus rei herbariae locupletissimus*, which was published in 1787 in Vienna. The lengthy title to *Phytanthoza-iconographia*, written in German and French in the first edition of the work published in 1735 in Regensburg and in German and Latin in the second edition, is of interest and deserves to be paraphrased here (*Phytanthoza* is a made Greek word meaning 'flowering plants'):

> The representation of about a thousand trees, saplings, plants, flowers, fruit and mushrooms grown in the four parts of the world, and which, by means of fine copperplates carefully engraved, appear in their true colors and with great realism using a technique hitherto unknown and here invented, printed by Johann Wilhelm Weinmann, councillor of the municipal office and first pharmacist of Regensburg. [There is, in addition,] an ample registry, written in the different, most widely read languages, that explicitly describes the species, qualities, forms, descriptions, and virtues of all the plants observed by the most celebrated naturalists, and gathered together by the most diverse authors who have to date written about them, with the economic support of Bartholomaüs Seuter and Johann Elias Ridinger and Johann Jacob Haid, famous artists and engravers of Augsburg.

The publishing history of this work is complex, as it was first printed in a series, consisting of fifty plates each, that appeared twice a year. Many botanists contributed to the text of *Phytanthoza-iconographia*. The first seventy-five botanical descriptions in Weinmann's herbal were written by Johann Georg Dietrich (1681–1737), who died before he could complete the project. His work was continued by his son Ludwig Michael Dietrich (1716–1769) and, after the death of Weinmann, the herbal was finished by Ambrosius Carolus Bieler of Regensburg (1693–1747). A number of artists were commissioned to make the preparatory drawings and engrave the plates. It is of interest to note that at the outset a young artist by the name of Georg Dionysius Ehret (1708–1770) was engaged for the period of one year to work on *Phytanthoza-iconographia*. This collaboration ended on an acrimonious note, however, for it appears that after producing five hundred botanical drawings Ehret asked to be paid, but Weinmann responded that their agreement had been for the delivery of one thousand drawings (see Calmann p. 16). With this excuse he dismissed the artist in 1728 with a much smaller remuneration than that for which he originally contracted. Ehert afterwards found a more satisfactory position with the banker Löschenkohl, depicting the plants in his private garden, and would go on to become one of the most celebrated botanical artists of his time (see No. 46 and *An Oak Spring Flora*, No. 50).

JOHANN WILHELM WEINMANN, *Phytanthoza-iconographia*, 1736. Frontispiece portrait of Johann Weinmann, volume 1

JOHANN WILHELM WEINMANN, *Phytanthoza-iconographia,* 1736. Five varieties of Lavender (*Lavandula*), volume III, plate 632

JOHANN WILHELM WEINMANN, *Phytanthoza-iconographia,* 1736. Two varieties of *Solanum: S. pensile* and *S. melongena,* volume IV, plate 933

The fame of Weinmann's work lay in the novelty of its illustrations. *Phytanthoza-iconographia* was the first botanical work to successfully apply the new technique of multi-color mezzotint engraving. This process had already been used in a limited fashion by the English professor John Martyn (1699–1768) in *Historia plantarum rariorum* (London, 1728–[1737]). This work contained plates engraved by Elisha Kirkall (*c.* 1682–1742), mostly based on drawings by Jacob van Huysum (1686–1740), and introduced the novelty of printing several colors from a single plate at one impression. Kirkall used the same technique for *Catalogus plantarum* published in 1730 by the Society of Gardeners (for color printing, see No. 27; *An Oak Spring Sylva*, No. 1, p. 5; and Henrey II.677–680).

The mezzotint process itself—which is capable of producing a painterly combination of rich blacks and subtle tones of grey—was invented in the mid-seventeenth century and developed largely in Holland. By the end of the 1660s, a market in such prints was established in France, Germany, and Holland. Mezzotint engraving was adopted enthusiastically in England after the arrival of a number of Dutch printmakers in the 1670s. By the eighteenth century, the method was so prevalent in Britain, especially for portraiture, that it was referred to as *la manière anglaise.* The Dutchman Johannes Teyler (Teijler) of Nijmegen (*fl.* 1650–1700) invented and patented (in 1688) a color-printing technique from a single plate. Works by Teyler's hand include the illustrations of plants and birds for *Opus Typocromaticum* (1683). This process, used for the *Phytanthoza-iconographia* mezzotints, was labor-intensive and demanded a high degree of skill. Every color had to be applied in its portion of the plate with great care to avoid the encroachment of one color upon another and the process had to be repeated for every impression taken. The production of more than 1,000 plates for *Phytanthoza-iconographia* was so time-consuming that, especially as the production went on, the work was sometimes carried out with less care and attention. This explains the fact that some illustrations are retouched with color and others entirely colored by hand. By itself, mezzotint engraving tended to produce somewhat blurry designs. In *Phytanthoza-iconographia* the engravers who contributed the most to the work, Bartholomaüs Seuter (1678–1754) and Johann Elias Ridinger (1698–1797), and later Johann Jacob Haid (1704–1767), had to skillfully combine the technique used for mezzotint engraving with etching to achieve a cleaner outline and a higher degree of definition.

Weinmann's imposing herbal bears some of the characteristics of the florilegium, a genre devoted to flower illustration that became immensely popular among collectors and the general public beginning in the seventeenth century. *Phytanthoza-iconographia* opens with an impressive allegorical frontispiece depicting Flora, Asclepius, and Time in a blue mezzotint engraving by Haid. The pages bearing the protracted title in Latin and German are followed by a blue mezzotint engraving of the author aged fifty-four. This is also prepared by Haid from a drawing by a member of the artistic Hirschmann family of Nuremberg known for their celebrated portraits. The frontispiece portrait of Ambrosius Bieler in the third volume is also the work of Hirschmann and Haid.

Phytanthoza-iconographia presents more than four thousand plants in alphabetical order, ranging from native to exotic species. Each plate is preceded by a detailed description, written in both German and Latin containing information on the form, characters, and pharmacological properties of the plant shown. Without question, however, its 1,025 engravings constitute the most outstanding aspect of Weinmann's book. In addition to Ehret, several other artists whose names we do not know contributed botanical illustrations to the work. These drawings are very fine and have been transformed into color engravings by Haid, Ridinger, and Seuter. The *mise en page* is skilful and pleasantly varied; some of the plates may show a single specimen with enlarged details, while in others a number of plants are arranged on the page.

In the first volume a series of aloe plants, perhaps the work of Ehret, are shown growing in elegant pots (Pls. 42–44, 46–50, and 58–61); equally fine is a series of drawings of *Brassicae* (256–267). In the second volume a handsome cacao plant is illustrated (277–278), while in the third volume the mandrake with its unusual root (708). Some of the engravings have been printed on larger sheets and folded, such as the waterlily (761), and a banana tree with the detail of a sectioned fruit (737). The illustrations in the fourth volume, which concludes with the an illustration of two species of *Zinziber* (*sylvestre* and *angustifolium*) (1025), appear to be the work of less talented artists.

37. François Alexandre Pierre De Garsault (1693–1778)

Description, Vertus et Usages de sept cents dix-neuf Plantes, tant Étrangeres que de nos Climats; et de cent trente-quatre Animaux, en sept cents trente Planches, Gravées en taille-douce, sur les desseins d'après nature, de M. De Garsault, par MM. De Fehrt, Prevost, Duflos, Martinet, &c. Et rangées suivant l'ordre du Livre intitulé Matiere Médicale De M. Geoffroy. Ouvrage utile à toutes Matieres Médicales, aux Artistes, aux Personnes charitables, & à tous ceux qui préparent eux-memes leurs Médicaments. [*vignette*]. A Paris, Chez P. Fr. Didot le Jeune, Quai Des Augustins, près du Pont S. Michel, à S. Augustin. [*double rule*] M. DCC. LXVII. Avec Privilege Du Roi.

[*Volume I, added copperplate illustrated title-page before plates, within a cartouche*]: Les Figures des Plantes et Animaux d'Usage en Medecine, décrits dans la Matiere Medicale de Mr Geoffroy Medecin, Dessinés d'après nature par M^r De Garsault, Gravés par Mrs Desehrt, Prevost, Duflos, Martinet &c. Niquet Scrip. [*within a compartment at foot*] A Paris Chez l'Auteur Rue St Dominique Porte St Jaques.

8° Five volumes, 21.3 × 14 cm. Volume I: π^2 (-2π3) a^8 A–E^8 F^4 (-F4) [*2*] *i–ii* iii–xiv *xv–xvi 1* 2–85 *86* [104 pp.]. Volume II: G–J^8 K–L^8 89–184 [96 pp.]. Volume III: M–R^8 (-R8) 185–*278* [94 pp.]. Volume IV: S–U^8 V^8 X–Y^8 (-Y8) 279–372 [94 pp.]. Volume V: Z–2E^8 2F^2 (-2F2) 375–472 [98 pp.].

PLATES: 729 total leaves of engraved and etched plates. Volume I: 118 leaves of plants and fungi numbered 1–118.

Volume II: 174 leaves of plants and fungi numbered 119–292. Volume III: 174 leaves of plants and fungi numbered 293–466. Volume IV: 177 leaves of plants and fungi numbered 467–643. Volume V: 86 leaves of insects, sea life, birds, and animals numbered 644–729.

BINDING: Contemporary mottled calf with gilt tooling on spines.

REFERENCES: Brunet II.1495; Hunt 585; Nissen 687; Plesch, p. 234; Pritzel 3,216; Stafleu & Cowan 1,961.

ONE OF THE MOST richly illustrated pharmacopoeias of the eighteenth century was the *Description, Vertus et Usages de sept cents dix-neuf Plantes*, compiled by Garsault and printed in Paris by Didot. Five volumes of engravings were published in 1764 and a volume of text in 1765 (*Explication abrégée de sept cent dix-neuf Plantes* ...). The six-volume set was re-issued in 1767, the last volume of which the Oak Spring Garden Library lacks. The title proclaims the genesis and aim of the work, which involved the preparation of 729 etchings of plants and 134 of animals from life by a number of artists working in Paris. This project, coordinated by Garsault, an artist, naturalist, and polymath who, as we may conclude from the *Avertissement*, originally conceived the idea for the work.

Garsault was indeed the artist of the preparatory drawings. From these drawings, copperplate etchings were made by various artists of a certain renown, including Benoit-Louis Prevost (1747–1804), Pierre Duflos (1742–1816), Carl de Fehrt (1723?–1774), and Angelique Martinet (*c.* 1731–1780). The text, which consists of brief descriptions of the plants and their pharmacological uses, was the work of Etienne François Geoffroy, a Paris physician and the author of an eight-volume works *Tractatus de materia medica* published in Latin in 1741 and then in French in 1757. Garsault notes in the *Avertissement* that Geoffroy's text provided the framework for his presentation of hundreds of plants and animals utilized in medicine. However, as we may read in the title, the work was intended not only for those interested in *materia medica*, but also '... aux Artistes, aux Personnes charitables, & à tous ceux qui préparent eux-memes leurs Médicaments.' Garsault explains how during the course of many years as an amateur naturalist he had been in the habit of making botanical and zoological drawings without any thought of their publication. It was only after the illustrious botanist Bernard de Jussieu (of the celebrated French family of botanists and physicians, Antoine, 1686–1758, Antoine Laurent, 1748–1836, and Bernard, 1699–1777) saw and praised them that he felt encouraged to bring together his work in a single publication for the use of the general public.

The published set is an admirable scientific undertaking. After the *Avertissement*, the botanical terms and characteristics of the flowers and fruits are defined for the reader. This is followed by an alphabetized list of medical terms used for illnesses. The plants in sections two through four, which are the indigenous, are arranged according to this alphabetized list of medical terms. Thus a practitioner could look up an illness and see which plants were appropriate. *Descript Medical*, is embossed on the spine of each the five volumes. The

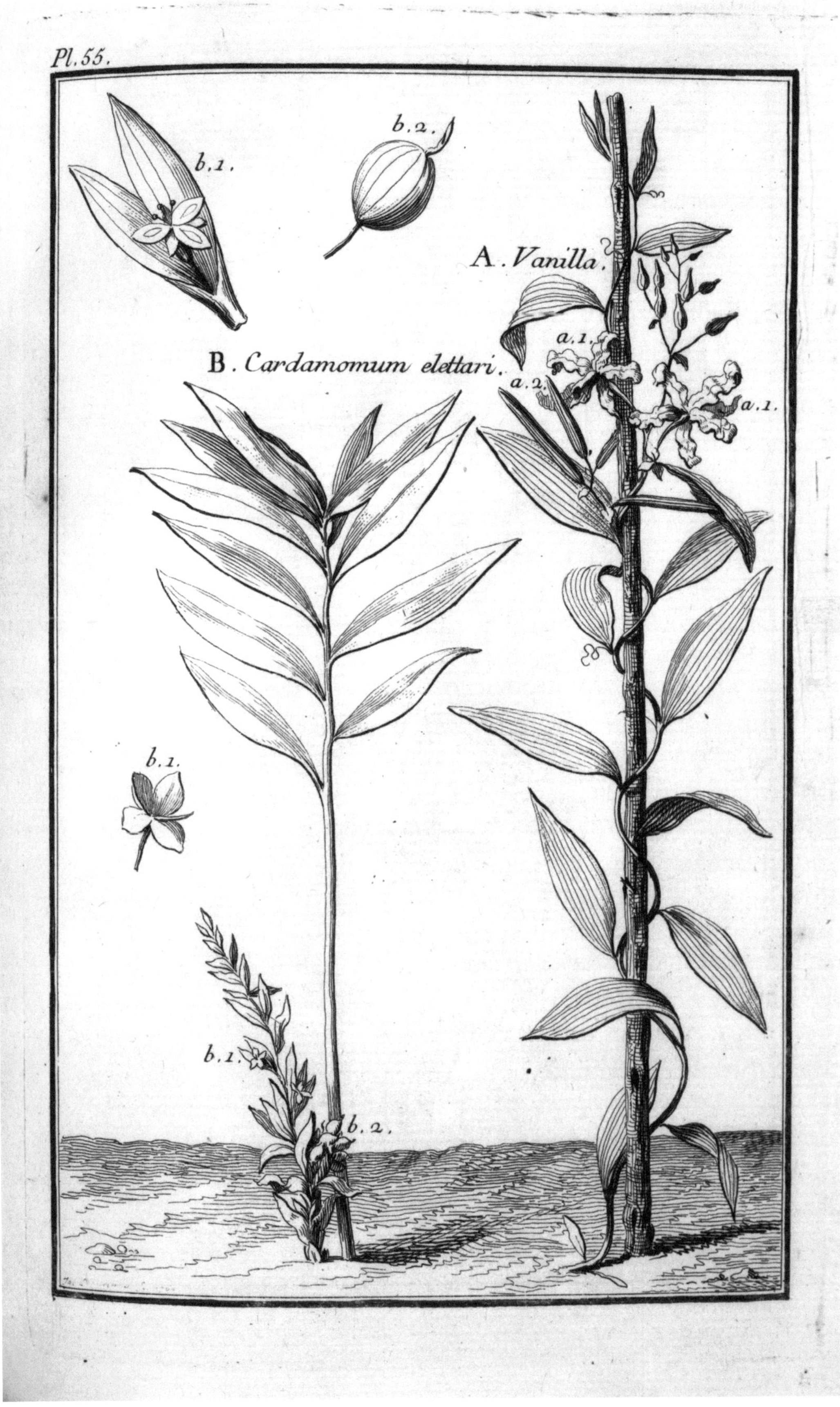

FRANÇOIS ALEXANDRE PIERRE DE GARSAULT, *Description, Vertus et Usages de … Plantes,* 1767. 'Vanilla' (*Vanilla planifolia*) and 'Cardamomum elettari,' True cardamom or Green cardamom (*Elettaria cardamomum*), volume 1, plate 55

first volume is devoted to exotic plants, defined as those plants and trees originating from other countries. Many exotic species were able to be included, for they were being cultivated, and hence were accessible for study, in the plant collections of the Jardin des Plantes, the Petit Trianon, and the private garden of M. Bombarde. Garsault dedicated volumes two through four to the study of those plants that grow in 'our climate.' The entire fifth volume is dedicated to descriptions and medical uses of animals, which include insects, seashells, and such exotics as a rhinoceros, with the last entry being an illustration of a man and women, in which the healing properties of urine are discussed. Most of the animals were drawn from life. Although some were copied from illustrations by well-known artists, interestingly, others were portrayed on the basis of eyewitness accounts by experts, such as the naturalist and traveller Michel Adanson (1727–1806).

Garsault portrayed each plant in its entirety, scrupulously recording not only the details of obvious interest (leaves, flowers, etc.), but also its general aspect, an element that he felt many botanical artists tended to neglect, even though it could be extremely helpful in identifying a plant. Many are shown growing from the earth in a more or less naturalistic setting and therefore with their roots hidden from view. Garsault evidently considered the root of the plant to be of secondary importance. Sometimes two varieties of the same species are presented in a single illustration.

Based on his personal experience Garsault offers some technical advice to artists interested in the genre of scientific illustration. Garsault begins with the paints to be used; he recommended mixing the colors with gum, avoiding the use of white (which tended to darken over time), and employing brushes of various size ranging from quite broad to extremely fine. Finally, he gratuitously admonishes artists to exercise great care and not allow their colors to extend beyond the outlines of the drawing. The *Avertissement* closes with the affirmation that those engaged in the decorative arts (printing textiles, etc.) would find in this work an endless source of naturalistic motifs.

The illustrations by Garsault are quite pleasing in their simplicity and are remarkably varied. In the botanical drawings he often included details of the leaf, flower, seed, and fruit. In some cases the illustrations are shown sectioned, documenting his in-depth studies of their structure. The animals in many cases have been portrayed against a landscape background, the artist using as his model the authoritative *Histoire naturelle générale et particulière avec la description du Cabinet du Roi* by Georges Louis Leclercq de Buffon (1707–1788), which was illustrated with a large number of extremely beautifully engraved plates. The French naturalist's work was published to great acclaim in instalments, beginning in 1749 and concluding well after the author's death in 1804.

In *Description, Vertus et Usages* each plant and animal is identified by its Latin name. The accompanying text is quite succinct, concentrating above all on the medicinal properties

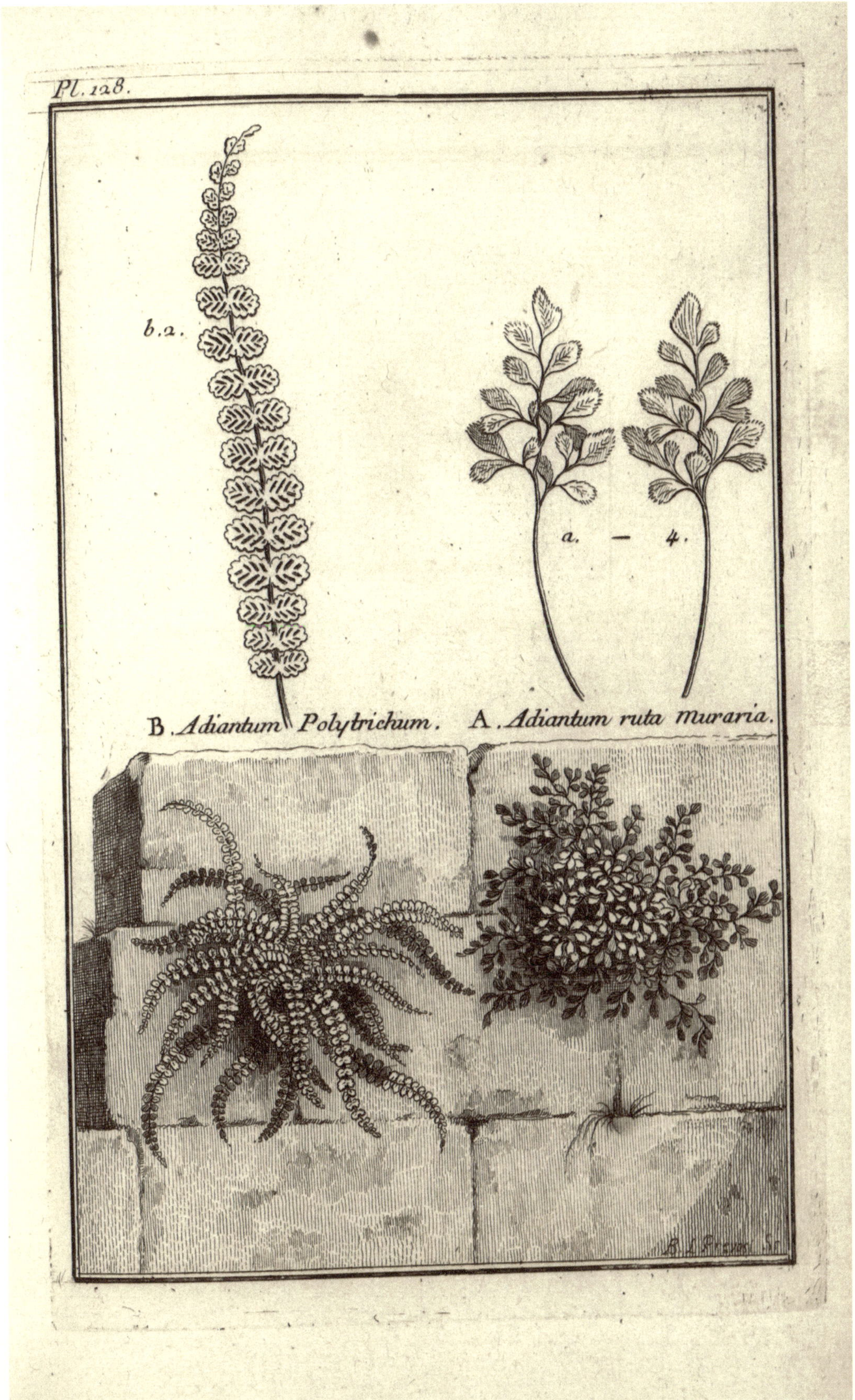

FRANÇOIS ALEXANDRE PIERRE DE GARSAULT, *Description, Vertus et Usages de … Plantes*, 1767. 'Adiantum ruta-muraria,' Wall-rue (*Asplenium ruta-muraria*), and 'Adiantum Polytrichum,' Maidenhair Spleenwort (*Asplenium trichomanes*), volume II, plate 128

FRANÇOIS ALEXANDRE PIERRE DE GARSAULT, *Description, Vertus et Usages de … Plantes,* 1767. 'Lactuca sativa,' Garden lettuce (*Lactuca sativa*), volume III, plate 314

of the plant or animal (occasionally with a recipe or two), which are described in clear and simple language. For example, in the chapter on *Pinus sativa*, a tree native to Langedoc and Provence (Pl. 65), we read that the leaves, flower and fruit 'qui naissent à part en divers endroits des memes branches, & commencent par un embrion qui devient un fruit gros comme un poing ... est pectoral, adoucissant: la dose est, depuis une demi-once, jusqu'à une once en émulsion' (I, pp. 49–50).

The plants, as he took care to underline, were drawn 'd'après nature avec la plus grande exactitude, comme on fait un portrait, chaque plante dans le lieu ou elle croit naturellement.' With regard to the larger species such as trees and shrubs, a few have been portrayed in their entirety, but Garsault generally focused on a single branch with its flowers or fruit. Plate 39, Volume I, illustrates the red coral, the author not specifying whether this was a plant or an animal; his observations would have been of interest, since in this period there was still some confusion regarding the exact nature of this organism.

This vast *encyclopédie* of botany and zoology, with its qualities of clarity, elegance, and utility, was a quintessential product of the Enlightenment, reflecting the rich scientific and artistic culture of the period.

38. John Edwards (1742 – *after* 1812)

The British Herbal, Containing One Hundred Plates of The most beautiful and scarce Flowers and Useful Medicinal Plants Which blovv in the open Air of Great Britain, Accurately colored from Nature, with their Botanical Characters, and A short Account of their Cultivation, &c. &c. By John Edwards. The Whole corrected according to the latest Editions of Botany. [*double rule*] London: Printed for the Author; And sold by J. Edmonson, Painter to Her Majesty, in Warwick-Street, Golden-Square; and J. Walter, at Homer's Head, Charing-Cross. M, DCC. LXX.

2° 46 × 28.2 cm. *i–ii 1* 2–50 [*2*] [54 pp.].

PLATES: 100 engraved and etched leaves of plants, all hand-colored.

BINDING: Contemporary calf that has been re-backed in brown leather.

PROVENANCE: Bookplate of Lady Boston; inscribed vertically across title-page in purple-colored pencil and in ink at top of title-page 'J. Williams Ashwau.'

REFERENCES: Blunt & Stearn, p. 221; Dunthorne 104; Henrey II.17–22, and No. 675; Nissen 578; *An Oak Spring Flora*, No. 64; Plesch, pp. 216–7; Pritzel 2,620; Stafleu & Cowan 1,624.

In 1770 an imposing volume in folio containing one hundred illustrations of plants was published in London as *Edwards's Herbal....* The book was printed at the expense of its author, John Edwards, an artist who happened to be working in London at the same time as the painter Edward Edwards (1738–1806), whose work was reproduced as prints. This circumstance was destined to generate some confusion, and the work was re-issued in the same year as *The British Herbal....* The original edition was issued in monthly parts and many of the plates are dated beginning in January of 1769. The work would be republished in 1775 by the bookseller Samuel Hopper with the title *A Select Collection of One Hundred Plates: Consisting of ... Exotic and British Flowers which Blow in our English Gardens.*

Although John Edwards categorized his work as an 'herbal' in the title, it mainly presents a collection of horticultural species, reflecting the broader definition that the term 'herbal' came to assume during the course of the eighteenth century. This volume represents an example of a new genre that enjoyed an immense vogue beginning in this period. Books such as Edwards' were lavishly illustrated with drawings of native and exotic flowering species, and these works inspired admiration, not only for the variety and beauty of their flowers, but also for the brilliant technical skill of the draftsmen. One exotic species presented in *The British Herbal*, for instance, is the 'Maedia' (*Dodhecanteon maedia*—Plate XLI). This species according to the author, was native to Virginia but was now cultivated all over England.

It may nevertheless be observed that, as the title specifies, next to these horticultural varieties, *The British Herbal* includes a number of 'useful medicinal plants,' such as the *Dictamus rubro flore* (Dittany with red flowers, No. 19). In the text that accompanies this illustration Edwards observes: 'The Roots of this plant are used in medicine, and estimed cordial and cephalick, resisting putrefaction and poison, and are useful in malignant and pestilential distempers, as also in epilepsies.' Furthermore, in keeping with the traditional format of the herbal, many of the illustrations include enlarged details of the flowers and seeds in the lower part of the page. Small letters appear next to these, which correspond to the descriptions of the anatomical parts of the flower. Each plant is labelled with its Latin and English names and is accompanied by a short text based on the botanical descriptions of Joseph Pitton de Tournefort (1656–1708) and Carolus Linnaeus (see No. 46).

John Edwards was born in Brentford, and lived and worked in London until about 1778, when he moved to Surrey. He was an artist and engraver who specialized in botanical illustration (see *An Oak Spring Flora*, No. 64, pp. 245–247) and the designing of ornamental motifs for textiles, especially cotton calico, whose manufacture was based, in part, in Surrey. Between 1763 and 1812 he had the opportunity to show his work at exhibits organized by the Royal Academy and the London Society of Artists, of which he was a member, and on numerous other occasions in London. Edwards also produced *A Collection of Flowers, Drawn after Nature & Disposed in an Ornamental & Picturesque Manner*, a florilegium issued in parts, in which many of the flowers that were illustrated in *The British Herbal* appear once again, although contradistinguished by a more markedly decorative style.

Most of the plates in *The British Herbal* were drawn and engraved by Edwards himself. Only a handful were executed by other such less-well-known figures as Ignace Fourgeron

JOHN EDWARDS,
The British Herbal, 1770.
Gas plant or Burning
bush (*Dictamnus albus*),
plate 19

and William Darling. Darling was an artist of whom we know only that his work was exhibited in 1762 at the Society of Artists. After they were printed, Edwards' illustrations were carefully colored by hand.

In contrast to the elaborate compositions in *A Collection of Flowers*, the botanical drawings in Edwards' *The British Herbal* are models of clarity and objectivity, soberly arranged in the center of the page with only modest concessions being made to the exuberant decorative taste of the period. Despite their undeniable elegance, the over-riding purpose of these works was didactic and scientific.

39. ANDREAS FRIEDRICH HAPPE (1733–1802)

Herbarium pictum sive Icones CCCCII. plantarum officinalium et hortensium ad Vivum depictarum a Dno: def: Happe. [*drawn rule*] Berolini. Quarto Volumine continentur

36.1 × 22.6 cm. 403 watercolors of plants, moss, and lichen, some heightened with gouache and some with white, numbered 1–402, each drawing inscribed with a Latin and French name, plus the Linnaean nomenclature. Volume 1: 1–100. Volume 2: 101–200. Volume 3: 201–300. Volume 4: 301–402. A drawing of a rose in a different hand with an inscription below in ink measuring 14 × 9.8 cm mounted between sheets 310 and 311.

BINDING: Contemporary green paper-covered boards; Volumes 2–4 have paper labels on spine: 'Herbier peint d'après nature'; paper cut-out cartouche labels on covers of Volumes 1–3 inscribed: 'herbier artificiel colorié—par de Happe composé de 402 figures'; Volume 4 label is left blank.

REFERENCES: There is no bibliography on this specific work, which is still unedited. On Happe, see Cleveland 564 and 611; Hunt II, No. 659, p. 421; Nissen 782–787; Pritzel 3,776–3,779; Stafleu & Cowan 2,384.

ANDREAS FRIEDERICH HAPPE—naturalist, entomologist, pharmacist, and botanical painter—was born in Ashersleben in 1733 and died in Berlin in 1802. Although the bibliographic evidence regarding his activity as an artist is uncertain and contradictory, he was quite prolific and we know that he was affiliated with the Academy of Science of Berlin as 'botanicus et rerum naturalium pictor.' Many works have been attributed to Happe, including several manuscripts of botanical paintings and the published works *Flora cryptogamica depicta* (1783), *Botanica pharmaceutica* (1785[–1806]), and *Flora depicta: aut plantarum selectarum icones ad naturam delineatae* ([1783–]1791[–1792]) containing 317 engravings produced in Berlin between 1783 and 1792, as well as the entomological work *Abbildungen der Schmetterlinge* (1783–1785), a preparatory manuscript of which is conserved in the Natural History Museum, London.

To these works must be added the four-volume *Herbarium pictum* conserved in the Oak Spring Garden Library, an impressive collection of 403 watercolors of medicinal and

ANDREAS FRIEDRICH HAPPE, *Herbarium pictum, c.* 1780. 'Le Dragonier,' Dragon's blood (*Calamus draco*), volume II, folio 149

ANDREAS FRIEDRICH HAPPE, *Herbarium pictum, c.* 1780. 'Papaver hortense,' Opium poppy (*Papaver somniferum*), volume III, folio 272

horticultural plants that was perhaps realized in the same decades as his published works. Even this extensive corpus, however, provides little information on the late-eighteenth-century naturalist-artist, apart from confirming that he had a thorough knowledge of pharmacology.

The plants in *Herbarium pictum* are arranged in alphabetical order by their Latin names, with a corresponding index provided at the beginning of the work. Each painting has been executed in opaque and transparent watercolors and is labelled at its base with the Latin and French names of the plant depicted. The drawings, consecutively numbered, appear on the recto side of each page. In Volume I, handwritten observations in French appear on the facing page to every illustration, the author providing once again the names of the plant in Latin (using the Linnaean nomenclature) and in French, followed by an ample description of its *proprietés* or medicinal and culinary uses. In the subsequent volumes the illustrations are not accompanied by notes, the author apparently not having been able to devote himself further to this task.

In his descriptions Happe cites—in addition to the undisputed authority Carolus Linnaeus—the botanist and voyager Joseph Pitton de Tournefort, the Swiss naturalist and botanist Caspar Bauhin, and Bernard de Jussieu. He provides specific information on the properties and uses of each of the plants in his herbal, writing for example on 'La petite Angélique sauvage' (*Angelica sylvestris*): 'Elle a des grandes vertues, on lui en suppose mème contre la gautte prise en substance. Elle est cordiale, sudorifique, vulnéraire ... Elle préserve de la peste, soit en jettant de la poudre de ses racines sur son habit, soit en approchant de ses narins....' (I, 20). On 'la Carline' (*Carlina acaulis*) he notes: 'Le medecin emploie les racines comme aperitives. On peut manger le réceptacle de la fleur, comme le cul des artichauds et des cardons' (I, 80). All of the plants presented are indigenous European species, with the exception of a cacao portrayed with its fruit in the fourth volume (IV, 362).

Most of the works in the *Herbarium pictum*, although drawn with scrupulous accuracy, are characterized by a certain rigid academicism. On occasion, however, Happe excelled himself, creating works of genuine elegance, such as in the drawing of *Cucurbita tagenaria* (II, 134); another illustrating *Papaver hortense* (III, 272) with its flower, a closed bud, and a large seed capsule; and the refined portrayal of *Trifolium pratense* (IV, 377).

Therefore we would be less severe in our judgement of this naturalist-artist than Pritzel, who wrote: 'Icones, etiam temporis respectu habito, satis mediocres 78 fasciculis annis 1788–1806 prodierunt' (These paintings, which must be considered rather mediocre considering the period in which they were produced, appeared in 78 issues during the years 1788–1806). From this passage we may infer that the drawings in Happe's herbal were also published in a series of issues for the general public.

40. Joshua Webster (1711–1803)

Webster's Distribution of English Medicinal Plants. Manuscript, England, circa 1780

37.3 × 24.3 cm. 226 watercolor/gouache drawings of plants on recto and verso, leaves numbered 1–230 in a contemporary hand; oil-on-paper frontispiece of an angel appearing before a man and a dog in a landscape; plus two leaves of text including a table of contents, and an index consisting of two leaves; plus a loosely inserted twelve-page sewn index booklet written in purple ink, 17.9 × 11.2 cm. Watercolor of a moth, 7.3 × 9.6 cm, on front paste-down.

BINDING: Contemporary vellum boards with painted frames within which are pasted marbled paper.

Joshua Webster was a member of the College of Physicians of London who practiced medicine close to the capital city, first in the town of St. Albans, Hertfordshire, and then in Chigwell, Essex. In 1745 he invented 'Cerevisia Anglicana,' a beer rich in Vitamin B that became an extremely popular drink during the second half of the eighteenth century. Webster eventually retired and moved to Chelsea, where he died at the age of ninety-two, retaining a remarkable vigor to the end of his days, which he attributed to the salutary effects of this beverage. In two works, *A true and brief account with Directions for Use of the Cerevisia Anglicana of syrupated English Diet Drink ...* (London, 1799) and *Practical Observations on Preservation of Health* (London, 1792), he extols the virtues of his invention without revealing the secret of its preparation. Webster was a figure of some prominence and counted among his friends the American scientist and diplomat Benjamin Franklin (1706–1790) and the Scottish poet James Thompson (1700–1748).

This one-of-a-kind manuscript conserved at the Oak Spring Garden Library opens to a watercolor of a moth on the left-hand, paste-down endleaf and, on the right, a frontispiece painting. Executed in opaque gouache with an ornate gold border, it depicts a shepherd with an injured leg sitting beneath a tree; a radiant angel by his side gestures toward the pastoral landscape with one hand, while in the other is held a banner that reads *Hic san[itas]* (Here is healing).

The manuscript herbal *Webster's Distribution of English Medicinal Plants* demonstrates that the London physician was also a knowledgeable herbalist and gifted artist. The purpose of this work, as the author notes in his brief introduction, was to demonstrate 'in a more useful way' the virtues of England's indigenous flora, as well as of those plants that had been acclimatized after many years of cultivation in England. These plants 'are frequently prescribed by Physicians, or have by a long course of practice & experience in private families, obtained a character for their usefulness in the healing art (p. 1).'

The plants are not arranged in a haphazard manner, as Webster explains, but in a precise order conceived by him and similar to the one used by a certain Doctor Quincy in his *Pharmacopea*. The method that Webster chose for his manuscript, he thought to be:

JOSHUA WEBSTER, *Webster's Distribution of English Medicinal Plants*, c. 1780 Stinging Nettle (*Urtica dioica*), with two mosquitoes (*Culicidae*), folio 70

> the best, and most useful, & at the same time to coincide pretty nearly with my own scheme; setting down such Plants only as are of real & known efficacy, arranging them according to their medicinal virtues, without regarding any scientific or botanical system of Class or Genus, being assured that a well chosen collection of Plants and Herbs, separated from those which are not so perfectly understood, nor so generally known or distinguished, with the Figures or Drawings from the real Plants themselves; must appear to every one the most rational and at the same time the most useful! (p. 1)

He grouped his plants in twelve 'sections,' some of which were further subdivided into 'classes' based on their curative properties and the conditions they were supposed to alleviate. The classes listed are as follows: 'Nervous, Strengthness, Stomachius, Balsamius, Diuretics, Diaphoretis, Emetis, Cathartics, Sternutatories, Narcotics, Coolers, and Topics.' A handwritten index is appended to the end of the manuscript along with a small booklet containing a list of the botanical illustrations.

Another note informs us that the drawings in this manuscript were selected from the many thousands executed by the author over the years. This observation shows how assiduously Webster applied himself, documenting his studies in the form of botanical illustrations. The paintings, usually of a single species, appear on recto and verso sides of every sheet. The left-hand margin of each drawing has been carefully ruled off and filled with the author's notations, which identify the section and class of the plant, its Latin name and growing season, and conclude with an extensive comment on the species' medicinal properties.

Classed among the 'Nervous simples,' Lily-of-the-Valley (*Convallaria maialis*) (p. 4), is noted for its pleasing scent, and Webster affirms that the flowers are considered to be 'Cephalic, useful in cold disorders of the Head … and Epilepsies,' and then furnishes a recipe designed to treat these disorders, which calls for a certain number of the stalks. Regarding the 'Mistletoe of the Oaks' (*Viscus quercinus*) (p. 117), classed among the 'Vulneraries,' Webster states that although the plant has given rise to various fanciful legends over the centuries, it possesses genuine healing properties. In addition to being an excellent remedy for nervous pains and hypochondria, he notes that he had used Mistletoe with success to treat convulsions in young patients.

While his classification system may seem somewhat eccentric, as a scientific work Webster's herbal is certainly of historical interest. It definitely merits a place in the history of botanical illustration due to the quality of its drawings, unique presentation, and noted botanical properties. More than two hundred botanical drawings are executed in vivid tempera colors over a lightly pencilled preparatory sketch. The artist has portrayed each plant with great realism, its various parts (often including the roots) elegantly arranged on the page. Particularly accurate and refined are the drawings of the 'Angelica' (p. 21), 'Sweet Fennel' (p. 23), and a superb common 'Asparagus' (p. 164).

JOSHUA WEBSTER,
Webster's Distribution of English Medicinal Plants,
c. 1780. 'Sweet Fennel'
(*Foeniculum vulgare*),
folio 23

JOSHUA WEBSTER, *Webster's Distribution of English Medicinal Plants*, *c.* 1780. 'Asparagus vul. Common Asparagus' (*Asparagus officinalis*), folio 164

We do not know if Webster ever intended to publish this work. What is clear is that it served as a tool for his study of medicinal plants. Sometimes the space in the left-hand margin of the drawing was left blank or contained only a few annotations, awaiting further additions as Webster pursued his daily studies. Webster's work follows the tradition of the manuscript herbal, which had been used by physicians, herbalists, and pharmacists since the Middle Ages to record and pass on their knowledge.

41. Nicholas Culpeper (1616–1654)

Culpeper's Complete Herbal, to which is now added, Upwards of One Hundred Additional Herbs, With a Display of their Medicinal and Occult Qualities; physically applied to the Cure of All Disorders incident to Mankind. [*decorative rule*] To which are now first annexed his English Physician Enlarged, and Key to Physic, With Rules for Compounding Medicine according to the true System of Nature. Forming a Complete Family Dispensatory, and Natural System of Physic. To which is also added upwards of Fifty Choice Receipts, Selected from the Author's last Legacy to his Wife. [*rule*] Embellished with Engravings of upwards of Four Hundred different Plants, With Other Subjects to Illustrate the Work. 'The Lord hath created Medicines out of the Earth; and he that is wise will not abhor them.' – Ecc. xxxviii. 4. [*double rule*] London: Published by Thomas Kelly, No. 17, Paternoster Row. And Sold by most Booksellers and Venders of Publications in the British Empire. J. Haddon, Printer, Tabernacle Walk. [*rule*] 1828.

4° 28.6 × 22.3 cm. 40 parts (some signatures unopened): (1): frontispiece, plate 1, *A*^{2} B–C^2, *i–iii* iv–vi *1* 2–6 [12 pp.]; (2): plate 2, D–F^2 7–18 [12 pp.]; (3): plate 3, G–I^2 19–30 [12 pp.]; (4): plate 4, K–M^2 31–42 [12 pp.]; (5 & 6): plates 5 and 6, N–R^2 43–62 [20 pp.]; (7 & 8): plates 7 and 8, S–Y^2 63–82 [20 pp.]; (9 & 10): plates 9 and 10, Z–2D^2 83–102 [20 pp.]; (11 & 12): plates 11 and 12, 2E^2–2I^2 103–122 [20 pp.]; (13 & 14): plates 13 and 14, 2K^2–2O^2 123–142 [20 pp.]; (15 & 16): plates 15 and 16, 2P^2–2T^2 143–162 [20 pp.]; (17 & 18): plates 17 and 18, 2U^2–3A^2 163–182 [20 pp.]; (19 & 20): plates 19 and 20, 3B^2–3F^2 183–202 [20 pp.]; (21 & 22): plates 21 and 22, 3G^2–3L^2 203–222 [20 pp.]; (23 & 24): plates 23 and 24, 3M^2–3Q^2 223–242 [20 pp.]; (25 & 26): plates 25 and 26, 3R^2–3X^2 243–262 [20 pp.]; (27 & 28): plates 27 and 28, 3Y^2–4C^2 263–282 [20 pp.]; (29 & 30): plates 29 and 30, 4D^2–4H^2 283–286 *287*–302 [20 pp.]; (31 & 32): plates 31 and 32, 4I^2–4N^2 303–322 [20 pp.]; (33 & 34): plates 33 and 34, 4O^2–4S^2 323–342 [20 pp.]; (35 & 36): plates 35 and 36, 4T^2–4Z^2 343–362 [20 pp.]; (37 & 38): plates 37 and 38, 5A^2–5E^2 363–375 *376*–382 [20 pp.]; (39 & 40): plates 39 and 40, 5F^2–5*K*^{2} (+ χ1) 383–398 [*6*] [22 pp.].

PLATES: Engraved frontispiece portrait with stipple engraving in Part 1 and 40 hand-colored engraved and etched plates.

BINDING: Original wrappers, parts 1, 2, 3, 4, 5 & 6, and 39 & 40 are printed; beginning with parts 5 & 6, all subsequent parts are doubled in one wrapper.

PROVENANCE: Bookplate of Paul Mellon.

REFERENCES: Anderson, p. 186; Arber, pp. 261–63; Blunt & Raphael, pp. 141, 182–183; Henrey 1.82–88 (note: Henrey lists 55 17th-century editions); Potterton; Rohde, pp. 163–168.

THE SON OF A CLERGYMAN, Nicholas Culpeper (1616–1654) was a famous pharmacist, herbalist, alchemist, and astrologer who owned a garden of medicinal plants at Spitalfields, London. He studied at Cambridge University and then became apprenticed to a master apothecary in London, having cultivated since his youth an interest in medicine and botany. Afterwards he built up a busy practice and was quite successful, even though he never obtained the title of physician, winning the respect of the general public and the love of his poorer patients through his generous efforts on their behalf.

According to Culpeper, through the providence of God simple remedies for every ailment could be readily prepared from plants growing on English soil. He was therefore vehemently opposed to the treatment of patients with exotic (and costly) nostrums—a fad promulgated by many of his colleagues—and made it his mission to educate the public by writing books that explained matters of health and medicine in uncomplicated language.

Combining notions from the classical tradition of Dioscorides with contemporary theories of herbal medicine, astrology, and alchemy, his works provide us with a detailed and fascinating picture of how medicine was practiced in seventeenth-century England. Culpeper was particularly interested in the areas of ophthalmology, gynaecology, and obstetrics; he even wrote *A Directory for Midwives,* and suggested well in advance of his time that hygiene played an important role in the healing process.

In 1649 Culpeper published an unauthorized translation from Latin into English of the *Pharmacopœia Londinesis* as *A Physicall Directory or a Translation of the London Dispensatory made by the College of Physicians in London,* to which he added an opening treatise on plants that included a number of herbal recipes. His aim was to furnish the layman with the knowledge needed to prepare homemade remedies as a defense against the meretricious and self-serving claims of supposedly reputable doctors and pharmacists. This work was condemned by the London College of Physicians, which—to protect its vested interests—insisted that all medical knowledge be communicated exclusively in Latin. That Culpeper's initiative was greatly appreciated by the general public was shown by the fact that *A Physicall Directory* became an immediate best seller.

In 1652 Culpeper published *The English Physitian, or, An Astrologo-physical Discourse of the Vulgar Herbes,* a text that remained in print for more than 350 years, serving as a practical manual for generations of physicians in England and America. The author's name became so well known that many books were spuriously attributed to him by other publishers. In 1789 Ebenezer Sibly (1751–*c.* 1800) published *Culpeper's English Physician; and Complete Herbal,* a work that enjoyed a similar success and went through many re-printings; indeed, it is still consulted by herbalists today.

The Oak Spring Garden Library owns a copy of the 1828 edition of Culpeper's herbal, which was published in London by Thomas Kelly. Kelly specialized in the printing or reissuing of affordable editions of celebrated works, such as *Buffon's Natural History* (issued in ninety-four instalments at the price of sixpence each), and manuals and popular works ranging from *Dr. Dickson's Live-Stock & Cattle-Keeper's Guide* to *The Life of our Lord and Saviour Jesus Christ* by Reverend J. Fleetwood (published in forty parts and sold for sixpence each). His edition of *The Complete Herbal* (which includes three other works by the same author), printed in instalments modestly priced at ninepence each that were distributed throughout Great Britain, testifies not only to the enduring fame of Culpeper's work, but also to the modifications that it underwent over the centuries.

The Complete Herbal opens with an engraved portrait of the author based on that which originally appeared in the first edition of *A Physicall Directory* (cf. Arber, p. 263). The bust of Culpeper is framed by a horoscope and floral garland; beneath is an engraving of his house at Spitalfields. The 'Epistle to the Reader' contains the celebrated allegorical passage: 'I consulted with my two brothers, Dr. Reason and Dr. Experience, and took a Voyage to visit my mother Nature, by whose advice, together with the help of Dr. Diligence, I at last obtained my desire: and, being warned by Mr. Honesty, a stranger in our days, to publish it to the world, I have done it.' A letter addressed to the author's beloved wife, Alice, follows, which closes with an affectionate salutation to her, his children, and the 'Arts and Sciences which I so dearly loved.'

This herbal presents a series of plants in alphabetical order, beginning with 'Amara Dulcis' (identifiable as the woody nightshade or felonwort, *Solanum dulcamara*). The text provides a botanical description, information on the plant's natural habitat and medicinal properties, and a recipe, which in the case of the nightshade is followed by the observation: 'You may drink a quarter of a pint of the infusion every morning. It purges the body very gently ... and when you find good by this, remember me.'

An adept in astrological medicine, Culpeper catalogued the plants that were under the dominion of the sun, the moon, and the planets, the heavenly bodies that could influence the onset, course, and cure of a disease. He furthermore was an exponent of the doctrine of signatures and the existence of 'sympathies and antipathies' between different elements and diseases, notions that were shared by some of the most eminent naturalists of the period, including Giovanbattista della Porta (see No. 51) and William Coles (who on the other hand strongly criticized Culpeper's practice of astrological medicine, see No. 34).

In his discussion of Melancholy, one of the connotative illnesses of his period (and of later centuries, albeit under different names), Culpeper writes that his sovereign cure for the malady, Melancholy Thistle,

> is under Capricorn, and therefore under both Saturn and Mars, one rids melancholy by sympathy, the other by antipathy. Their virtues are but few, but those not to be despised; for the decoction of the thistle in wine being drank, expels superfluous melancholy out of the body, and makes a man as merry as a cricket.... Dioscorides saith, The root borne about one doth the like, and removes all diseases of Melancholy. Modern writers laugh at him; *Let them laugh that win:* my opinion is, that it is the best remedy against all melancholy diseases that grow; they that please may use it.

NICHOLAS CULPEPER, *Culpeper's Complete Herbal*, 1828. Frontispiece portrait and image of his home, part 1

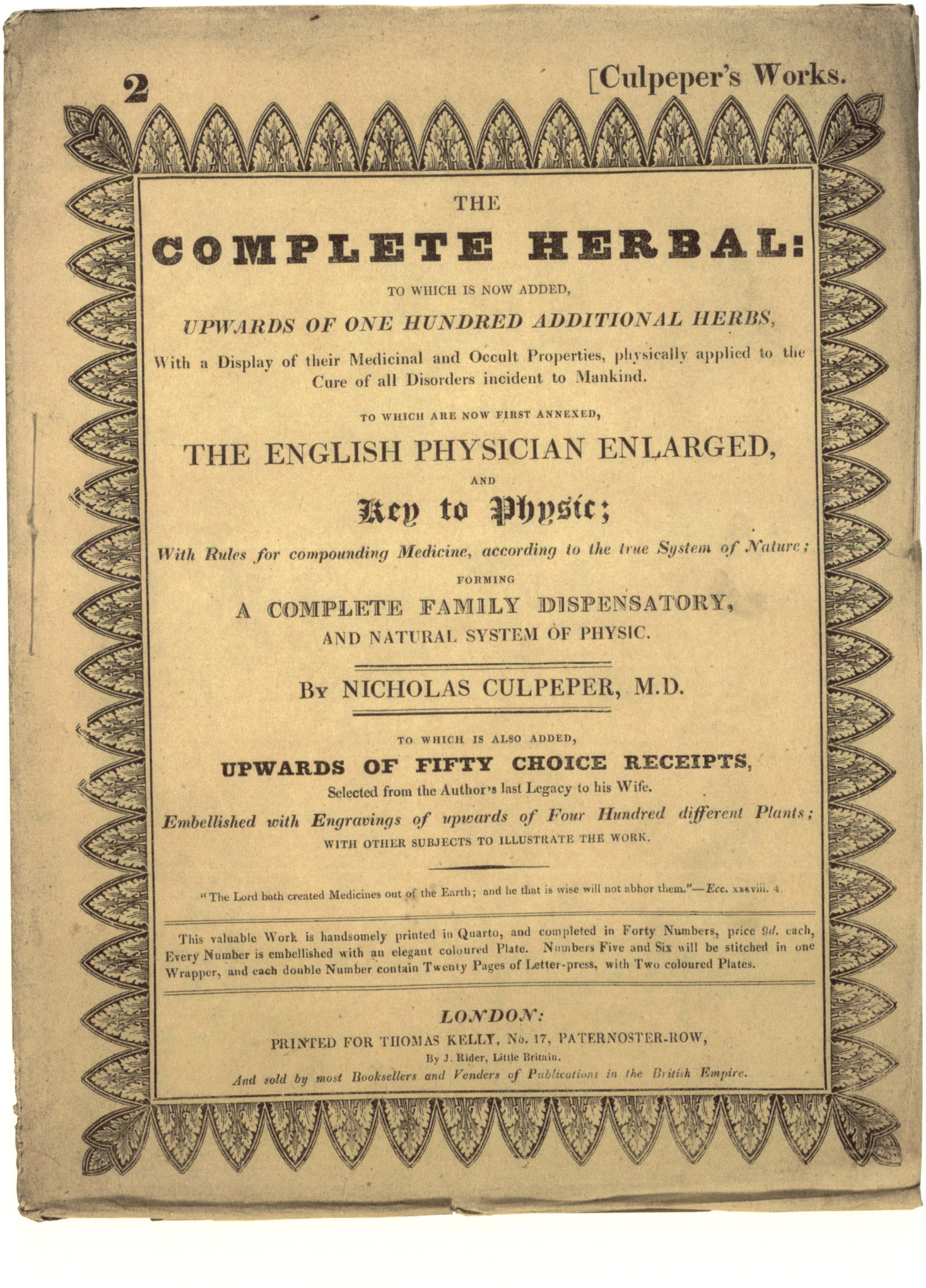

2 [Culpeper's Works.

THE

COMPLETE HERBAL:

TO WHICH IS NOW ADDED,

UPWARDS OF ONE HUNDRED ADDITIONAL HERBS,

With a Display of their Medicinal and Occult Properties, physically applied to the Cure of all Disorders incident to Mankind.

TO WHICH ARE NOW FIRST ANNEXED,

THE ENGLISH PHYSICIAN ENLARGED,

AND

Key to Physic;

With Rules for compounding Medicine, according to the true System of Nature;

FORMING

A COMPLETE FAMILY DISPENSATORY,

AND NATURAL SYSTEM OF PHYSIC.

By NICHOLAS CULPEPER, M.D.

TO WHICH IS ALSO ADDED,

UPWARDS OF FIFTY CHOICE RECEIPTS,

Selected from the Author's last Legacy to his Wife.

Embellished with Engravings of upwards of Four Hundred different Plants;

WITH OTHER SUBJECTS TO ILLUSTRATE THE WORK.

"The Lord hath created Medicines out of the Earth; and he that is wise will not abhor them."—*Ecc.* xxxviii. 4.

This valuable Work is handsomely printed in Quarto, and completed in Forty Numbers, price 9*d*. each, Every Number is embellished with an elegant coloured Plate. Numbers Five and Six will be stitched in one Wrapper, and each double Number contain Twenty Pages of Letter-press, with Two coloured Plates.

LONDON:

PRINTED FOR THOMAS KELLY, No. 17, PATERNOSTER-ROW,

By J. Rider, Little Britain.

And sold by most Booksellers and Venders of Publications in the British Empire.

NICHOLAS CULPEPER, *Culpeper's Complete Herbal,* 1828. Printed wrapper, part 2

NICHOLAS CULPEPER, *Culpeper's Complete Herbal*, 1828. 'Common Alder,' Alder (*Alnus glutinosa*); 'Garden Arrach,' Orache (*Atriplex hortensis*); 'Alehoof,' Ground Ivy (*Glecoma hederacea*); 'Agrimony' (*Agrimonia eupatoria*); 'Water Agrimony' (*Bidens tripartite*); 'Alexander' (*Angelica* sp.); 'Amaranthus Flower Gentle' (*Amaranthus hypochondriacus*); 'Anemone' (*Anemone* sp.); 'Asparagus' (*Asparagus officinalis*); 'Alacost' (*Lysimachia* sp.?); 'Angelica,' Wild Angelica (*Angelica sylvestris*) or (*A. archangelica*); 'Allheal,' an unidentified member of *Lamiaceae* family. Plate 1

In Kelly's edition of *Culpeper's Complete Herbal*, each fascicule is accompanied by a single colored plate. The first issue, which breaks off abruptly at the end of the second line of Culpeper's description of 'The Black alder-tree,' contains a modest engraving of twelve plants, not all of which are discussed in the text. They are depicted in simple, roughly colored outline drawings, and each is labelled with its common name. Culpeper's herbal treatise continues for another nineteen issues, concluding with the 'Yarrow, called Nose-Bleed, Milfoil and Thousand-leaf.'

In a further twenty fascicules, Kelly presents other works by the English physician: *Directions for making Syrups, conserves, & ...* in fifteen chapters; *The English Physician and family Dispensatory* containing recipes for various electuaries, decoctions and medicines; and, in closing, *A key to Galen's method of Physic*, in which Culpeper discusses the medicinal plants known to the ancient Romans, their active principles, and how these could be extracted.

In the Oak Spring Garden Library's collection is another edition of *The Complete Herbal*, published by Thomas Kelly in 1865.

V

HERBALS OF THE BOTANICAL GARDENS AND PRIVATE GARDENS OF EUROPE

42. JOACHIM CAMERARIUS *the younger* (1534–1598)

[*Title-page in red and black*]: Hortus medicus et philosophicus: In quo plurimarum stirpium breves descriptiones, novae icones non paucae, indicationes locorum natalium, observations de cultura earum peculiares, atque insuper nonnulla remedia euporista, nec non philologica quaedam continentur. Autore Ioachimo Camerario Reipub. Norimberg. Medico D. Item Sylva Hercynia: Sive catalogus plantarum sponte nascentium in montibus & locis plerisque Hercyniæ Sylvæ quæ respicit Saxoniam, conscriptus singulari studio à Ioanne Thalio Medico Northusano. Omnia nunc primum in lucem edita. [*Printer's device*]. Francofurti ad Mœnum. M.D.LXXXVIII. Cum Gratia & Privilegio Cæs. M [*lower right-hand corner of title-page is missing*].

[*Colophon, 2H3, verso*]: Impressum Francofurti ad Moenum, apud Iohannem Feyerabend, impensis Sigismundi Feyerabendij, Heinrici Dackij, & Petri Fischeri. [*Printer's device*] M.D.LXXXVIII.

4° 19.4 × 15.3 Three parts in one volume, each with separate title-page. A–B⁴ a–z⁴ A–R⁴ 2A–2H⁴ *i–xvi* 1–184 *1–2* 3–133 [5] (53 as 55, 89 as 79) *I–XLVII I–IX* [*6*] [400 pp.]

PLATES: 56 woodcuts in Part III.

BINDING: Contemporary vellum with its leather ties missing.

PROVENANCE: Duplicate from the Munich University Library with release stamp on verso of title-page.

REFERENCES: Blunt & Stearn, pp. 86–87; Camerarius; Hunt 156; Nissen 311; Pritzel 1,439.

JOACHIM CAMERARIUS, who was born in Nuremberg in 1534, was not only a physician and botanist but also an erudite scholar with a profound knowledge of classical culture. His father, Joachim Camerarius the Elder (1500–1574)—who had adopted the name Kammermeister (Chamberlain) in place of the family name Liebhard, and Latinized it to Camerarius—was an active figure in German intellectual circles during the first half of the sixteenth century. Learned in Latin and Greek, the elder Camerarius translated many classical Greek texts into Latin and some of them into German. He was furthermore a good friend of Albrecht Dürer and translated some of his writings into Latin, thus facilitating their dissemination all over Europe. In addition to his learning he was a deeply pious man and played an active role in the Protestant Reform movement. The young Joachim therefore received an elite education, and was able to benefit from the company of his father's cultivated friends, among them the writer, reformer, and educator Philipp Melanchthon (1497–1560), of whom his father had written a biography and at whose home he often stayed.

Despite his classical education, Joachim decided to become a physician and frequented classes at the universities in Leipzig, Wittenberg, and Breslau before completing his formation at the celebrated faculties of medicine in Padua and then Bologna, where he received his degree in 1562. During his period in Italy, Camerarius had the opportunity to meet a number of eminent physicians and naturalists, including the anatomist Gabriele Falloppia of Modena (1523–1562); the surgeon Girolamo Fabrici d'Acquapendente of Padua (1533–1619); the physician and botanist Prospero Alpini (see No. 15); athe Flemish herbalist Jodocus de Goethuysen (*c.* 1535–1595), who spent the greater part of his life in Tuscany and Italianized his name to Giuseppe Casabona; and the naturalist Ulisse Aldrovandi. With many of these men he established life-long friendships, exchanging letters, information, and botanical specimens on a regular basis.

Camerarius then returned to Nuremberg, where in 1564 he was appointed official physician and director of the town's 'Collegium Medicum.' Once he had settled down after his peripatetic student days, Camerarius was reluctant to travel except on rare occasions, such as when he was invited by Wilhelm IV, the Landgrave of Hesse, to come to Kessel and design a new plan for his garden. Determined to continue his botanical studies, Camerarius acquired the garden of medicinal plants lying outside the city walls that had belonged to the pharmacist Jörg Öllinger. In his 'suburbanus hortus' Camerarius cultivated many rare plants, some of them sent to him by Italian colleagues or the Flemish botanist Carolus Clusius, who was then director of the Imperial Botanic Gardens in Vienna (see Nos. 9 and 14). This free intercourse between men of science, which took little heed of national boundaries or social rank, reflected the dimension of sociability that was one of the most significant characteristics of the relationship between intellectuals from the Renaissance to the Baroque periods.

According to Camerarius, the serious study of botany had to begin with the works of the great Italian naturalist Pietro Andrea Mattioli (1500–1577, see No. 7), whose texts he therefore translated into German, so that they would be more accessible to his colleagues. In 1586 he produced *De Plantis Epitome*, an abridged version of *I Discorsi,* and in 1590 a complete edition entitled *Kreuterbuch*. Both of these were printed by Sigmund Feyerabend in the city of Frankfurt. In 1588 Feyerabend also published Camerarius' first original work, *Hortus medicus et philosophicus*, which was bound together with *Sylva Hercynia*, a short treatise on the flora of the Harz Mountains in eastern Germany by the physician Johann Thal (1542–1583).

In the title to his work Camerarius links the two realms of medicine and philosophy, and in his preface addressed to the learned reader (*studioso lectori*) the physician explains that he considered it opportune to supplement his scientific discussion with various remarks about philology (*quaedam philologica*); these consisted for the most part of classical citations designed to complement the more specifically 'botanological' material in his herbal. His aim was to write a work whose appeal would extend beyond the limited circle of fellow specialists to a broader audience of cultivated scholars.

Camerarius' perspective was shared by many of his fellow naturalists, from Conrad Gesner (1516–1565) and Ulisse Aldrovandi to Mattioli himself, all of whom in the prefaces to their works expressed the desire to connect their knowledge of nature with other disciplines relating to the study of Antiquity. The humanist education that Camerarius had received in his youth would also influence his second work, dedicated to the symbology of plants and animals (see No. 52).

III.

Alcea arborescens. Pag. 9.

Plantulam hic exhibemus, qualis apud nos enata fuit ex semine. Florem vero communicauit CL. Vir D. Rauvvolfius.

JOACHIM CAMERARIUS, *Hortus medicus et philosophicus*, 1588. 'Alcea arborescens' (*Althaea* sp.), Aa3

In his preface Camerarius furthermore declares that nature could be compared to a splendid 'museum' whose herbs, plants, and trees far surpassed in beauty the statues and paintings of Antiquity. Thus he was inspired to write this book describing the prodigious variety of plants and flowers that, in addition to their beauty, embodied 'vires salutares hominum' (salubrious virtues for mankind). Notwithstanding its philosophical dimension, however, *Hortus medicus et philosophicus* was a true herbal based on the author's direct study of botanical specimens. It catalogues in alphabetical order the plants growing in his garden with succinct descriptions that include a list of their healing virtues, one of the author's convictions being that the vast array of nostrums prescribed by herbalists, physicians, and charlatans could be reduced to a handful of genuinely effective remedies based on medicinal plants.

Camerarius provides information on the form, the names in different languages, the provenance, and the habitat of each plant. He continues with advice on how it was to be cultivated (taking into account the cold climate of his native city) and its medicinal properties, citing those classical and modern authors in whose texts either a description or an illustration of the plant could be found. Frequently he inserted personal observations regarding the places where he had seen the species growing *in situ* or had gathered a specimen of it, for the most part during his sojourn in Italy as a student. Thus he notes that *Cistus mas* could be found growing spontaneously along the beaches of Pisa and Porto Ercole, while *Cistus foemina* was native to the hills of Euganei near Padua (p. 42). In addition to specific places, Camerarius mentions various colleagues, such as his botany professor at the university in Padua, Luigi Anguillara (*c.* 1512–1570) (p. 97), and those to whom he was indebted for the gift of rare or unusual plants, in particular Carolus Clusius of Leiden and Giuseppe Casabona, herbalist to the Grand-Duke of Tuscany (p. 96).

Like many of his fellow botanists, Camerarius devoted considerable time and attention to the documentation of his studies by means of botanical illustrations. For this purpose he acquired not only a large collection of drawings and engravings that had belonged to the pharmacist Öllinger, but he also acquired the estate of Conrad Gesner (1516–1565). Gesner, a Swiss scholar and naturalist, whom his contemporaries dubbed 'the Pliny of his Age,' published an important work on zoology, *Historia animalium*, and had many drawings and woodblocks commissioned for a projected, but never completed, treatise *Historia plantarum*. Camerarius hoped to use these woodblocks for a more ambitious book of his own, *Paralipomenon rei herbariae*, on which he had already begun work as he affirms in the introduction to *Hortus medicus et philosophicus*, but this project, like Gesner's, was never finished.

Hortus medicus did include about thirty of Gesner's illustrations, while the remainder were carved by the great Swiss engraver Jost Amman (1539–1591), based on preparatory drawings by Camerarius' nephew Joachim Jungermann (1561–1591). The illustrations do

XXXI.

Nasturtium Indicum. Pag.105.

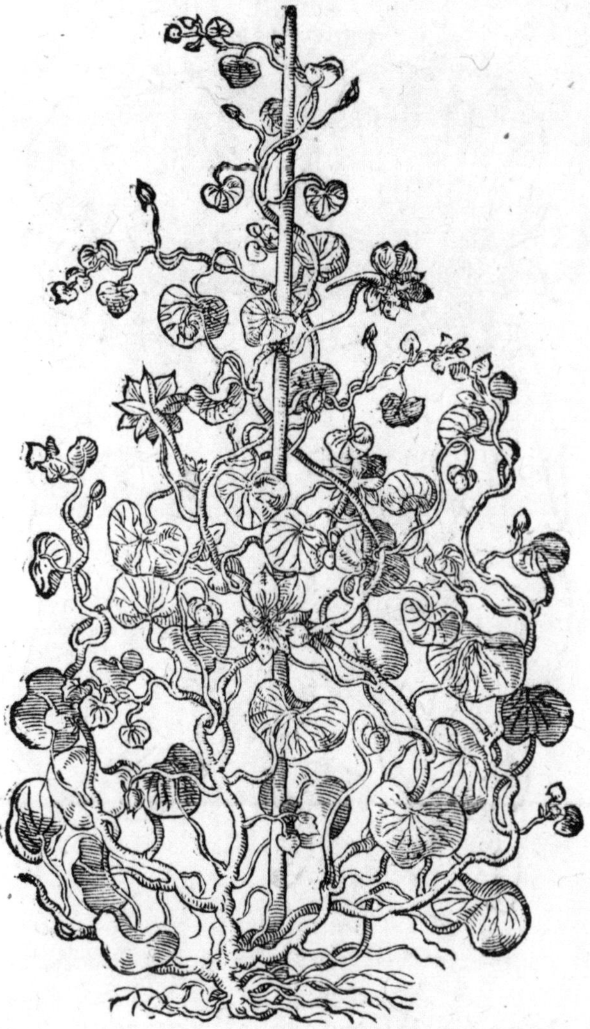

Hæc est vera imago eximiæ huius plantæ, qualem ad viuum ab industrio pictore exprimendam curauimus.

Ee Flores

JOACHIM CAMERARIUS, *Hortus medicus et philosophicus*, 1588. 'Nasturtium Indicum' (*Tropaeolum majus*), Ee

not appear next to their descriptions in the text, but have rather been grouped together in separate section entitled *Icones accurate nunc primum delineatae praecipuarum stirpium, quarum descriptiones tam in Horto quam in Sylva Hercynia suis locis habentur.* These botanical illustrations are characterized by a *mise en page* of great freedom; often the entire plant is shown, with its roots or details of the flowers, leaves, or seeds added to help in its identification. At the top of the page the Latin name of the plant is given, as well as a page reference to allow the reader to look up the corresponding description in the text. Beneath each illustration is a brief comment, the author on occasion observing that the specimen had been cultivated in his own garden, from which we may infer that it was probably drawn from life by Joachim Jungermann. Such was the case for the illustration of *Alcea arborescens* (Pl. III), grown 'apud nos ... ex semine' (in our garden from a seed). The notes beneath the illustration of *Aphyllanthes anguillarae* (*Globularia vulgaris*) (Pl. VII) suggest that this engraving probably came instead from the collection of Conrad Gesner. Among the exotic species illustrated in the herbal, *Quamoclit* (Pl. XL) and *Aloe americana* (Pl. V) are particularly significant because, as the author comments, although they had already been described by Andrea Cesalpino, they had never before been illustrated in a botanical text. With regard to the drawings of *Nasturtium indicum* (Pls. XXXI and XXXII), the author states specifically that they were the work of 'his excellent artist,' i.e., Joachim Jungermann.

Further confirmation of Camerarius' direct interest and involvement in the production of botanical illustrations has emerged with the recent, fortunate discovery of a florilegium that was compiled by him around the year 1589, *The Camerarius Florilegium* (Christie's, 20 May 2002). This collection of 473 botanical drawings constitutes an important record of the indigenous and exotic plants being cultivated in Germany at the end of the sixteenth century. The significance of this manuscript becomes even greater if one takes into account its links with another work, Basil Besler's *Hortus Eystettensis* (1613) (see *An Oak Spring Flora,* No. 11, pp. 52–57), an extraordinary florilegium of the plants that were being cultivated in the garden of the Prince-Bishop Johann Conrad von Gemmingen at Eichstätt. Camerarius not only supervised the construction of the prelate's garden, but also gave his patron suggestions regarding the plants that might be cultivated there. Although the splendid botanical drawings in *The Camerarius Florilegium* are unsigned, it is quite possible that they were the work of Joachim Jungermann. The young artist had already contributed to the illustration of *Hortus medicus et philosophicus* and his work had been warmly praised by such eminent botanists as Carolus Clusius and Giuseppe Casabona; indeed the latter invited Jungermann to accompany him on his herborizing expedition to Crete.

The Camerarius Florilegium is today conserved in the library of the University of Erlangen, which also houses the naturalist's papers. His correspondence furnishes ample evidence of Camerarius' ties with physicians and scientists all over Europe, the respect with which he was regarded, and the positive reception that was accorded his works. In another letter, conserved in the Arents Collection at the New York Public Library, Camerarius conveys his thanks to the geographer Abraham Ortelius (1527–1598) for his 'generous and erudite letter' commenting on *Hortus medicus et philosophicus*.

43. Guy de la Brosse (1586?–1641)

Description du Iardin Royal des Plantes Medecinales, estably par le Roy Louis le Juste, à Paris. Contenant le Catalogue des Plantes qui y sont de presens cultivées, ensemble le Plan du Iardin. Par Guy de la Brosse, Medecin ordinaire du Roy, & Intendant dudit Iardin. [*device of the coat of arms of Louis XIII*]. A Paris, [*full measure rule*] M. DC XXXVI.

4° 22.5 × 16.7 cm. A–N⁴ O² *1–2* 3–52 *53*–107 *108* (23 as 32, 83 as 38) [108 pp.].

PLATES: Two folded, etched plates, the first one is signed 'Federic Scalberge pict. sculp. et fec Anno, 1636.'

BINDING: Contemporary vellum, with modern gilt red leather label on spine; inscribed pressmark on front cover 'B. A. H. 114.'

PROVENANCE: Blue oval French library stamp (illegible) on title-page.

REFERENCES: BM Cat. (*Nat. Hist.*) III.1040; Evelyn; Hamy; Pritzel 1,187.

In 1633 Guy de la Brosse—botanist, chemist, and physician to Louis XIII—was appointed by the superintendent of the state finances Charles de Buillon to construct a 'jardin medical,' in compliance with the king's wish to see a botanical garden established in Paris that would rival those of the other intellectual capitals of Europe. This decision was much awaited, for the pharmacists of Paris had been petitioning the king since 1620 to build a garden similar to the one at the university in Montpellier.

Guy de la Brosse was without a doubt the most qualified person to oversee this project. Son of the physician Isaïe de la Brosse, he completed his studies at the illustrious faculty in Montpellier where he received the title of doctor of medicine. In 1619 he was appointed personal physician to Henri II de Bourbon, the Prince of Condé. During this period he published various works, including a treatise on the plague, *Traicté de la peste* (1623), and a botanical text influenced by Aristotelian philosophy, *De la nature, vertu, et utilité des plantes* (1628), which he dedicated to the prime minister, Cardinal Richelieu (1585–1642), who had become his protector. A follower of the Swiss alchemist Paracelsus, de la Brosse frequented the circles of free thinkers in Paris and was a friend of the philosopher-scientists Marin Mersenne (1588–1648) and Pierre Gassendi (1592–1655).

In the same year that he published *De la nature, vertu, et utilité des plantes*, de la Brosse was appointed one of the *médecins ordinaires* to Louis XIII and shortly afterwards received the royal commission to establish a garden of medicinal plants. He immediately drew up

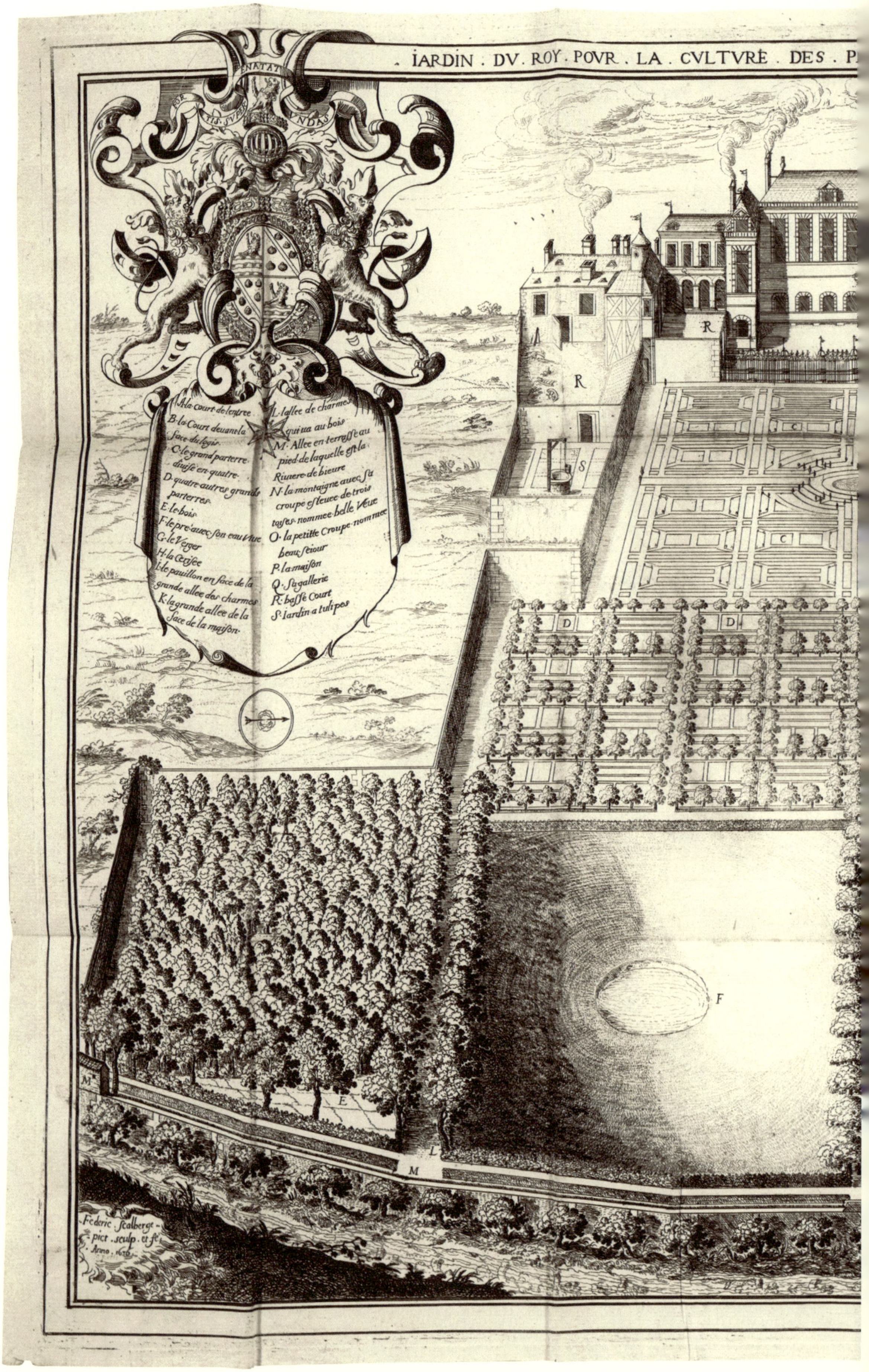
IARDIN . DV . ROY . POVR . LA . CVLTVRE . DES . P
A. la court de lentree
B. la Court deuant la face du logis
C. le grand parterre diuise en quatre
D. quatre autres grands parterres
E. le bois
F. le pre auec son eau viue
G. le Verger
H. la Cerisee
I. le pauillon en face de la grande allee des charmes
K. la grande allee de la face de la maison
L. lallee de charmes qui va au bois
M. Allee en terrasse au pied de laquelle est la Riuere de bieure
N. la montaigne auec sa croupe esleuee de trois toises nommee belle Veue
O. la petitte Croupe nommee beau seiour
P. la maison
Q. sa gallerie
R. basse Court
S. Iardin a tulipes
Federic Scalberge pict. sculp. et fe. Anno. 1636

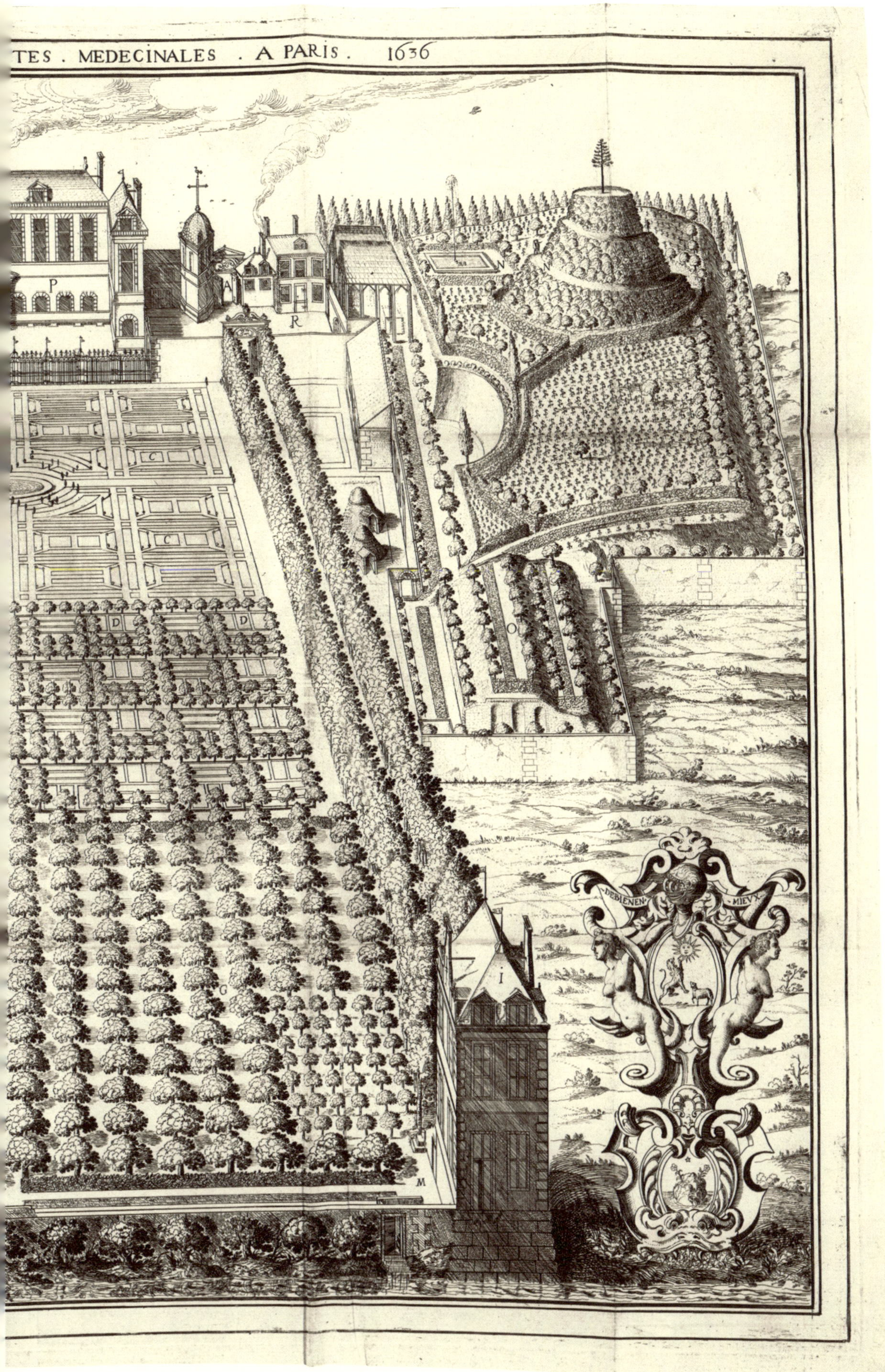

GUY DE LA BROSSE,
Description du Jardin Royal des Plantes, 1636.
Folding plate of the 'IARDIN DU ROY POUR LA CULTURE DES PLANTES MEDECINALES, A PARIS, 1636'

and published a *Dessein d'un jardin royal pour la culture des plantes médicinales à Paris* (1628), which was followed by an *Avis pour le jardin royal des plantes médicinales* (1631). His proposal generated some opposition, particularly among the physicians at the Sorbonne, who feared that the new facility might compete with their own institution, diluting its influence and power. Plans went forward notwithstanding and, after a suitable site had been chosen, de la Brosse began to supervise the planting of the new garden in 1634.

Work advanced quickly and de la Brosse was soon able to publish a *Description du Jardin Royal des Plantes*. This he dedicated to Charles de Buillon who, as the superintendent in charge of appropriations for the public works sponsored by the king, had found the large sums necessary to complete his project. The physician was quick to justify the royal investment: '... entre tant d'ouvrages pieux, necessaires, utils & de decoration ... il n'y a aucun où toutes les pensées des hommes concurrent, que l'establissement du Jardin Royal des Plantes Médicinales.' He was convinced that, in addition to its scientific importance, the new garden had great social and political significance, pointing out that ever since the first flower beds had been planted two years earlier: 'la porte est constantement ouverte aux Nations estrangères qui les sont venu à le visiter & admirer.'

While it was true that Paris already had 'un jardin des Plantes Médicinales, sinon Royal, au moins stipendié par le Roy,' built some years earlier and tended by 'sieur Robin, Herboriste du Roy,' as de la Brosse pointed out this was a rather modest plot, 'n'ayant jamais ecedé trointes toises de terre, & pour le quelle, & et pour la pension, il n'y avait que quatre cens livre par an' (p. 14). The four hundred livres per year received by Jean Robin and his son Vespasien represented a totally insufficient sum for the upkeep of an important medicinal garden (see No. 18).

Construction was finally completed, but the new garden would be officially inaugurated only in 1640, an occasion which de la Brosse would mark by the publication of another short work, *L'ouverture du jardin royal de Paris*. The verdant space, which occupied some 'cinquante arpens' in the Faubourg Saint Victor, did not constitute a mere 'vain ornament à la France & à Paris ... mais un très necessaire et utile Escole, de la matière Médicinale.' Three doctors, including the chief physician at court, Jean Héroard, were engaged to lecture on the plants growing there and on the remedies that could be prepared from them, either directly or after chemical manipulation (p. 19). These lessons were open to the public and were conducted in French, in contrast to the Sorbonne whose professors continued to teach in Latin and retained the exclusive right to confer degrees.

In *Description du Jardin Royal des Plantes*, de la Brosse explains how the garden came to be built and provides a detailed description of its layout; the text is furthermore illustrated with two large folded engravings. The first, which is signed 'Federic Scalberge pict. sculp. et fec. Anno, 1636,' reproduces a watercolor by Frédéric Scalberge (d. 1640), a Flemish artist

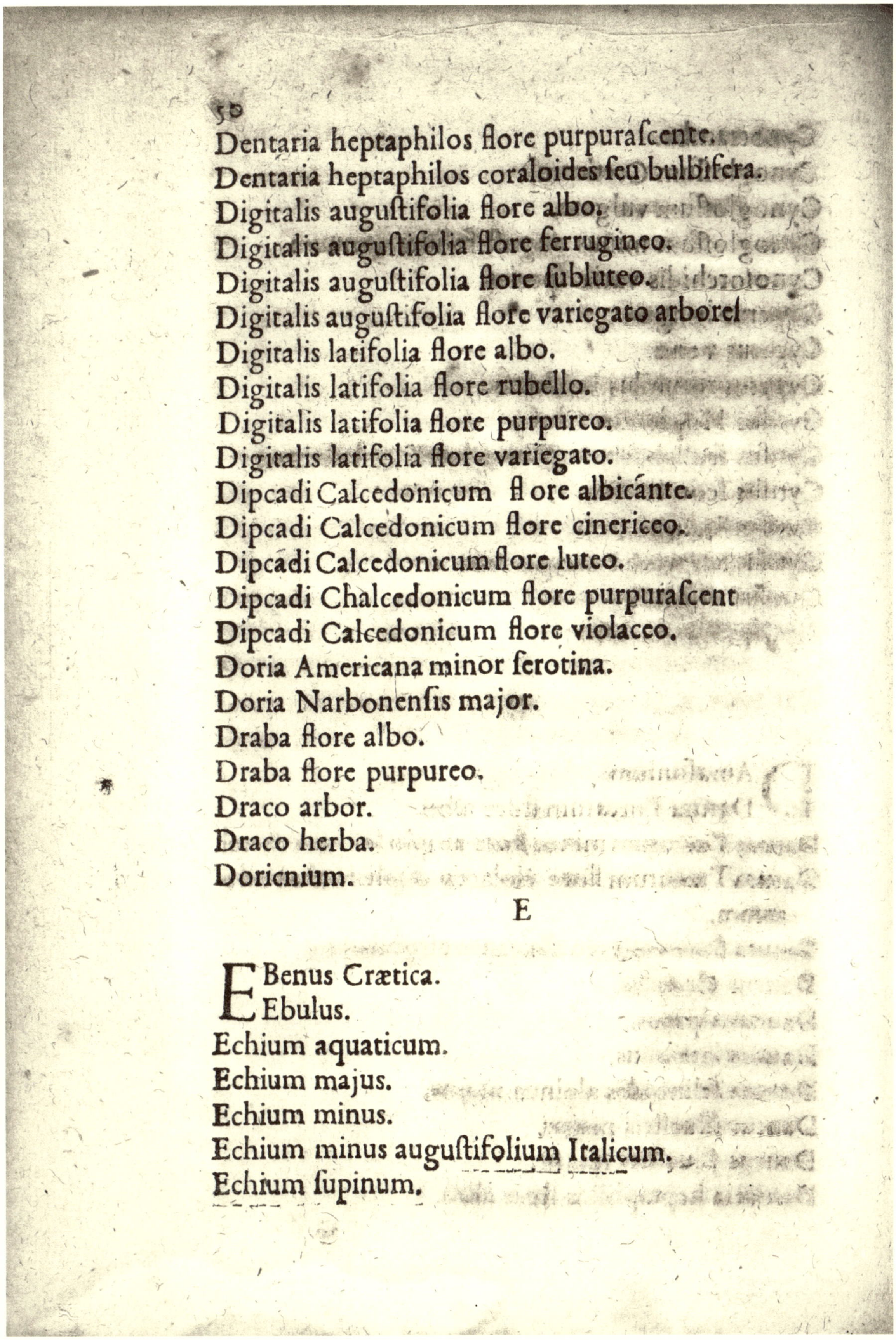

Dentaria heptaphilos flore purpuraſcente.
Dentaria heptaphilos coraloides ſeu bulbifera.
Digitalis auguſtifolia flore albo.
Digitalis auguſtifolia flore ferrugineo.
Digitalis auguſtifolia flore ſubluteo.
Digitalis auguſtifolia flore variegato arborel
Digitalis latifolia flore albo.
Digitalis latifolia flore rubello.
Digitalis latifolia flore purpureo.
Digitalis latifolia flore variegato.
Dipcadi Calcedonicum flore albicánte.
Dipcadi Calcedonicum flore cinericeo.
Dipcádi Calcedonicum flore luteo.
Dipcadi Chalcedonicum flore purpuraſcent
Dipcadi Calcedonicum flore violaceo.
Doria Americana minor ſerotina.
Doria Narbonenſis major.
Draba flore albo.
Draba flore purpureo.
Draco arbor.
Draco herba.
Doricnium.

E

EBenus Cratica.
Ebulus.
Echium aquaticum.
Echium majus.
Echium minus.
Echium minus auguſtifolium Italicum.
Echium ſupinum.

GUY DE LA BROSSE, *Description du Jardin Royal des Plantes*, 1636. One leaf from the index, G1, listing some plants cultivated in the garden

who was active in Rome between 1627 and 1633 (the original painting is conserved in the library of the Musée Nationale d'Histoire Naturelle in Paris). The engraving presents a panorama of the magnificent garden, including its buildings, pavilions and wooded groves, its orchards, the tulip garden, the garden of Alpine plants, and so on. The layout of each was dictated by the requirements of the plants in terms of exposure to the sun and elements. The second engraving illustrates the 'grand parterre' of medicinal plants, which was laid out in accordance with the rigorous quadrangular ground plan of the botanical gardens of the period.

The sophisticated design of the Jardin Royal was admired by the English polymath John Evelyn (1620–1706) who, after visiting it in 1640, wrote: '[it] may truely be esteemed an excellent pattern for our Imitation' (Evelyn, p. 404). De la Brosse could now vaunt the fact that Paris had a botanical garden equal to those of Padua, Pisa, and Leiden, and indeed with a larger and more complete collection than many of the older gardens. A handful of critics complained that there were too many plants, taking particular issue with the large number of ordinary species being cultivated there. The author replied crisply that native plants counted no less than the most exotic ones—all were plants endowed with specific virtues. It would be scientifically incorrect to limit one's interest to what came from far away, neglecting things closer to home, and therefore even a king's garden should not consist only of rare plants.

Following the text, which provides invaluable information on the origins of the Jardin Royal des Plantes, is a catalogue listing in alphabetical order two thousand plants that could be found growing in the garden just two and half years after its construction. In addition to common European species there were in fact many from foreign lands, such as the *Draco arbor*, *Gnaphalium Americanum*, *Polygonatum Americanum*, *Sorbus Americana*, and several varieties of the tobacco plant. Many horticultural varieties and exotic fruit trees appear in the list as well, their opulent flowers and rare fruit evidently destined for the king's table. An expanded and updated edition of this catalogue was printed in 1641.

Pritzel has observed (No. 1,183) that there is an extremely rare collection of forty-eight botanical etchings with a manuscript title-page conserved in the British Museum of London, *Reliquiae operis historici plantarum in horto regio Parisiensi educatarum, a Guidone de la Brosse; ab Abrahamo Bosse aeri incisarum*. On one sheet is an annotation by Antoine de Jussieu stating that the work of de la Brosse had been interrupted and all of the copperplates destroyed. Providentially, Guy Crescent Fagon, de la Brosse's successor, managed to recover a certain number, and Fagon's colleagues, de Jussieu and Sébastien Vaillant (1669–1722), managed to have twenty-four copies of the completed etchings printed for distribution among their friends. It may be hypothesized that the great Flemish artist Abraham Bosse (1602–1672), author of many illustrations for the monumental project of the Académie

Royale des Sciences *Mémoires pour servir à l'Histoire des Plantes* (see *An Oak Spring Flora*, Nos. 43 and 44), made these precious engravings of some of the plants in the king's garden, which after the French Revolution came to be known simply as the Jardin des Plantes.

44. DENIS DODART (1634–1707)

Mémoires pour servir à l'histoire des plantes. Dressez par M. Dodart, de l'Academie Royale des Sciences, Docteur en Medecine de la Faculté de Paris. [*Royal device*]. A Paris, De L'Imprimerie Royale. [*rule*] M. DC. LXXVI.

2° 53.8 × 39.9 cm. π^1 $\tilde{a}^1$ $\tilde{e}^1$ A–3V[1] *i–vi 1* 2–131 *132* [138 pp.].

PLATES: Frontispiece signed 'S. Le Clerc in. et f.' and 'Goyton ex.,' 39 leaves of plants, most are by Nicolas Robert (1614–1684), three head- and tail-pieces, two are signed by Le Clerc, all engraved and etched.

BINDING: French mottled calf with gilt tooling on spine and gilt rules on front and back covers.

PROVENANCE: Bookplate of Belton House and The R[t] Hon[ble] John L[d] Brownlowe, Baron Charleville.

REFERENCES: Blunt & Stearn, pp. 121–122; Hunt 343; Nissen 502; *An Oak Spring Flora*, pp. 167–170; Plesch, pp. 203–2; Pritzel 2,341; Sitwell & Blunt.

THE IDEA of publishing a monumental *Histoire des Plantes* was conceived in 1667 and promoted by Claude Perrault (1613–1688)—architect, physician, scientist and one of the foremost intellectuals at the court of Louis XIV. Nicolas Marchant, botanist to the king's uncle Gaston d'Orléans (1608–1660), who was himself a passionate gardener and student of the botanical sciences, agreed to write the text, whilst the gifted botanical painter Nicolas Robert (1614–1684), who had worked for many years for Gaston, was designated to prepare the drawings. Abraham Bosse (1602–1676) and Louis Chatillon (1639–1734) were then to translate these drawings into engravings. By 1692 a total of 319 botanical illustrations had been completed; a three-volume work containing these engravings without letterpress was printed in 1701, but never placed on sale, and surviving copies are extremely rare (see *An Oak Spring Flora*, pp. 167–170).

When in 1673 Denis Dodart—Docteur Régent at the Faculty of Medicine in Paris and counsellor and physician to the king—was elected to the prestigious Académie Royale des Sciences, founded just seven years earlier, he became an energetic supporter of Perrault's original project. Through his efforts, in the manner of a prelude to the complete work, in 1676 a magnificent folio volume containing thirty-nine of Nicolas Robert's drawings—*Mémoires pour servir à l'Histoire des Plantes*—was published by the Imprimerie Royale (although the date 1765 appears in the colophon).

Dodart takes no personal credit for the *Mémoires*. His *Avertissement* begins with the declaration, 'Cet livre est l'ouvrage de toute l'Académie' and, to emphasize this point, the work opens with a magnificent frontispiece depicting Louis XIV on an official visit to the Académie (see frontispiece). In his preface, after explaining the aims of the project and noting that an herbal should provide accurate descriptions based on the study of real specimens, Dodart turns to a consideration of the botanical illustration. In his view, a correct drawing must show the plant in its natural dimensions; when this was not possible the artist could divide his portrayal into two sections. Sometimes it was also necessary to show certain parts, such as the flower, fruit, or seeds, separately in enlarged detail. Since it was important to convey a realistic idea of the plant's general appearance, the roots should be drawn as well.

Dodart noted that, while color was an essential element in the portrayal of a plant, printing in color was not possible, coloring by hand was quite expensive, and artists did not always possess the necessary skill for this exacting task. Therefore, the botanical etching should be prepared with great care, exploiting the capacity of this technique to suggest subtle gradations in hue through the use of various tones of black and gray. Finally, it was indispensable that the artist draw his plants *d'après nature*.

This lengthy excursus shows how much careful thought the author had devoted to the problem of the botanical illustration, reflecting an interest that was shared by many men of science and their patrons in this period. Louis XIV himself was persuaded by his minister Jean-Baptiste Colbert (1619–1683) to expand the royal collection of *vélins* (botanical and zoological drawings executed on fine vellum) and to nominate Nicolas Robert 'peintre ordinaire de Sa Majesté pour la miniature.'

Dodart then focuses his attention on more specifically scientific issues regarding the plants, their cultivation, and above all their virtues. The latter topic continued to interest readers and could not be neglected, even if the *Mémoires* had been conceived as an expression of the entire scientific and artistic culture in France under the reign of the Sun King. Significantly, Dodart affirmed that the properties of a plant should be determined 'par l'experience,' i.e., based on their verifiable effects rather than on accepted tradition, and expressed his doubts concerning della Porta's doctrine of signatures (see No. 51).

One novel and quite 'modern' aspect of this work is its discussion of the chemical reaction or the alterations that plant materials may undergo when heated, dissolved in water, or combined with other substances. The author evidently judged this topic to be of great importance for he treats it at some length, bringing to bear the weight of his immense expertise. The vignette preceding Dodart's account of the history of the project for the *Mémoires*—which was drawn and etched by Sebastien Leclerc the Elder (1637–1714), the author of the celebrated opening engraving of Louis XIV—is quite significant in this regard. It shows a group of academics watching a chemistry experiment, a scene denoting the most advanced research then being conducted in the field of botany and the author's rejection of a scientific tradition based on the passive acceptance of knowledge conveyed from the past, in a radical departure from most herbals of the period (see p. xlii).

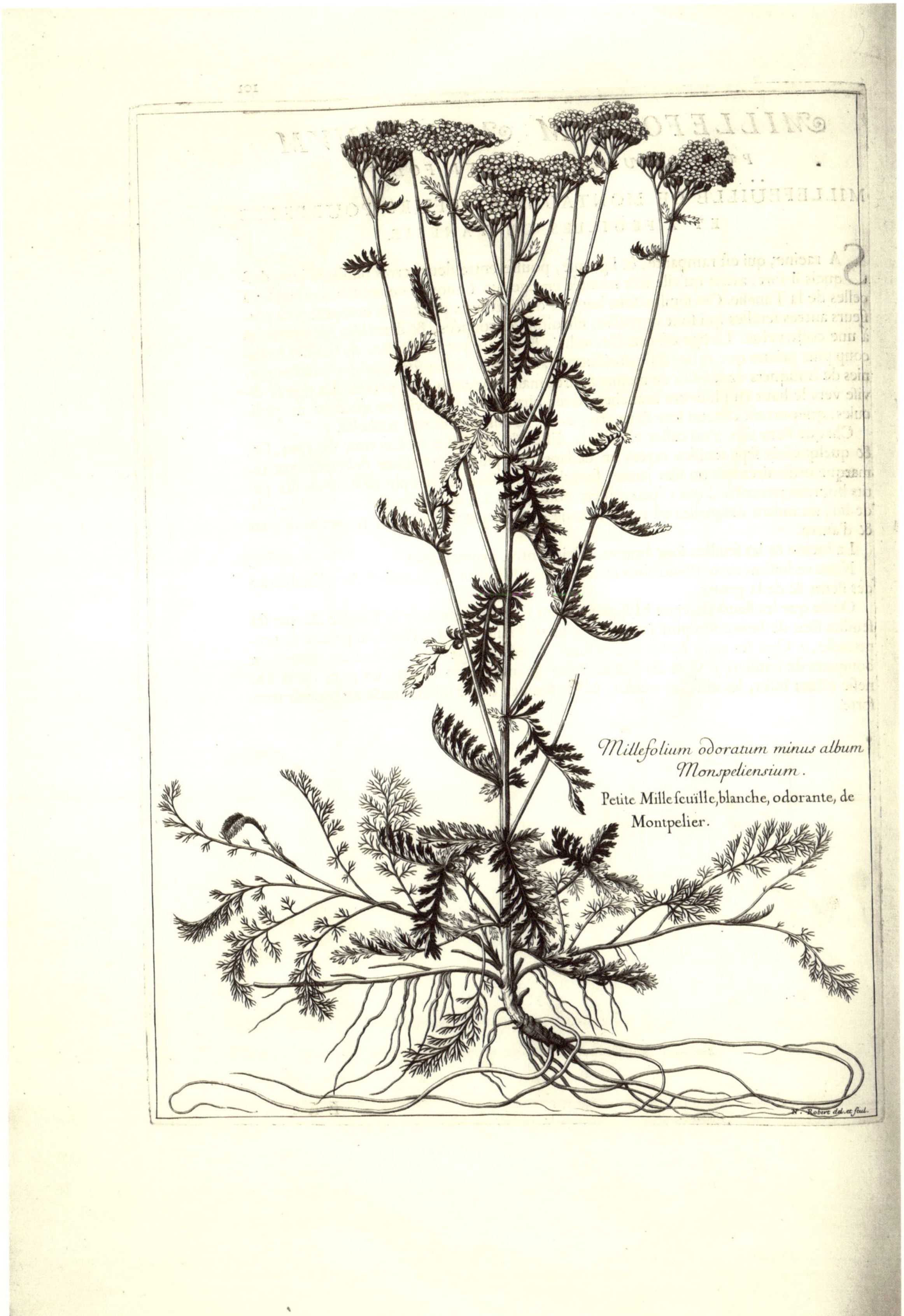

DENIS DODART,
Mémoires pour servir à l'Histoire des Plantes,
1676. 'Millefolium odoratum,' Yarrow (*Achillea millefolium*), EEe verso

What sets the *Mémoires* apart, however, are its illustrations, which are truly exceptional. They portray a series of unusual or exotic flowering plants arranged in alphabetical order by their Latin names, from *Angelica acadiensis* to *Urtica altera pilulifera*. The Latin name of the plant, followed by its common name in French, appear at the base of each engraving. Most of the illustrations were drawn and etched by Nicolas Robert, sometimes with the collaboration of the engraver Abraham Bosse. The artist no doubt benefited from the influence of Dodart in obtaining the prestigious commission to illustrate the academy's *Histoire des Plantes*, although he already had considerable experience as a botanical painter and his talent was sufficient to speak for itself. A charming flower book painted by him in 1645 for a wealthy patron, *Ghirlande de Julie*, attracted great admiration and praise. He then passed into the service of Gaston d'Orleans at the Château de Blois, where he began the series of large botanical paintings on vellum that would constitute his greatest achievement and the core of the royal collection.

The artist's unparalleled technique and sensitivity to natural phenomena emerge clearly in the thirty-nine etchings published in Dodart's *Mémoires*. Every one of them is a work of art of exceptional virtuosity, Robert demonstrating his ability to capture the delicate unfurling of a new leaf, an inflorescence of tiny flowers (as in his illustration of *Millefolium odoratum*), but also the opulence of more imposing species such as *Clematis americana*, the mass of stems and magnificent leaves of *Virga aurea mexicana*, and the rough stalks of *Scolymus chrisantemus*. Sometimes he shows a plant springing regally from the earth (*Rapuntium americanum*), while in others the roots are portrayed in meticulous detail, surprising us with the intricate beauty of their ramifications (*Sanicula sive Cortusa indica*).

A truly worthy monument to the reign of the Sun King, we cannot but agree with Wilfrid Blunt, who described the *Mémoires* as 'a landmark in botanical illustration' (p. 122).

45. Jan Commelin (1629–1692) *and* Caspar Commelin (*c.* 1668–1731)

[*Printed in red and black*] [*Volume I*]: Horti Medici Amstelodamensis Rariorum Tam Orientalis, quàm Occidentalis Indiæ, aliarumque Peregrinarum Plantarum Magno studio ac labore, sumptibus Civitatis Amstelodamensis, longà annorum Serie collectarum, Descriptio et Icones Ad vivum æri incisæ. Auctore Joanne Commelino Urbis Amstelodamensis (dum viveret) Senatore. Opus Posthumum. Latinitate donatum, Notisque & Observationibus Illustratum, à Frederico Ruyschio, M. D. Botanices Professore, &c. & Francisco Kiggelario. [*Printer's device*]. Amstelodami, Apud P. & J. Blaeu. nec non Abrahamum à Someren. M. DC. XCVII.

[*Printed in red and black*] [*Volume II*]: Horti Medici Amstelaedamensis Rariorum Tam Africanarum, quàm Utriusque Indiæ, aliarumque Peregrinarum Plantarum Magno studio ac labore, sumptibus Civitatis Amstelædamensis longa annorum Serie collectarum, Descriptio et Icones Ad vivum æri incisæ. Auctore Casparo Commelino M. D.

Et horti Medici Botanico. Pars Altera. [*Printer's device*]. Amstelædami, Apud P. & J. Blaeu. nec non Viduam Abrahami à someren. M. DCCI.

2° 39.3 × 25.4 cm. Two volumes. Volume I: π^8 $\star^2$ $2\star^2$ A–5P^2 *i–xxiv* 1–145 *146*–195 *196*–197 *198*–220 [4] (220 numbered pages interleaved with 220 unnumbered pages that contain 110 engraved plates) [468 pp.]. Volume II: π^8 $\star^2$ A–5X^2 *i–xx* 1–13 *14*–17 *18*–19 *20*–21 *22*–23 *24*–25 *26*–27 *28*–37 *38*–73 *74*–95 *96*–97 *98*–99 *100*–139 *140*–143 *144*–149 *150*–157 *158*–175 *176*–177 *178*–187 *188*–189 *190*–201 *202*–203 *204*–209 *210*–213 *214*–215 *216*–219 *220*–221 *222*–224 [4] (62 as 61, 65 as 66, 66 as 65, 186 as 182) (224 numbered pages interleaved with 224 unnumbered pages that contain 113 engraved plates) [472 pp.].

PLATES: Volume I: frontispiece, printer's device on Latin and Dutch title-pages, two coat-of-arms of 'D. Joannes Huydecoper,' and 'D. Joannes Commelinus,' 110 leaves of plants, all engraved and hand-colored. Volume II: frontispiece, printer's device on Latin and Dutch title-pages, three coats-of-arms of 'D. Franciscus De Vroede,' 'D. Joannes Huydecoper,' and 'D. Gerbrandus Pancras,' all hand-colored and heightened with silver and gold, 113 leaves of plants, all engraved and hand-colored.

BINDING: Late 18th- or early 19th-century tree calf with gilt tooling on covers and spines.

REFERENCES: Blunt & Stearn, pp. 153–4; Cleveland 304; Hunt 399; Nissen 389; Plesch 180–181; Pritzel 1,833; Stafleu & Cowan 1,187; Wijnands.

ONE OF THE MOST sumptuous botanical texts ever published in Holland was *Horti Medici Amstelodamensis,* which appeared in Amsterdam between 1697 and 1701. Its two magnificently illustrated volumes dedicated to exotic plants from India, Africa, and the New World constitute a veritable monument to the Amsterdam Physic Garden.

The authors—Jan and his nephew Caspar—were members of a Protestant family of printers and publishers by the name of Commelin (originally Commelyn or Commelijn), who had been forced to leave their native town of Douai in northern France and seek refuge from religious persecution in the Low Countries. Jan Commelin, a prominent merchant in pharmaceutical plants and other medicinal substances, eventually rose to become *commissarius* (director) of the *Hortus Medicus.* He filled Amsterdam's Botanical Garden with tropical plants discovered and sent back from Holland's colonies in the West Indies, Ceylon, the cape of South Africa, and the Malabar and Coromandel coasts of India. Some of these specimens were the first of their kind to be introduced in Europe.

Jan Commelin was the author of the original Dutch text of *Horti Medici Amstelodamensis,* but died in 1692 before it could be completed. It was edited and published posthumously by the celebrated physician and *Professor Botanicus* of the garden, Frederik Ruysch (1638–1731), together with a leading apothecary Frans Kiggelaer (1648–1722), who supplemented the text with their own information set apart by asterisks. Commelin's nephew Caspar, a physician and professor of botany who was appointed botanist of the *Hortus Medicus* in 1696, undertook to complete the work, writing the text to the volume on the flora of Africa, *Horti Medici ... Africanarum.* Caspar was nominated 'prælector exoticis' of the *Hortus* in 1701, lecturing on exotic plants, while Ruysch continued as lecturer on indigenous species. The text throughout is arranged in parallel columns of Latin and Dutch.

Volume I opens with an elegant frontispiece modelled on an iconographic conceit that recurs in many of the scientific texts published between the end of the seventeenth and

JAN AND CASPAR COMMELIN, *Horti Medici Amstelodamensis Rariorum … Plantarum*, 1697–1701. Frontispiece, volume 1

JAN AND CASPAR COMMELIN, *Horti Medici Amstelodamensis Rariorum … Plantarum*, 1697–1701. Jan Commelin's coat-of-arms, volume I

the first decades of the eighteenth centuries, in which a female figure is shown seated in state and accepting homage in the form of lavish gifts. In *Horti Medici Amstelodamensis* she represents the personification of the city of Amsterdam; at her feet are botanical tributes from the continents of Asia, Africa, and the Americas, while a row of tropical palm trees forms an elegant arcade behind her.

The frontispiece is followed by two title-pages—one in Latin and one in Dutch—and two imposing coats-of-arms, the first belonging to the family of Joan Huydecoper van Maarseveen, who served with Jan Commelin as co-director of the botanical garden, and the second to Jan Commelin himself, *commissarius dum viveret* (for life). A note addressed to the reader cites Holland's outstanding tradition of botanical study, embodied in such illustrious figures as Carolus Clusius and Rembert Dodoens, and underlines the importance of the founding of the *Hortus Medicus* (in 1682 by a decree of the city council) as confirmation not only of this tradition but also of Amsterdam's position as the capital of the Low Countries and a trading power.

The second volume, which was entirely prepared by Caspar Commelin, opens with a frontispiece depicting a figure from Antiquity (perhaps Theophrastus or Dioscorides) lecturing in Amsterdam's Botanical Garden to a group of students and colleagues, some of whom are engaged in taking notes. In the background extends an orderly range of flowerbeds and greenhouses where the garden's collection of exotic plants was cultivated. This is followed by two title-pages, and the coats-of-arms of three prominent directors of the *hortus*: the senator François de Vroede, Joan Huydecoper, and Gerbrand Pancras.

Horti Medici Amstelodamensis is illustrated with 220 large engravings of exotic species, presented in alphabetical order. Carefully composed and drawn with great realism, they are striking in their visual impact. Often the plants are shown growing out of the ground in a naturalistic setting, with details such as seeds or flowers scattered around them or a colorful butterfly hovering above. If a species was too large to fit on the page, it might be portrayed in sections—first the base and then, beside or in front of it, the flowering stem. The artist sometimes took liberties or displayed a certain ingeniousness in his compositions, for example taking a cut stem and 'planting' it in the ground to make it appear more life-like, or adapting the curve of a stem or leaf to fit his composition, giving priority to artistic considerations rather than scientific accuracy.

The original watercolor paintings on which these engravings were based may be found in the library of the Botanical Garden of Amsterdam, conserved in eight volumes of the famous *Moninckx Atlas*. These contain the drawings of 420 plants from the garden commissioned by Jan and Caspar Commelin beginning in 1686. They were in large part the work of Jan Moninckx (d. 1714), who completed about 270 paintings between 1687 and 1708 with the help of his daughter Maria (1673?–1757). Two other artists collaborated

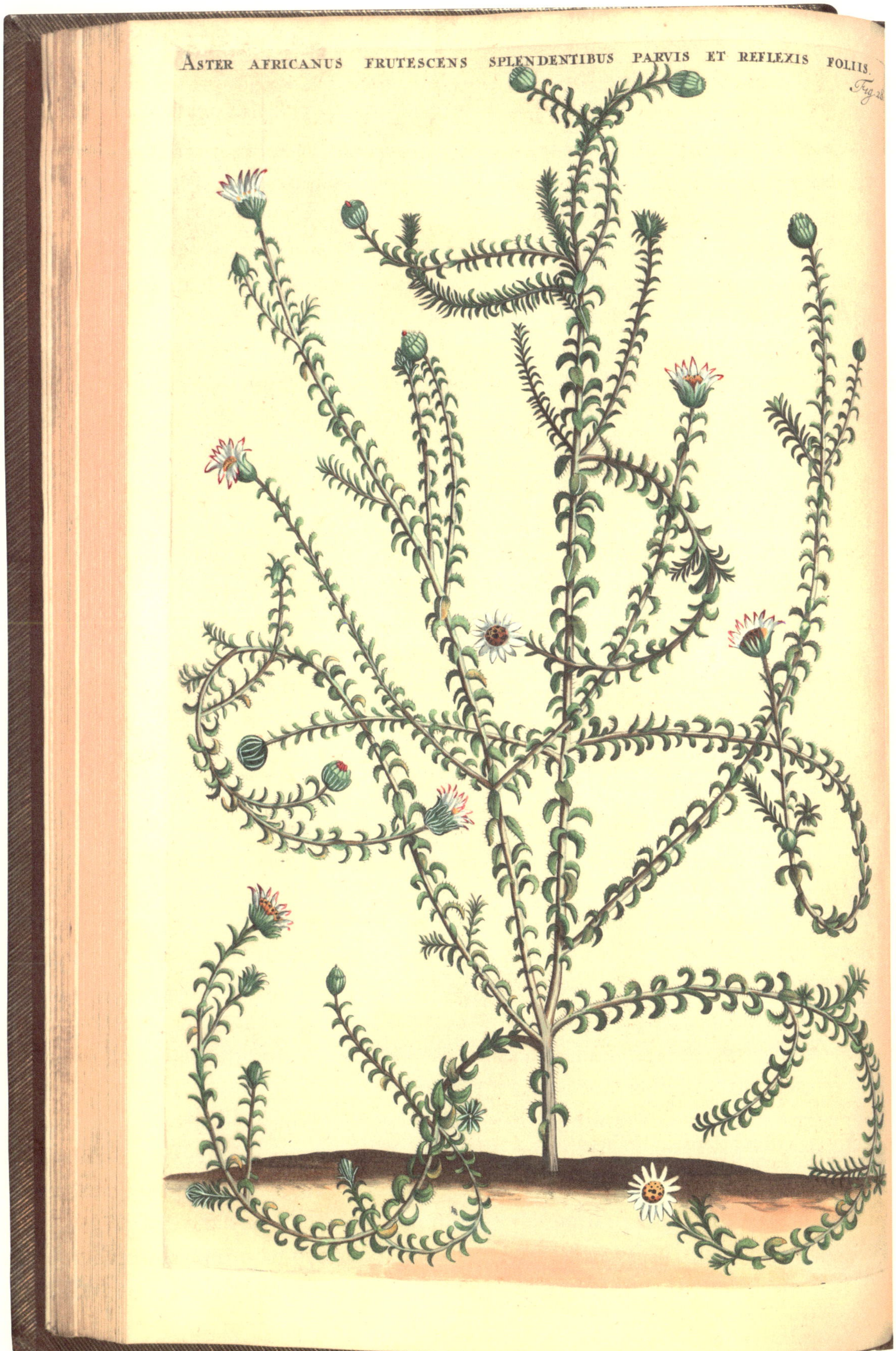

JAN AND CASPAR COMMELIN, *Horti Medici Amstelodamensis Rariorum … Plantarum*, 1697–1701. 'Aster Africanus,' possibly a member of *Polyarrhena* or *Felicia*, South African Compositae, volume II, plate 28

on the project—Johanna Helena Herolt (b. 1668), daughter of the botanical painter Maria Sibylla Merian (see *An Oak Spring Flora*, No. 101, pp. 380–384), and Alida Withoos (1659/60–1730), the sister of Mattias Withoos (see *An Oak Spring Flora*, No. 21, pp. 87–89).

The engravings vary greatly in quality; the second volume, for example, contains a fine series of aloe plants, whereas other works appear somewhat rigid or artificially composed, such as the 'Frutex africanus aromaticus' (Pl. LIX) and the 'Aster africanus frutescens' (Pl. XXVIII), whose stems curve in improbably perfect circles. Particularly beautiful is the *Planta lactaria Africana* (Vol. I, Pl. XVII).

The principal aim of this herbal was to document the collection of exotic species being cultivated in the botanical garden of Amsterdam. The Oak Spring Garden Library copy of *Horti Medici Amstelodamensis* has been colored by hand with great skill, transforming this exceptional printed book into a genuine work of art.

46. Carolus Linnaeus (Carl von Linné) (1707–1778)

[*Printed in red and black*]: Hortus Cliffortianus Plantas exhibens quas In Hortis tam Vivis quam Siccis, Hartecampi in Hollandia, Coluit Vir nobilissimus & Generosissimus Georgius Clifford Juris Utriusque Doctor, Reductis Varietatibus ad Species, Speciebus ad Genera, Generibus ad Classes, Adjectis Locis Plantarum natalibus Differentiisque Specierum. Cum Tabulis Æneis. Auctore Carolo Linnæo, Med. Doct. & Ac. Imp. N. C. Soc. [*engraved and etched vignette*] [*rule*] Amstelaedami. 1737.

2° 41.9 × 25.1 cm. π^4 $\star^2$–$6\star^2$ $7\star^2$ ($7\star^2$ as $9\star^2$) $8\star^2$ ($8\star^2$ as $7\star^2$) $8\star^2$ A–3M² 4G–6P² (-6L2) *i–xxxiv* I–X 1–231 *232* 301–501 [*17*] [494 pp.].

PLATES: Frontispiece (signed 'J. Wandelaar inv. et fecit 1738'), vignette on title-page (signed 'A. vander Laar inv. et fecit 1737'), and 36 leaves of plants, by Georg D. Ehret (1708–1770) and Jan Wandelaar (1690–1759), all engraved and etched.

BINDING: Contemporary speckled Cambridge calf with blind tooling on front and back covers; rebacked with original gilt tooled leather.

PROVENANCE: Bookplate of Prince of Liechtenstein and Rachel McMasters Miller Hunt (1882–1963).

REFERENCES: Calmann, pp. 45–51; Cleveland 387; Dunthorne 186; Hunt 504; Nissen 1215; *An Oak Spring Flora*, p. 187; *An Oak Spring Garland* 17; Pritzel 5,408; Sitwell & Blunt, pp. 31, 115; Stafleu & Cowan 4,719.

According to Hunt (II, p. 173) the three works *Hortus Cliffortianus*, *Genera plantarum* (1737), and *Species plantarum* (1753), in which the Swedish botanist Carolus Linnaeus (1707–1778) laid out his principles for the naming and classification of plants, constituted one of the most fundamental contributions to the botanical literature of all time. *Hortus Cliffortianus* offered Linnaeus his first opportunity to compile a catalogue of plants in accordance with his own system of scientific criteria. This project was sponsored by the wealthy Anglo-Dutch banker George Clifford (1685–1760), who owned a garden with a large collection of rare plants (acquired through his connections with the Dutch East

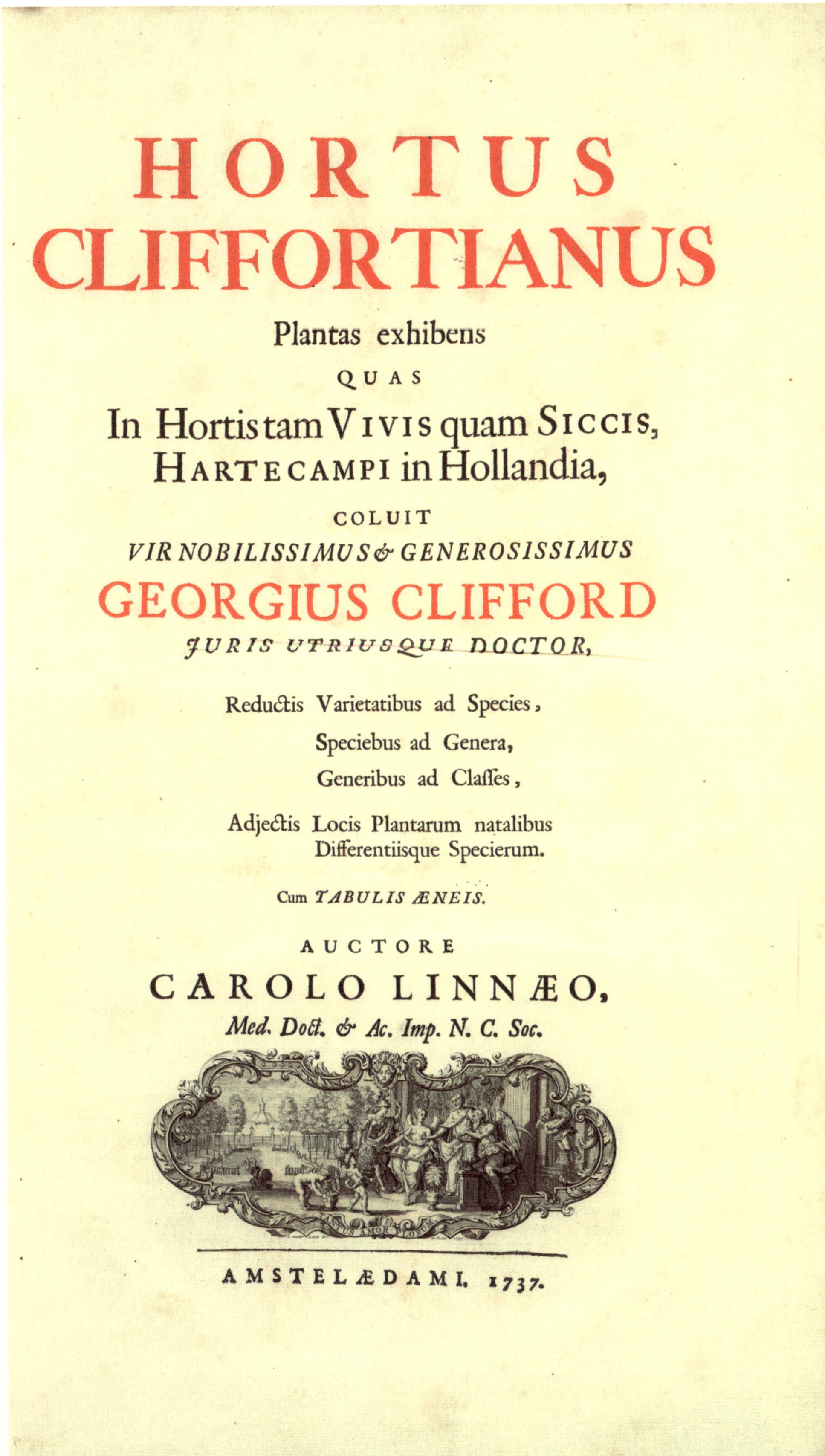

HORTUS
CLIFFORTIANUS
Plantas exhibens
QUAS
In Hortis tam VIVIS quam SICCIS,
HARTECAMPI in Hollandia,
COLUIT
VIR NOBILISSIMUS & GENEROSISSIMUS
GEORGIUS CLIFFORD
JURIS UTRIUSQUE DOCTOR,

Reductis Varietatibus ad Species,
Speciebus ad Genera,
Generibus ad Classes,

Adjectis Locis Plantarum natalibus
Differentiisque Specierum.

Cum *TABULIS ÆNEIS.*

AUCTORE
CAROLO LINNÆO,
Med. Doct. & Ac. Imp. N. C. Soc.

AMSTELÆDAMI. 1737.

CAROLUS LINNAEUS,
Hortus Cliffortianus,
1737. Title-page

India Company) in Hartekamp near Haarlem. It was at Clifford's home that the brilliant scientist met the gifted botanical artist Georg Dionysius Ehret (1708–1770), an encounter that would lead to a remarkably fertile collaboration. As William Stearn noted, this event 'marks the beginning of a new era in botanical illustration and foreshadows the golden century of great flower-book production which extended from 1760 to 1860' (Introduction to the facsimile of *Genera Plantarum*).

Hortus Cliffortianus opens with an elegant frontispiece depicting a crowned figure, perhaps the personification of The Netherlands, with Apollo on her right and figures representing the continents of Europe, Africa, and Asia bearing floral tributes on her left. In the background is a garden filled with exotic plants and presided over by a herm of Janus (or perhaps Asclepius), on whose pedestal is an image of the serpent Oroborus biting its tail, a symbol of eternity. This work was drawn and etched by Jan Wandelaar (1690–1760), a somewhat well-known Dutch artist, and bears the date 1738, indicating that the herbal took two years to complete.

Linnaeus dedicated his work to Clifford, whose enlightened patronage did so much to further the botanical sciences in this period. The author's gratitude is expressed in the closing words of the acknowledgement that follows the title-page:

> Floreat hic Tuus Hortus, reddatque fructus non minus gratos, quam aliorum Magnatum suis Patronis & Tibi erit impensus labor, sumptus & cura soluta! Floreas in Tuo Horto in Nestoreos annos! Floreas ex tuo horto post fata, per saecula! Vivas, qui supervixeris UNICUS BOTANIORUM MAECENAS
>
> (May your garden flourish and bear fruit no less appreciated than are to their protectors the gardens of other great men and, thanks to you, [our] exertions and expenses will be recompensed and [our] worries resolved. May you flourish in your garden to the venerable age of Nestor. May your garden flourish after you and for centuries to come. For as long as you live, you will remain the GREATEST PATRON OF BOTANISTS).

This dedication is signed with the author's name and the words 'Ex Musaeo Cliffortiano 1737.'

Linnaeus then addresses a word to the 'Botanist reader,' explaining how he had structured the descriptions in this catalogue of the plants in his patron's garden; first the Species was identified, then its Synonyms, Variety, Specific Names, the Places where it could be found, and finally an accurate description was provided. Furthermore, the plants were grouped according to their provenance—Africa, the Americas, Asia, and southern Europe.

The first two plates are devoted to the wide variety of leaf forms that could be found in nature, first the 'Folia simplicia' and then the 'Folia composita' and 'Folia determinata,' presented in simple outline drawings well suited to their didactic purposes. The next two

CAROLUS LINNAEUS,
Hortus Cliffortianus,
1737. *Kaempferia* sp.,
plate 3

plates, of the *Kaempferia* and the pepper plant, were drawn and engraved by Jan Wandelaar, while the remaining thirty illustrations were almost all drawn by Ehret and etched with great virtuosity by Wandelaar. Of particular interest are the illustrations of two *Cliffortiae* (Pls. 30 and 31), a plant that Linnaeus named in honor of his patron, and an elegant work portraying the *Hura* or sandbox tree (Pl. 34), a tall species native to Mexico and Central America. Behind the drawing of the plant appears a full-plate, enlarged detail of the leaf, realized with a thin, ghosted line.

The illustrations, which have not been colored by hand in this copy, are of exceptional quality—beautifully composed and drawn by Ehret, who had no rival as a botanical artist, and meticulously translated into etchings by Wandelaar. In some cases small numbers or letters have been placed next to specific parts of the plant, referring the reader to the notes that appear at the foot of the page after the plant's Latin name. This practice reflects the rigorous scientific approach of the *Hortus Cliffortianus*, which must be considered not only an outstanding work of art, but also a genuine milestone in the history of the botanical sciences.

47. Giovanni Battista Morandi (*fl.* 1717–1751)

[*Floral border surround with coat-of-arms at bottom. Within ribbon before the title*]: Divinitas, Utilitas, Et Delectatio. Collectio Plantarum quae in diversis Europae, Asiae, Africae, et Americae locis nascuntur graphice delineatarum, expressarumque à Joanne Babtista Morandi Mediolanensi Anno MDCCXXXVII una cum varijs illarum nominibus juxta celeberrimos Rei Herbariae Scriptores. T.

45.7 × 30.3 cm. 1–108 leaves, numbered in modern hand, with drawings of plants, plus manuscript text.

PLATES: 85 leaves of plants, 12 leaves with indexes of plants surrounded by a landscape border, all in pen-and-ink and gouache.

BINDING: Modern green morocco, bound by Zaehnsdorf, London, England.

REFERENCE: Tongiorgi Tomasi, 2003 b.

THE MILANESE ARTIST Giovanni Battista Morandi is known for having been the first officially appointed painter at the Botanical Garden of Turin. He was unusually qualified for this position, because to his artistic activity he joined a considerable experience in specifically scientific activities. In 1744 he published a botanical text, *Historia botanica practica*, which was so well received that it was reprinted in 1761. The original manuscript is conserved at the Natural History Museum in London, and on its title-page the author describes himself as a 'botanico-galenico-pittore' (botanist—follower of Galen—artist).

The Botanical Garden of Turin was constructed near the Castello del Valentino in 1729 at the behest of Victor Amadeus II of Savoy, and its first director was Giovanni Bartolomeo Caccia (?–1749), the professor of botany at the university's faculty of medicine.

GIOVANNI BATTISTA MORANDI, *Collectio Plantarum*, 1737. Title-page with specimens of *Gagea, Viola, Vinca, Knautia, Scilla*, folio 1

For his teaching and research activities Caccia regularly commissioned botanical drawings from artists, a function that Morandi would officially fill from 1732 to 1741. When he arrived in the capital city, Morandi was well versed in this genre of painting. In the short treatise *Risposta del cavalier Giovambattista Morandi* published in 1743—as a parentheses to an acrimonious debate set off by his work *Historia botanica practica* among the herbalists of Milan—he mentions that twenty years earlier, in 1723, he had already compiled a *Miscellanea Plantarum*, and that in 1725 in Rome he produced a volume of *Fiori Ortensi* at the request of an illustrious patron, Cardinal Pietro Ottoboni. Furthermore, conserved at the Collegio Alberoni in Piacenza are three volumes of botanical drawings dating from 1731 that include exotic species from Malabar.

We know that during his tenure at the Botanical Garden of Turin, Morandi completed no less than three hundred botanical paintings that were assembled in three volumes: *Exterarum, et rariorum plantarum, quae in Horto Regio Acadamia Taurinensis excoluntur, imaginae ad vivum expressae a Joanne Baptista Morandi Mediolanensi.* This valuable collection was stolen from the library of the Botanical Garden in 1979 and has never been recovered.

Exterarum, et rariorum plantarum may be considered one of the first body of works in a vast collection of botanical paintings projected by the *prefetto* Caccia and continued by his successors at the Botanical Garden over a period of more than one hundred years—the *Iconographia Taurinensis*. This titanic undertaking, which began in 1752 and was completed in 1868, saw the realization of 7,500 paintings in gouache and watercolor that were bound in sixty-four imposing volumes. The modest beginnings of this remarkable collective effort, in which many artists participated, may be traced back to another botanical painter, Giovanni Ignazio Francesco Peyroleri, who worked at the Botanical Garden of Turin until 1732, first as an *erbolaio* (herbalist) and then as a draftsman. He was succeeded by Morandi who, as we have seen, was appointed as the first official botanical artist at the garden.

There is no doubt that the experience acquired in Turin was invaluable to Morandi for it allowed him to perfect the skills needed for this highly specialized genre. His own account in *Risposta* of his earliest attempts are quite revelatory for they show that from the very outset he adopted a rigorously scientific approach to the portrayal of botanical specimens: 'Questa pianta fu da me in Torino fedelemente delineata, e dipinta dal vero col suo fiore, secondo la sua naturale grandezza, ed ancora fresca di sugo l'ho incisa in diversi modi, e le feci l'Anatomia' (This plant was faithfully sketched by me in Turin, and painted from life with its flower, according to its natural size, and still fresh and full of sap I cut it in different ways, and studied its Anatomy).

When he returned to Milan, Morandi continued to devote himself energetically to his art, producing many *corpora* of paintings: 1,200 works completed between 1725 and 1748 and acquired in 1756, after the artist's death, by the Biblioteca Ambrosiana of Milan, are

GIOVANNI BATTISTA MORANDI, *Collectio Plantarum*, 1737. Index leaf, folio 16, with Morandi's signature

today bound in five volumes; the thirteen volumes of *Plantarum Icones*; and the two volumes of *Miscellaneae Plantarum* now in the Biblioteca del Museo Civico di Storia Naturale of Milan. The honorary decoration of 'Cavaliere Milanese' was conferred upon the artist, circa 1725.

The engraved plates that illustrate Morandi's book *Historia botanica practica* were the product not only of his brush, but also of his—not always entirely successful—essays in the art of engraving. This is how he recounts his experience in *Risposta*:

> Ogn'Erba ha una sua figura, la quale fù prima da me dipinta dal vero secondo la sua natural grandezza, indi l'ho delineata, e ridotta ad una grandezza intelliggibile, e non sapend'io per anco incidere, mi raccomandai tanto a Dio benedetto, ed al mio grande Protettore S. Giuseppe, che senza Maestro in brevissimo tempo imparai ad intagliare in Rame, che dippoi colle mie mani per maggior sicurezza l'ho incisa in rame, con appresso la figura di un Fiore, un Frutto, o di una foglia di esso Fiore, ò un seme alla grandezza naturale, perchè con maggior facilità si possa distinguere il genere, e la Specie di quell'Erbe da adoperarsi nella medicina...
>
> Every herb has its picture, which was first painted by me from life in its natural dimensions, then afterwards I drew it, and reduced it to an intelligible size, and since I did not know the least thing about engraving, I entrusted myself entirely to my blessed God, and to my great protector Saint Joseph, and in this way without the benefit of a Teacher in a very short time I learned to engrave in Copper, after which with my own hands for greater reliability I engraved [the herb] in copper, with beside it the figure of a Flower, a Fruit, or a leaf of that Flower, or a seed shown life-sized, so that with greater facility one may distinguish the genus, and the Species of that Herb that is to be used in medicine ....

Sources, furthermore, mention that the artist also produced an illustrated manuscript in thirteen volumes for the prestigious Collegio Ghislieri of the University of Pavia, of which all trace has been lost *ab antiquo*. It is known, however, that it bore the title *Collectio Plantarum quae in diversis Europae, Asiae, Africae, Et Americae locis nascuntur graphice delineatarum, expressarumque À Joanne Babtista Morandi Mediolanensi Anno MDCCXXXVII una cum varijs illarum nominibus juxta celeberrimos Rei Herbariae Scriptores*, which coincides exactly with the title that appears on the first page of the manuscript at the Oak Spring Garden Library. The possibility that this manuscript constitutes the first volume of the lost work is supported by the presence of the coat of arms of the Collegio Ghislieri on the elegant title-page. The title is framed by undulating stems of flowers that start beneath the escutcheon and gracefully meander up the sides to the top where they intertwine with a scroll bearing the words 'Divinitas, Utilitas et Delectatio' (Divinity, Utility, and Delight). It is not unlikely that the escutcheon, with its diagonal bands of red and yellow crowned

GIOVANNI BATTISTA MORANDI, *Collectio Plantarum*, 1737. Blood lily (*Haemanthus* sp.), folio 30

with regalia and papal and military insignia at the foot of the page, was that of the author himself, as Morandi was awarded the decoration of 'Cavalier Milanese.' In addition, it is known that the artist had finished the manuscript for the Collegio well before 1744, the year in which *Historia botanica practica* was published. The author mentions the thirteen-volume herbal in his *Historia*, and the date 1737 which appears in the Oak Spring Garden Library manuscript would be entirely consistent with this attribution.

The manuscript contains eighty-five botanical drawings, and twelve index leaves which mark off the twelve sections into which the herbal has been divided: the plant genera *Genista, Spartium, Viola, Lilium, Ephemeron, Linaria, Scabiosa, Scilla,* and *Liliastrum*. On each drawing the name of the plant is written in elegant calligraphy, and the parts shown in enlarged detail have been carefully labelled. The author also evidently consulted the works of many of the greatest authorities in the area of botany, for there are repeated references to Caspar Bauhin, Jean Bauhin, Konrad Gesner, Carolus Clusius, Joseph Pitton de Tournefort, Hermann Boerhaave, and others. Each plant is portrayed with a cut stem and beside it details showing particulars of the roots (to which the artist dedicates special attention), fruit, and seeds. The gouaches and watercolors have been applied in dense, emphatic brushstrokes that entirely cover the brown ink underdrawing.

A certain number of drawings occupy a double-page, in many cases so that the artist could portray the natural dimensions of the plant: *Lilium album* (fols. 35^v and 36^r); *Lilium album* (fols. 39^v–40^r), which was cut from another sheet and glued on; *Scabiosa alpina* (fols. 66^v–67^r); *Scabiosa alpina* (fols. 69^v–70^r); *Scabiosa pyrenaica* (fols. 78^v–79^r); *Scilla radice alba* (fols. 90^v–91^r); *Scilla vera* (fols. 92^v–93^r); *Scilla africana* (fols. 94^v–95^r); *Scilla vulgaris* (fols. 96^v–97^r); and *Liliastrum* (fols. 100^v–101^r).

Collectio Plantarum is distinguished by a most pleasing particularity: each group of plants is preceded by a title-page bearing a long scroll that lists the species that are illustrated in that section. The scroll is presented *à trompe l'oeil* against a varied series of backgrounds: a landscape vista with buildings and rivers, or framed by the wall of what might be a garden, or against a background of architectonic elements, such as sections of columns or marble fragments. The twelve title-pages have been signed by the artist, who reiterates each time that his plants were painted from life (*fideliter, ad vivum*). Indeed, the natural proclivity for research and teaching that made up an essential element in Morandi's character wedded itself harmoniously with his artistic propensities.

The artist's taste for decorative effects found elegant expression in the drawing of a stem of *haemanthus*, African Blood Lily, with its crown of fruit, set off by a fluttering blue ribbon (fol. 30^r). The same decorative approach characterizes the two-volume painted herbal *Miscellaneae Plantarum* dating from 1744 cited above, in which the artist chose to depict more than one species on each page, and by a manuscript presenting a fanciful alphabet constructed from botanical elements that was sold on the antique market some years ago and whose present whereabouts are unknown.

A portrait in watercolors of Giovanni Battista Morandi can be found in the Iconoteca Bassiana at the Department of Biology of the University of Bologna.

48. Giorgio Bonelli (1724–1782)

[*Printed in red and black*] [*Volume I*]: Hortus Romanus juxta systema Tournefortianum paulo strictius distributus a Georgio Bonelli Monregalensi in Subalpinis, Publico Medicinae Professore: Specierum nomina suppeditante, praestantiorum, quas ipse selegit, adumbrationem dirigente Liberato Sabbati Maevaniensi in Umbria Chirurgiae Professore, et Horti Custode; Adjectis unicuique volumini Rariorum Plantarum Tabulis C. aere incisis Tom. I. [*Engraved image*] Romae MDCCLXXII. Sumptibus Bouchard Et Gravier. [*rule in red and black*] Ex Typographia Pauli Junchi. Superiorum Permissu.

[*Printed in red and black*] [*Volume II–V*]: Hortus Romanus secundum systema J. P. Tournefortii a Nicolao Martellio Aquilano Medico Doctore Linnaeanis characteribus expositus adjectis singularum plantarum analysi ac viribus: [*red double rule*] Species suppeditabat ac describebat Liberatus Sabbati Mevanias ejusdem Horti Custos[.] et Chirurgiae Professor[.] Accedunt Tabulae Centum propriis plantarum coloribus expressae. Tom. II [–V]. [*Engraved image*]. Romae MDCCLXXIV [–DCCLXXVIII]. Sumptibus Bouchard, Et Gravier. [*rules in red & black*] Ex Typographia Pauli Junchi Superiorum Permissu.

[*Printed in red and black*] [*Volume VI–VIII*]: Hortus Romanus secundum systema J. P. Tournefortii a Nicolao Martellio Aquilano inter Archigimnasii Romani Professores Botanicæ Practicæ Lectore ejusdemque Horti Præfecto Linnæanis Characteribus expositus adjectis singularum plantarum analysi ac viribus. Species suppeditabat ac describebat Constantinus Sabbati Mevanias ejusdem Horti Custos, et Chirurgiæ Professor[.] Accedunt Tabulæ Centum propriis plantarum coloribus expressae. Tom. VI [–VIII]. [*Engraved image*]. Romæ MDCCLXXX [–MDCCXCIII]. Sumptibus Bouchard, Et Gravier. [*double rule in red & black*] Ex Typographia Pauli Junchi [–(Volume VIII): Joannis Zempel]. Superiorum Permissu.

2° 54.9 × 37.6 cm. Eight volumes. Volume I: π^1 $2\pi^1$ $\star^1$ $2\star^1$ a–d^1 B–P^1 *i–viii* *I–VIII* 3–30 [44 pp.]. Volume II: π^1 $2\pi^1$ a–d^1 A–L^1 *i–vi* *I–VI* 1–22 [34 pp.]. Volume III: π^1 $2\pi^1$ a–b^1 A–I^1 *i–iv* I–IV 1–18 [26 pp.]. Volume IV: π^1 $2\pi^1$ $3\pi^1$ A–L^1 *i–vi* 1–22 [28 pp.]. Volume V: π^1 $2\pi^1$ c^1 A–K^1 *i–vi* 1–20 [26 pp.]. Volume VI: π^1 $2\pi^1$ a^1 A–I^1 *i–vi* 1–18 [24 pp.]. Volume VII: π^1 $2\pi^1$ A–I^1 *i–iv* 1–18 [22 pp.]. Volume VIII: π^1 $2\pi^1$ A–H^1 *i–iv* 1–15 *16* [20 pp.].

PLATES: Double-page etching in Volume I entitled 'Prospectus Horti Romani,' printed in green, and signed 'Andrea de Rossi fe.'; five engraved and etched portraits of cardinals in volumes 1–4 and 6; 800 hand-colored leaves of plants (100 in each volume) with blue and purple wash borders, engraved by Maddalena Bouchard, daughter of the publisher, after drawings by Cesare Ubertini, all engraved, etched, and hand-colored. Each volume's dedication begins with an etched initial.

BINDING: Contemporary Italian quarter-bound vellum and geometric-pattern printed boards; in marbled paper sleeves.

PROVENANCE: Robert de Belder; bookplates of Ewald Schlundt in each volume, and of Arpad Plesch, 1899–1974, in Volume I.

REFERENCES: Cleveland 509; Dunthorne 45; Hunt 629; Nissen 200; *An Oak Spring Flora*, No. 28; Pirotta & Chiovenda; Staffleu & Cowan 634.

THE MONUMENTAL HERBAL *Hortus Romanus* is devoted to the plants that were being cultivated in the Botanical Garden of Rome during the last decades of the eighteenth century. This garden was created three centuries earlier at the behest of Pope Nicholas V (1447–1455), but was moved to different sites within the walls of the Vatican before Pope

Alexander VII (1665–1667) ordered in 1660 that a garden on a much larger scale worthy of the prestige of the Holy See be constructed on Janiculum Hill, where it could also serve as the teaching garden of the city's university.

Under the direction of Giovanni Battista Trionfetti (1656–1708), an illustrious botanist who added more than three thousand species to its collection, the renown of the Vatican's garden spread to every corner of Europe. A period of relative inactivity followed as series of lesser-known figures were appointed to succeed Trionfetti, but the situation changed once again with the arrival of Pope Benedict XIV (1740–1758), who took a renewed interest in the garden and designated Francesco Maratti (1748–1777) as its *prefetto* (director). Maratti was a member of the Benedictine abbey of Vallombrosa, which was during this period one of the most active centers of scientific culture in Europe. He was followed by Niccolò Martelli (1734–1829) of Aquila, whose tenure lasted from 1777 to 1803 and who succeeded Giorgio Bonelli as editor of the *Hortus Romanus*.

Indeed, although authorship of the *Hortus Romanus* has generally been ascribed to Bonelli, the herbal took twenty years to complete and he was actually responsible for only the first volume. The greater part of the work could not have been realized without the participation of a series of botanists, beginning with Liberato Sabatti (1714–1779) from the Umbrian city of Bevagna, and his son Costantino, both of whom served as directors of the Botanical Garden and during their respective tenures planted many new species that are described and illustrated in the herbal.

The moving spirits behind the *Hortus Romanus* were not the botanists at the garden, but a French publisher Giovanni Bouchard (d. 1790)—who published *Carceri* (1745) by Giovanni Battista Piranesi—and his partner Giovanni Giuseppe Gravier, who also bore the considerable expense of its publication. Bouchard and Gravier had both emigrated to the Eternal City from the province of Dauphiné in southeastern France, and together set up a bookshop in Via del Corso that was much frequented by the intellectuals of Rome. It was they who penned the opening dedication addressed to Pope Clement XIV, whose portrait—an etching in blue ink—follows the title-page to the first volume. All eight volumes were printed by Paolo Junch at his typographic works in Rome.

Very little is known about Giorgio Bonelli, except that he was a physician born in Mondovì, Piedmont, who studied in Turin under the illustrious botanist Giovanni Bartolomeo Caccia. Bonelli moved to Rome in 1757, where it appears that he taught medicine for a period at the university; his only known published work is *Memoria sull'oglio di ricino* (Some notes on ricin oil), which appeared in Rome in 1782.

The first volume of *Hortus Romanus* begins with an introduction in which the author Bonelli explains that he was invited to participate in this project by its publishers, Bouchard and Gravier. They had already decided that the artist Cesare Ubertini should execute the

HORTUS
ROMANUS
JUXTA SYSTEMA TOURNEFORTIANUM
PAULO STRICTIUS DISTRIBUTUS
A
GEORGIO BONELLI
MONREGALENSI IN SUBALPINIS,
PUBLICO MEDICINAE PROFESSORE:
SPECIERUM NOMINA SUPPEDITANTE,
PRAESTANTIORUM, QUAS IPSE SELEGIT, ADUMBRATIONEM DIRIGENTE
LIBERATO SABBATI
MAEVANIENSI IN UMBRIA
CHIRURGIAE PROFESSORE, ET HORTI CUSTODE;
ADJECTIS UNICUIQUE VOLUMINI RARIORUM PLANTARUM
TABULIS C. AERE INCISIS
TOM. I.

ROMAE MDCCLXXII.
SUMPTIBUS BOUCHARD ET GRAVIER.

EX TYPOGRAPHIA PAULI JUNCHI.
SUPERIORUM PERMISSU.

GIORGIO BONELLI,
Hortus Romanus,
1772–1793. Title-page, volume 1

1 Ingressus ad superiorem Horti partem.
2 Aedes, in quibus Theatrum, Musaeum, et Cubicula Custodis.
3 Ostium Theatri.
4 Ostium Musaei.
5 Musaeum, et Cubicula Custodis.
6 Clathri.
7 Muri Hortum sepientes.

8 Gradus lateritii pro Pl. Exoticis.
9 Muri Sessiles pro elegantioribus Pl.
10 Ambulacra.
11 Hydrophyllacium pro Aquaticis.
12 Fontes.
13 Stylobatae pro majoribus fictilibus.
14 Quadrae cum ambulacris, semitis, et pulvinis, ubi aluntur Stirpes.

Methodo T
14. ad 35.
36 Pergulae
37 Aquæductu
38 Ingressus a
39 Quadrae,
Classium
40 Muri sessi

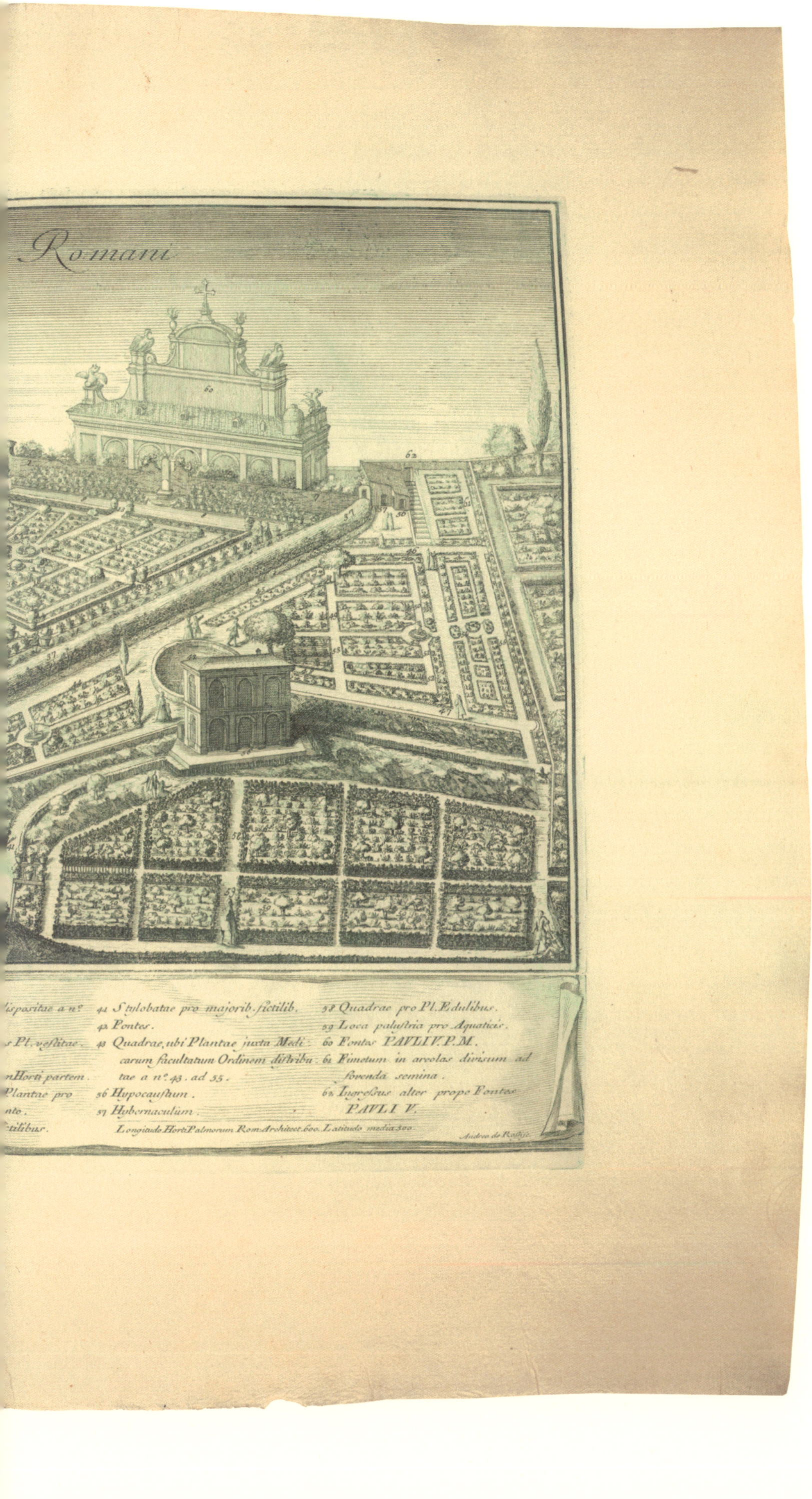

GIORGIO BONELLI,
Hortus Romanus, 1772–1793. 'Prospectus Horti Romani,' volume I

drawings, on the basis of which Bouchard's daughter Maddalena (who was enrolled in the Accademia dell'Arcadia under the name of Cleonice Aphrodisia) would prepare the etchings for the printed work. From the introduction it would also appear that the decision to classify the plants in the herbal using the system of Tournefort was made not by Bonelli, but by the two French editors. When Niccolò Martelli—who was a much more competent botanist than his predecessor—joined the project, he insisted that the more up-to-date and scientific binomial system of Linnaeus be adopted instead.

Not surprisingly, this long and complex work had a somewhat involved publishing history; the first five volumes appeared in 1772, while the remaining three were printed in 1780, 1784 and 1793, respectively. Complete sets are rare. It is furthermore uncertain whether all eight hundred plates were the work of Ubertini and Bouchard, the artists originally engaged on the project.

Hortus Romanus met with a cool reception on the part of experts, who found it to be extremely weak from a scientific standpoint; for one thing, most of the species included were quite well known or even commonplace and therefore of little interest to the botanist. In addition, they judged the illustrations to be stiff, unrealistic, and poorly executed. The first volume of *Hortus Romanus juxta systema Tournefortianum* nevertheless includes a fine etching by Andrea De Rossi, 'Prospectus Horti Romani,' which, spread over two pages and printed in green ink, offers a bird's eye view of the garden situated on the slope of Janiculum Hill. The title-page is decorated with a charming vignette of unknown authorship showing flower beds and terracotta vases filled with blooming plants, from what was perhaps an actual corner of the garden.

Bonelli's introduction is followed by a short account of the garden, 'Horti Romani Brevis Historia,' in which the author explains that the Vatican decided to move its garden to Janiculum Hill to make it more accessible to visitors, and underlines the important contributions made by some of its more distinguished *prefetti*, from Michele Mercati (1541–1586), Andrea Bacci (1524–1600), Castore Durante (1529–1590), Johannes Faber (1575–1629), and Pietro Castelli (*c.* 1575–1657), to the aforementioned Trionfetti and Maratti. It may be noted that Castelli was also the author, with Tobia Aldini, of a splendid flower book, *Exactissima descriptio rariorum quarundam plantarum*, dedicated to the most famous private gardens in Rome (see *An Oak Spring Flora*, No. 28, pp. 113–116).

In each volume, the text or *Enumeratio Plantarum*, consisting of succinct descriptions in Latin of one hundred different plants with detailed notes on their medical applications, is presented first. This is followed by one hundred botanical etchings, some of them enhanced with stippling and, in this copy, colored by hand. Each illustration is surrounded by a simple linear frame colored in blue, beneath which appear the names of the plant in Italian and French. The herbal begins with two etchings of the mandrake, followed by the

GIORGIO BONELLI, *Hortus Romanus*, 1772–1793. Rhubarb (*Rheum palmatum*), volume 1, plate 34

GIORGIO BONELLI, *Hortus Romanus*, 1772–1793. Water-prince (*Nymphoides peltata*), volume II, plate 67

belladonna and other species well known for their healing 'virtues.' These are sometimes portrayed with their roots, or with enlarged details of their seeds or flowers, in keeping with the traditional format of the herbal. The subsequent volumes of the *Hortus Romanus* are similarly formatted, each beginning with a title-page printed in black and red, in most cases followed by the etched portrait of a Church official, and then the descriptions and illustrations of one hundred plants from the Vatican's garden.

In the title-page to the second volume, Bonelli's name has been replaced by that of Niccolò Martelli and it is announced that the authors would henceforth use the classification system of Linnaeus. This page is decorated with the vignette of a flowering garden, and is followed by an etching of the cardinal Francesco Gioacchino de Pierre de Bernis, signed in 1774 by Ignazio Benedetti. The text is once again presented in the form of an *Enumeratio Plantarum*, followed by the illustrations that show no substantial variations in iconography, with the exception of Plate 67, in which a *Ninfea minore* or waterlily is portrayed floating in a fountain rather than growing in its natural habitat.

The third volume opens with a title-page embellished with another view of the garden, and a portrait of Cardinal Francesco Saverio de Zelada. A brief *Monitum* or note by the author, instead of the usual *Enumeratio*, precedes the section of botanical illustrations, each framed in blue as in the preceding volumes. Some of the larger species, such as the 'Orobanche' in Plate 11 and the 'Acanto' in Plates 13 and 14, have been portrayed in sections in order to fit them on the page.

Volume IV follows the format of the first two volumes, and its etched portrait of Cardinal Bernardino De Giraud is signed by the engraver Antonio Fiori. Particularly impressive is the plate of the 'Brassica capitata alba' or Bolognese cabbage (Pl. 24). In Volume V the canonical portrait is lacking. In the sixth volume, the vignette on the title-page has been copied from Volume I. The portrait on the following page of the papal legate to Bologna, Cardinal Ignazio Buoncompagni, was drawn by Marco de Antoniis and engraved by Augusto Campanella.

In the seventh volume the name of Liberato Sabatti has been replaced by that of his son Costantino as the principal collaborator on the text. In both this and the eighth volume of the herbal the portrait has been omitted.

Despite the disparagement of critics, the botanical illustrations prepared by Cesare Ubertini and Maddalena Bouchard are quite worthy of Bouchard and Gravier's herbal. The engravings in the Oak Spring Garden Library copy of the *Hortus Romanus* were colored with somewhat less skill, however, by an artist who used a relatively limited palette of bright colors and often applied his paints quite thickly, making little attempt to capture nuances in hue or shading. As a result, many details of the stem, leaves and flowers present in the original etchings have been blurred.

49. Gaetano Savi (1769–1844)

Materia Medica Vegetabile Toscana del Dottor Gaetano Savi Professor di Fisica nell' Università di Pisa Firenze Presso Molini, Landi E C.º MDCCCV.

39 × 24.4 cm. π^2 1–14^2 *i–iv 1* 2–49 *50–51* 52–56 [60 pp.].

PLATES: 60 leaves of plants engraved by G. Canacci after drawings by B. Benvenuti, all engraved, etched, and hand-colored.

BINDING: Half speckled calf and brown paste paper boards.

REFERENCES: Nissen 1,733; Plesch, p. 397; Pritzel 8,061; Stafleu & Cowan 10,386.

Gaetano savi was born in the capital city of Tuscany on 13 June 1769 to a modest Florentine family and died in Pisa on 28 April 1844. He began his scientific studies at the Benedictine Abbey of Florence and in 1785 transferred to Pisa to pursue a degree in medicine. In 1791, in his capacity as 'aggregato' or assistant to the celebrated naturalist and geologist Professor Giorgio Santi (1746–1822), he was appointed 'Custode del Museo e del Giardino dei Semplici di Pisa' (Curator of the Museum and Botanical Garden of Pisa). During this period as a student, Savi formed close ties with another professor at the University of Pisa, Ottaviano Targioni Tozzetti, under whose guidance he developed an interest in botany and discovered his true vocation—'the study of nature.'

Savi received his degree in 1794 and for a period of one year, in 1797, served 'di mala voglia' (with great reluctance) as a 'Medico di turno dello Spedale' or hospital doctor. However, following his inclination for 'cose naturali' (natural sciences), he began to collaborate once again with Giorgio Santi, accompanying him on botanic expeditions around Tuscany. Savi contributed short botanical notes to Santi's published accounts of these expeditions, *Viaggi per la Toscana* (1795–1806).

In 1798 Savi published a two-volume work of his own on the indigenous plants of Tuscany—*Flora Pisana*—that was of striking modernity and won him great fame. The text was exemplary, for in his botanical descriptions the author applied the nomenclature and classification criteria (based on the reproductive structures of the plant) recently developed by the Swedish botanist Carolus Linnaeus (see No. 46). This book was followed by *Materia Medica Vegetabile Toscana* (1805), and then *Botanicum Etruscorum* (1808), which was written in Latin and describes roughly 1,509 species of plants native to the region of Tuscany.

'Salito in fama di egregio botanico' (with his fame as a distinguished botanist firmly established), in 1801 Savi was appointed to a professorship at the University of Pisa, first in physics, which he taught until 1810, and then in botany. He published a collection of his lectures, *Lezioni di Botanica* (1811), which was adopted as a textbook in many Italian universities and reprinted many times; a completely revised edition entitled *Istituzioni Botaniche* appeared in 1833.

When Giorgio Santi died, Savi was appointed to take his place as director of the Botanical Garden of Pisa and under his firm guidance the institution went through a period of expansion and modernization. This was a most fertile period for Gaetano Savi; he continued to pursue his scientific research and at the same time published many works whose wide-ranging scope was to popularize the subject of botany. For example, he wrote a history of the Botanical Garden, which emphasized its contributions to research and teaching. He also sought to encourage a broader interest in gardening through works written for amateur gardeners rather than botanists, in order that—as he put it—the native plants and trees might become with 'loro fiori, frutti e coll'ombra loro ... vero profitto nell'economia rustica e forestale' (their flowers, fruits and with their shade ... truly profitable within the [local] agrarian and forest economies).

Other works by Savi in this period include *Almanacco di giardinaggio* (Gardener's Almanac, 1822–1824); *Trattato degli Alberi della Toscana* (Treatise on the Trees of Tuscany, 1811), in which he discusses, among other topics, the problem of deforestation and its environmental consequences; and a three-volume *Flora Italiana* (1818–1824). This monumental work was embellished with 120 illustrations prepared under the direction of the author by Tuscan artists including the draftsmen Baldassarre Benvenuti and Antonio Serantoni (1780–1837) and the engravers Giuseppe Canacci (active during the first half of the nineteenth century) and Giuseppe Ciuti. He wrote numerous other *memorie* or essays, among which we may mention his *Notizie sul Té* and *Notizie sul Caffè* ('Notes on Tea' and 'Notes on Coffee') both published in the journal *Nuovo Giornale dei Letterati* in 1835. In recognition of his distinguished career, at the first national meeting of Italian scientists sponsored by Grand Duke Leopoldo II of Tuscany and held in October 1839 in Pisa (which had been designated the scientific capital of Italy for the occasion), Savi was elected to preside over the 'Sezione di Botanica e Fisiologia Vegetabile.'

Materia Medica Vegetabile Toscana constitutes Gaetano Savi's first extended work and is a significant contribution to nineteenth-century botanical-pharmaceutical studies, describing the indigenous medicinal plants of Tuscany together with their properties. The volume was dedicated to a benefactress with a profound interest in botany—Elena Amici Mastiani Brunacci, who would later become lady-in-waiting to the sister of Napoleon Bonaparte, the Princess of Lucca and Piombino (and later Grand Duchess of Tuscany), Elisa Baciocchi (1777–1820).

In the introduction to *Materia Medica Vegetabile*, Savi briefly reviews the history of 'phytotherapy' (the treatment of illnesses with remedies based on plants), beginning with Theophrastus and Dioscorides, continuing with the contributions of eminent Italians such as Cesalpino, and concluding with the latest reforms introduced by Tournefort and Linnaeus. His appraisal of the situation in his day was somewhat pessimistic, for, as he notes, very few of his contemporaries seemed to be interested in the study of the pharmacological properties of plants.

With this work Savi was therefore seeking to fill a lacuna and to 'rendere qualche servigio ... ai Pratici e agli Speziali' (render a useful service to Practitioners and Pharmacists). He furthermore declared that:

> ... io non posso abbastanza raccomandare agli Speziali e ai Medici che procurino di conoscere bene le piante officinali, sapendo per prova che non vi è da fidarsi agli erbaioli i quali ingannano per ignoranza, e talvolta ancor per malizia, se conoscono di poterlo fare impunemente
>
> (I cannot sufficiently recommend to Pharmacists and Physicians that they make every effort to acquire a thorough knowledge of medicinal plants, knowing from experience that one cannot place any faith in herbalists, who deceive us through ignorance, and sometimes even out of malice, when they know that they may do so with impunity, p. 2).

Materia Medica Vegetabile Toscana is divided into fourteen sections, each concentrating on a specific medicinal property—astringents, digestives, and so on. Within each section the author enumerates a series of plants with their given properties and furnishes a description of each one—its appearance, provenance, and 'qualità medicamentose' (medicating properties). He also lists the 'preparazioni più usitate che voglion farsi, e [...] le dosi nelle quali si possono amministrare' (the most common preparations that can be made ... and the doses in which they may be administered). Each description is preceded by the scientific name of the plant (adopting the system of Linnaeus), its common name, and the number of the corresponding illustration. The volume concludes with an index of the plants arranged alphabetically. The text is illustrated with an impressive series of sixty botanical engravings; only Savi's *opus magnum*, *Flora Italiana*, would surpass this work in terms of the number and quality of its illustrations. In his introduction the author underlines that here, 'si trovano le figure delle specie indigene adoprate presso di noi, fatte è vero, senza gran lusso d'incisione, ma non equivoche, e capaci di darne una chiarissima idea' (one may find illustrations of the indigenous species used in our region, presented—it is true—without elaborate engravings, but they are sufficiently accurate and capable of giving a most clear idea [of the plants], p. 3).

Most of the botanical drawings were prepared by Baldassarre Benvenuti, an artist who had studied in Pisa under Giovan Battista Tempesti (1729–1804) who was well known as a painter of portraits and religious scenes. He also had already collaborated with Giorgio Santi on the illustration of his *Viaggi per la Toscana* (1795–1805). All of the engravings were executed by Giuseppe Canacci. Other artists contributed to the work as well; the illustrations of *Aconitum napellus* L. (Pl. 50) and *Veratrum album* L. (Pl. 53) can be ascribed

GAETANO SAVI,
Materia Medica Vegetabile Toscana, 1805. Ramping fumitory (*Fumaria capreolata*), Common fumitory or Earth Smoke (*Fumaria officinalis*), plate 1

to Felice Pontecchi, while Gaetano Gaglier was the author of the drawing of *Gentiana asclepidea* L. (Pl. 40). These illustrations, as Savi took pains to underline, were intended primarily for teaching and study purposes. Nevertheless, from an artistic point of view they are characterised by an elegant *mise en page* and a *silhouette* of extreme clarity; the sharp, decided outlines reflect the neoclassical style of the period and testify to the superb draftsmanship of the artists involved.

In some of the engravings more than one species is shown for the purposes of comparison; in others a single plant is portrayed in its entirety, sometimes with the enlarged detail of a flower, leaf, or bulb. Benvenuti and Canacci have devoted particular attention to the rendition of chiaroscuro effects through the use of fine parallel lines or crosshatching, and the images are quite convincing and full of life. The copy of Savi's herbal in the Oak Spring Garden Library has furthermore been colored by hand with exceptional skill.

50. John Lindley (1799–1865)

Flora Medica; A Botanical Account of all the more Important Plants used in Medicine, in different parts of the World. By John Lindley, PH.D. F.R.S. Professor of Botany in University College, London; Vice Secretary of the Horticultural Society, Etc. Etc. Etc. [*rule*] Certa feram certis autoribus; haud ego vates—. [*rule*] London: Printed for Longman, Orme, Brown, Green, and Longmans, Paternoster-Row. 1838.

8° 22.3 × 14 cm. A–2T⁸ *i–xvi* 1–656 [672 pp.]. 16 pages of advertisements at end.

BINDING: Original blind-stamped green cloth with 'Lindey's Flora Medica' stamped in gilt on spine, bound by Westley and Jarvis, London, with their label on rear paste-down.

PROVENANCE: Inscribed on front free endpaper: 'Joseph Pease Southend Darlington' and 'Edward Pease Darlington.' Bookplates of Darlington Gardener's Institute and Edward Pease, Greencroft West, Darlington. Label on top left corner of front paste-down: 'Montpellier Library Cheltenham' and inscribed in ink below it: 'R. E. Way.'

REFERENCES: *An Oak Spring Pomona*, No. 44, pp. 120–122; Pritzel 5,359; Stafleau & Cowan 4,652; Stern; Wheelwright, p. 161.

JOHN LINDLEY was one of the most outstanding, versatile, and dynamic botanists in nineteenth century England. He promoted the French Natural System (as opposed to what was viewed as Linnaeus' 'unnatural' system) of plant classification, wrote what many consider to be one of the best texts on the physiological principles of horticulture, *Theory of Horticulture* (1842), and earned the name of 'the father of orchid taxonomy' for his studies on the *Orchidaceae*. A tireless administrator, he rose to the position of vice-secretary and then secretary of the (Royal) Horticultural Society, and organized England's first flower shows under its auspices. He played an instrumental role in the expansion and improvement of the Royal Botanic Gardens, Kew; it was upon his recommendation that the

FLORA MEDICA;

A BOTANICAL ACCOUNT

OF ALL THE MORE

IMPORTANT PLANTS USED IN MEDICINE,

IN DIFFERENT PARTS OF THE WORLD.

BY

JOHN LINDLEY, PH.D. F.R.S.

PROFESSOR OF BOTANY IN UNIVERSITY COLLEGE, LONDON;
VICE SECRETARY OF THE HORTICULTURAL SOCIETY,
ETC. ETC. ETC.

Certa feram certis autoribus; haud ego vates —— .

LONDON:

PRINTED FOR

LONGMAN, ORME, BROWN, GREEN, AND LONGMANS,

PATERNOSTER-ROW.

1838.

JOHN LINDLEY,
Flora Medica, 1838.
Title-page

gardens, which had fallen into neglect, were turned over to the nation to serve as the Horticultural Society's headquarters, with collections that soon included more than 40,000 different plants, an herbarium of 6,500,000 dried specimens, and a library of more than 230,000 volumes. Lindley's own collection of 1,300 books forms the core of the superb Lindley Library at the Royal Horticultural Society.

Lindley was born in Catton, Norfolk, in 1799 to a nurseryman and expert pomologist, and died in Turnham Green in the county of Middlesex in 1865. He was so fortunate as to obtain an introduction to the illustrious botanist Sir Joseph Banks (1743–1820), who gave him access to the enormous library and unique collection of plants at his private residence in London. Lindley was thus able to frequent the house at Soho Square of this generous patron of the sciences, pursuing studies on the genus *Rosa* and meeting the people of eminence from all walks of life whom Sir Joseph entertained there, an experience that proved crucial to his career and to his formation as a scientist.

At the age of twenty-two Lindley began working on *Collectanea Botanica*, a sumptuous folio volume of forty-one botanical drawings, almost all by his own hand, that included two orchid from Trinidad. This marked the beginning of the botanist's life-long interest in the *Orchidaceae*; he conducted important work on the description, classification, and cultivation of this vast family of flowering plants and assembled a valuable herbarium of 7,000 specimens that he eventually left to Kew Gardens. In addition to his activities as administrator, horticulturist, taxonomist, and editor (he was one of the co-founders of the influential weekly paper of record *The Gardeners' Chronicle*), he was appointed in 1829 as the first professor of botany at the University College of London; he would later also teach botany at Cambridge. He served as director of the historic Physic Garden in Chelsea, and developed the Horticultural Society's Chiswick Garden, setting up a school of horticulture there.

A prolific author, Lindley wrote monographs and books on a wide range of subjects. In *The Genera and Species of Orchidaceous Plants* (1830–1840) he offered a basic classification of the family with short diagnoses and descriptions in Latin. This was followed by *Sertum Orchidaceum* (1837–1841), a magnificent folio volume embellished with forty-nine lithographed plates based on his own drawings. He was also the author of a fundamental text on the fruit trees of the British Isles, *Pomologia Britannica*, which was published with a splendid series of plates in 1841 (see *An Oak Spring Pomona*, No. 44, pp. 120–122). Believing, however, that the botanist's perspective should not be limited to a single family or part of the world, Lindley also conducted studies on the flora of Australia and produced a work on palaeobotany, *The Fossil Flora of Great Britain* (1831–1833).

In his *Flora Medica*, Lindley was one of the first botanists to compile a comprehensive herbal of all the medicinal plants known to man. In the preface he stresses '... the necessity that students should have access to a botanical account of the plants which furnish the substances used medicinally in different parts of the world,' in part because '... the very nature of the climate of tropical countries generally causes the properties of plants to be more concentrated than in northern latitudes.' The 1,350 entries are fully descriptive and note the properties ascribed to the plants by accepted authorities or, in the case of tropical species, by the indigenous peoples who used them. Lindley also discusses the relationships between various plants, in his view one of the keystones of the *materia medica.*

It must be observed nonetheless that Lindsey's primary interest was in plant classification rather than pharmacology, for even in *Flora Medica*—a work on medicinal plants—the botanical descriptions are considerably more detailed than the discussion of their properties. Furthermore, despite the author's own interest in and talent for botanical drawing, this work is not illustrated, because it was intended primarily to serve as a reference work for students and specialists.

VI

CURIOUS AND STRANGE HERBALS

GIOVANBATTISTA DELLA PORTA, *Phytognomonica*, 1588. Title-page

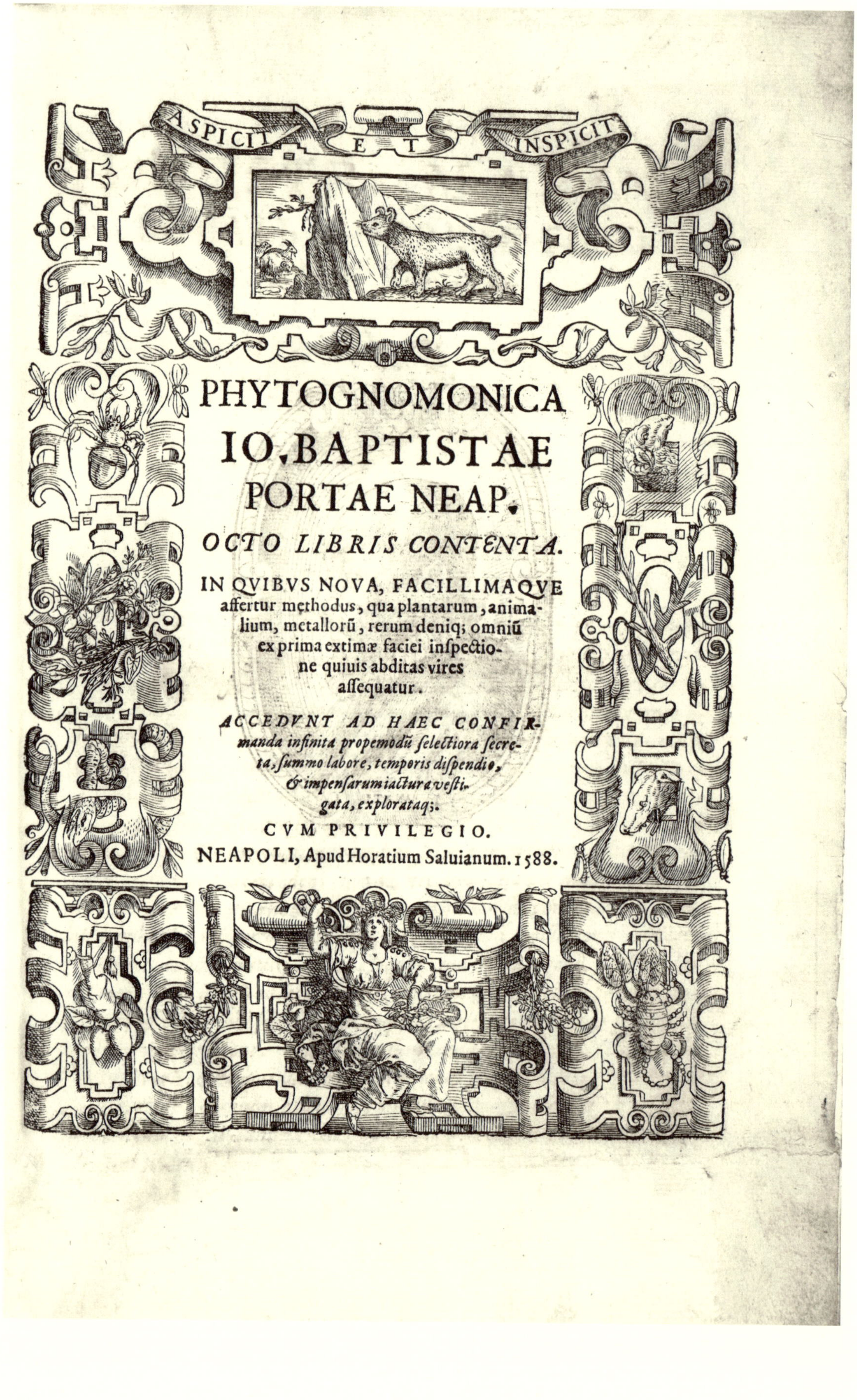
ASPICIT ET INSPICIT

PHYTOGNOMONICA
IO. BAPTISTAE
PORTAE NEAP.
OCTO LIBRIS CONTENTA.
IN QVIBVS NOVA, FACILLIMAQVE
assertur methodus, qua plantarum, anima-
lium, metallorũ, rerum deniq; omniũ
ex prima extimæ faciei inspectio-
ne quiuis abditas vires
assequatur.
ACCEDVNT AD HAEC CONFIR-
manda infinita propemodũ selectiora secre-
ta, summo labore, temporis dispendio,
& impensarum iactura vesti-
gata, explorataq;.
CVM PRIVILEGIO.
NEAPOLI, Apud Horatium Saluianum. 1588.

51. Giovanbattista della Porta (1535?–1615)

[*Within woodcut border*]: Phytognomonica Io. Baptistae Portae Neap. octo libris contenta. In quibus nova, facillimaque affertur methodus, qua plantarum, animalium, metalloru[m], rerum deniq[ue] omniu[m] ex prima extimæ faciei inspectione quiuis abditas vires assequatur. Accedunt ad haec confirmanda infinita propemodu[m] selectiora secreta, summo labore, temporis dispendio, & impensarum iactura vestigata, explorataq[ue]. Cum Privilegio. Neapoli, Apud Horatium Saluianum. 1588. [*In ribbon, above*]: Aspicit et Inspicit.

2° 31.6 × 21.5 cm. A–2R⁴ 2A–C⁴ *1–2* 3–6 *7–9* 10–320 [*24*] (300 as 301, 301 as 302) [344 pp.].

PLATES: Woodcut portrait of author on verso of title-page and 32 woodcuts of plants and their corresponding animals and body parts.

BINDING: Modern vellum with ties.

REFERENCES: Adanson, IX; Anderson, pp. 193–200; Arber, pp. 251–253; Gabrieli, p. 734; Hunt 158; Mortimer, *Italian* II.399; Nissen 463; Pirotta & Chiovenda 88; Pritzel 7,273; Séguier 151.

GIOVANBATTISTA DELLA PORTA was one of the most eclectic and encyclopaedic personalities to frequent the cultural circles of sixteenth-century Italy. He was a passionate student of natural philosophy, an alchemist, physicist, naturalist, agronomist, botanist, and man of letters whose boundless desire to learn was coupled with immense erudition and a remarkable spirit of observation. Della Porta was considered by his contemporaries to be a most perspicacious student of nature and diligent explorer of its secrets, and he occupies a prominent position in the history of science even if, when compared to such great scientists of the Renaissance as Galileo and Kepler, he remained tied to the magic-scientific tradition of the past. For instance, della Porta never fully grasped the revolutionary significance of experimental research, whose primary use in his view was to demonstrate the 'wonders' of the universe. His work, nevertheless, is quite fascinating, for it represents a bridge between medieval philosophical speculation and modern science, and his multi-faceted personality awaits more thorough study, particularly in the light of the many manuscripts and letters by his hand that still remain to be edited.

Born into an aristocratic family in Naples, della Porta dedicated himself quite early to the study of philosophy and the natural sciences, establishing a *Wunderkammer* of natural history specimens and curiosities in his city. While still in his early twenties, in 1558, he published *Magiæ Naturalis*, a collection of herbal prescriptions and reflections on medicine, handicrafts, optics, and other 'secrets of nature,' whose purpose was to demonstrate how magic could be used as an instrument to study the causes of natural phenomena. Thus, in contrast to the natural philosophy of Aristotle, whose scope was to explain the normal, everyday aspects of nature, *Magiæ Naturalis* sought to reveal its exceptional and wonder-working dimensions.

But how was it possible to delve into the most profound secrets of nature? According to della Porta it was Nature herself who provided subtle clues in the form of 'signatures' or visual likenesses, by which means the attentive observer could recognize the essential nature of a thing, be it a plant, animal, or mineral. After identifying these characteristics one might through careful study and experimentation discover the hidden affinities and contrasts between different elements. The promulgation of such a thesis was in reality quite perilous in the tense political and religious climate created by the Council of Trent, but della Porta pursued his researches undaunted and elaborated upon his ideas in further works. He even established an academy where eminent philosophers and scientists could meet and devote themselves to the study of natural phenomena, the Accademia dei Secreti della Natura.

Following the publication of *Magiæ Naturalis*, for the next twenty years della Porta spent much of his time travelling through Europe and establishing contacts with the most illustrious scientists of the period. Enlightened sovereigns such as the Hapsburg emperor Rudolph II (1576–1612), themselves fascinated by *le cose della natura*, were pleased to receive him at their courts and bestowed flattering tributes upon him. In 1579 he accepted the invitation of Cardinal Luigi d'Este to stay as his guest and pursue his work at the prelate's palace in Rome. There he devoted himself to experiments in optics and on the *lapis philosophalis*, the miraculous but elusive substance which, according to the philosophers of Antiquity, possessed the power to transform base substances into gold. The philosopher's stone formed the object of endless research on the part of alchemists, and became synonymous with studies into the secrets of nature.

In 1583 della Porta published two short treatises on agronomy, *Pomarium* and *Ulivetum*, in which he discussed the results of his experiments in the cultivation of fruit and olive trees on his estate. The following year saw the completion of *De Humana Physiognomonia* (although it would not be published until 1586), a work devoted to physiognomy, an art which had long formed an element in the repertoire of occult practices and which, analogous to the theory of signatures, hypothesized that the inner nature and disposition of a person could be read in his outward form and lineaments. Della Porta presented 'parti e degli costumi degli animali per li quali possiamo congetturare i costumi degli huomini' (the parts and customs of animals from which we may conjecture the customs of man) and discussed his system for reading and interpreting these signs. For example, a person with a leonine appearance might be expected to be strong, generous, and courageous, while a person whose features resembled that of a mastiff would tend to be despotic in nature, and someone with a face like a fox would be insidious and deceptive.

In this period an anonymous denunciation was brought against della Porta and the Tribunal of the Inquisition in Naples, in the wake of a papal bull denouncing the practice of the magical arts, commanded him to abstain from his studies of magic and from any further 'giudicii astronomici' (astronomical speculations). The Accademia dei Secreti della Natura was forced to close its doors from 1592 to 1598, and the publication of all of della

Porta's works was prohibited. This ban was reiterated several times during the remainder of della Porta's life, despite the fact that he was besieged with requests from printers all over Europe to publish further editions of his works.

In 1588, two years after it was written, della Porta ventured to publish his herbal *Phytognomonica*, and the next year saw the publication of a second expanded edition of *Magiæ Naturalis* in twenty books. The latter was an immediate success all over Europe, and more than twenty editions in Latin, as well as translations into Italian, French, German, Dutch, and English followed. In these works he energetically sustained his theories regarding the science of 'natural astronomy'—that natural phenomena were influenced by the movement of the heavenly bodies.

Thus we see that della Porta viewed the universe of nature as being bound together by a network of secret correspondences that could be unveiled through a study of analogies. This underlying order was particularly legible in the plant world and it was on this thesis that *Phytognomonica*—one of the most curious and fascinating herbals of the sixteenth century—was based. 'Plantarum cum homine multam habere similitudinem' (Great is the resemblance between plants and man), the Neapolitan scientist wrote, and in his work he sought to demonstrate how most plants (as well as many animals and minerals) possess 'signatures' that could enable one to deduce their inherent properties and apply them to the treatment of various illnesses. In particular, those plants (or animal organs) which resemble a specific human organ, *simpatizzano* (have a particular rapport) with that part of the body and could be used to cure the disturbances or diseases that might afflict it through a kind of 'imitative natural magic.' Therefore, according to *Phytognomonica* plants with yellow flowers might be used to purge a patient of bile, plants with a milky sap would stimulate the production of milk, 'bony' plants could be administered to treat bone ailments, and plants with bright red flowers (*quae flammeo sunt flore*) would reduce inflammation. Herbs whose leaves or flowers resemble a scorpion in form were clearly destined by nature to provide a remedy for scorpion bites, while plants whose stems are dappled like the skin of a snake, such as the *Arum* and the *Dracunculus*, 'contra eorum morsum valent' (are excellent remedies against their bite). The author also believed that an intimate relationship existed between plants and the stars and discusses at length the qualities of various 'lunar' and 'solar' plants.

The woodcut illustrations scattered throughout the work add considerably to its fascination. Depicted in soberly elegant outline are different plants that have been grouped together, not because they are related botanically, but because they present analogies in the form of their leaves, flowers, or fruit. A small number of animals are also illustrated. The visual component of *Phytognomonica* is further enhanced by a portrait of the author, shown as a dignified figure in his fifties, and by an elaborate title-page surmounted by a lynx, the wild feline known for its acute eyesight, and the motto 'Aspicit et Inspicit' (Look and Examine).

Phytognomonica was an enormous success and was reprinted in an expanded version in 1591 in Frankfurt; new editions appeared in Hanover in 1599 and in Frankfurt in 1608, in Rouen in 1650, and would continue to be produced during the entire course of the seventeenth century. The author wrote to the illustrious botanist Ulisse Aldrovandi about his book in a letter, dating from perhaps 1590:

> I giorni adietro mandai fuora un libro de fisionomia delle piante intitolato *Physiognomonica*: dove per nuovo metodo si investigano i vitij et virtù delle piante ... Sono molti in Alemagna et di colà me vien scritto che piace moltissimo, se l'ha visto, scrivermelo il suo parere che volendosi ristampar di nuovo in Venezia, possa riccomodarlo, e se costì non ve ne fussero, harò pensiero inviargliene uno.

> In the last few days I have sent out a book on the physiognomy of plants entitled *Physiognomonica* where, by a new method, the vices and virtues of plants may be investigated ... There are many in Germany ... who have written to me that [the work] pleases them exceedingly and, if you have seen it, write to me regarding your opinion on it in order that—since I wish to print it anew in Venice—I may correct it, and if it is not the case [that you have seen it] I can undertake to send you a copy. (Gabrieli, p. 734).

In addition to his scientific endeavors, it is pertinent to recall here that since his youth della Porta had cultivated a literary inclination, and he was the author of various theatrical pieces as well.

The Neapolitan philosopher played an important role in the activities of the Accademia dei Lincei from its very foundation in 1603 by the Roman nobleman Federico Cesi. This was the first scientific academy created in Italy and Cesi asked della Porta to organize a branch in Naples. The lynx was adopted by Prince Cesi as the chief symbol of the Academy and it may be hypothesized that della Porta's device served as the Academy's eponym, as it had appeared the title-page of his work *Phytognomonica* in 1588 and was reused on the second edition of *Magiæ Naturalis* in 1589. The celebrated scientist became a member of the Accademia in 1610, and his adherence brought additional prestige and authority to the institution.

It is interesting to note, in closing, the dispute that arose between della Porta and another member of the academy, Galileo Galilei, regarding the invention of the telescope. Della Porta had written a short treatise, *De Telescopio* (which lay unpublished until 1962), in which he described the various steps that went into the construction of this new instrument, learned by watching Dutch craftsmen who had already engaged for some time in the development of such optical devices. As Galileo demonstrated, however, the

BVVLSOS à maxilla dentes tabellula oculis subiecta exprimit, à dextris è dentarijs speciem vnam designauimus, putantes ad dentium similitudinem effigiandam suffecturam, à sinistris pineum conum desquamatum, vt è suis loculis iacentes, & exempti pytides, siue pineoli, & in medio mali punici cortices, & membranulam dissecuimus, vt acini eosdem simulantes in conspectum venirent.

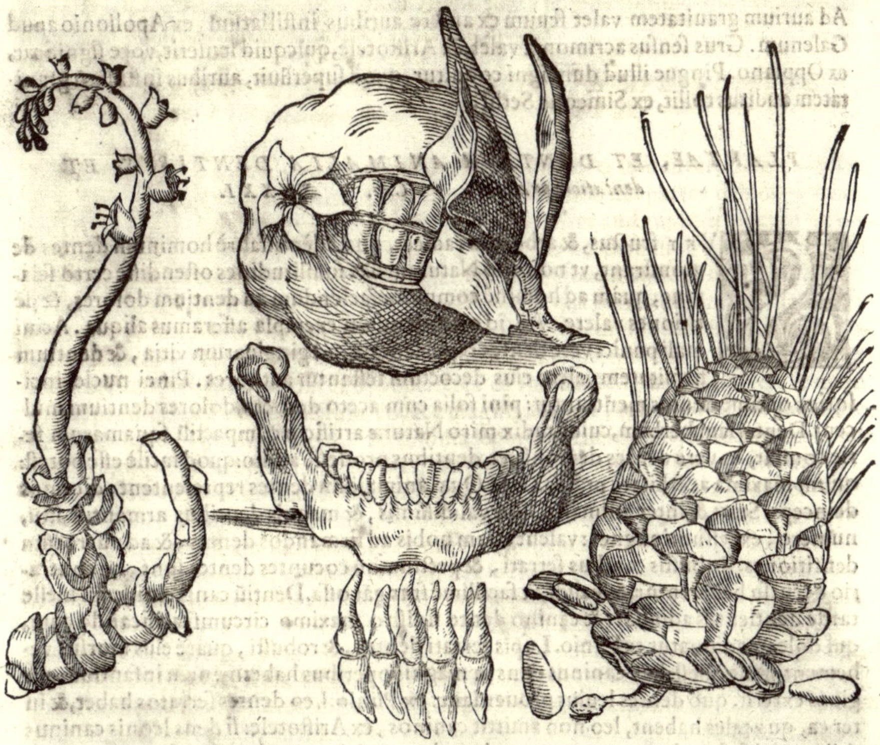

Canis galeus piscis dentes habet acutos, & ad latera recuruos, sic & glaucus: infantiũ gingiuis, dentitionibusq; multum conferunt vtriusq; piscis dentium cineres: horum etiam dentibus gingiuas tangi prodest, & è collo suspensi. Sus, elephas, & hippopotamus exertos habent dentes, eisq; ad auxilium, & robur datos: quibus non decidunt, ex Aristotele. Cerebrum illitum gingiuis puerorum valet ad faciles dentitiones. Sunt & robusti animalium rostri, validorum dentium vice, ad molendum cibum, & defensionem dati à Natura. Pici rostro quasuis arbores, & quercus tundunt, vt Græcis dricolaphi dicatur, ex Aeliano, pascuntur enim vermibus, & tam vehementer arbores excauãt, vt eas sternent, perduras amygdalas etiam tertio ictu pertundant, & nucleum erodant ex Aristotele: quorum rostrum ad collum suspensum, omnem tollit dentium dolorem. Coruini generis auibus prædurum rostrum datum est, ex Aristotele: dentium dolorem statim eo alligato abire tradit Plinius. Galli pugnaces, & valido sunt rostro, vt etiam lapillos rumpant: officulo gallinarum si quis gingiuas scarificet, proiecto officulo, statim dolor abit, ex Plinio, & sic de cæteris.

ALAE

GIOVANBATTISTA DELLA PORTA, *Phytognomonica*, 1588. Toothwort (*Lathraea squamaria*); Pomegranate (*Punica granatum*); *Pinus pinea*; all showing resemblance to teeth, page 138

'revolutionary' nature of the telescope did not lie in its discovery and the details of its construction, but rather in the uses to which it might be put in the innovative and 'critical' observation of natural phenomena, making it possible for man to finally pierce the secrets of the heavens.

52. Joachim Camerarius *the younger* (1534–1598)

[*Engraved title-page. Vol. I*]: Symbolorum & Emblematum ex Re Herbaria desumtorum centuria una collecta a Ioachimo Camerario Medico Norimberg. In quib[us] rariores Stirpium pro prietates historiae ac Se[n]tentiæ memorabiles no[n] paucæ breviter exponu[n]tur. An° Salut: CIↃ·IↃ·XↃ· [*Within circular frame at bottom*]: Deo Omnia florerebunt prospiciente [*On ribbon within circle*]: Spes [*monogram initials bottom right*]: HS.

[*Engraved title-page. Vol. II*]: Symbolorum & Emblematum ex Animalibus Quadrupedibus desumtorum centuria altera collecta. A Ioachimo Camerario Medico Norimberg. Exponuntur in hoc libro rariores tum animalium proprietes tum historiæ ac sententiæ memorabiles. An° Salut: CIↃ·IↃ·XↃV· [*Within circular frame at bottom*]: Cet Pietas Pura Pia.

[*Engraved title-page. Vol. III*]: Symbolorum & Emblematum ex Volatilibus et Insectis desumtorum centuria tertia collecta a Ioachimo Camerario Medico Norimberg. In qua multæ rariores proprietates ac historiæ et sententiae memorabiles exponuntur A° Salut. CIↃ·IↃ·XCVI· [*Within circular frame at bottom*]: Zephyris Aspirantibus.

[*Engraved title-page. Vol. IV*]: Symbolorum & Emblematum ex Aquatilibus et Reptilibus desumptorum centuria quarta a Joachimo Camerario Medi: Nor: coepta: absoluta post eius obitum a Ludouico Camerario JC^to^ Joach Fil. In qua itidem res memorabiles plurimæ exponuntur. An° Salut: CIↃ DCIV. [*Within circular frame at bottom*]: Oculos Sursum.

4° 18.8 × 14 cm. Volume I: A–Z^4 a–b^4 c^2 fols. *1–2* 3–60 (4 as 3) 65–68 61–64 69–102 [204 pp.]. Volume II: A–Z^4 a–c^4 fols. 1–50 *51* 52–91 *92* 93–103 *104* [208 pp.]. Volume III: A–Z^4 a–c^4 d^2 fols. [*2*] *1–2* 3–7 *8* 9–66 *67* 68–102 *103–104* (65 as 45) [212 pp.]. Volume IV:)(^{4} A–Z^4 a–b^4 c^2 fols. [*4*] 1–100 *101–102* [212 pp.].

PLATES: Engraved title-pages, 400 etched circular emblems, 100 in each volume, by Hans Sibmacher.

BINDING: Contemporary blind-stamped vellum with previous owner's initials and date, 'HHVE 1611' stamped on front covers; remnants of ties. Spines are inscribed 'Symbolorum et Emblematum.'

REFERENCES: Camerarius; Hunt 181; Nissen 312; Praz, pp. 295–296; Pritzel 1,441.

AFTER PUBLISHING *Hortus medicus et philosophicus* in 1588 (see No. 42), Joachim Camerarius dedicated himself to the preparation of another highly original work, *Symbolorum et Emblematum*, which was printed in Nuremberg beginning in 1590. Inspired by the prevailing vogue for emblems and devices, *Symbolorum et Emblematum* was divided into four *centuriae*, each presenting one hundred emblems or symbolic images based on botanical or zoological themes.

The origins of the 'impresa' (device) and the 'emblema' lie in the culture of Neoplatonism; their aim was to create a symbolic code made up of images (the *corpo* or body of the device) and words (a motto that formed its *anima* or spirit) that would encapsulate essential truths and suggest moral lessons, thus linking the visible and invisible worlds. The composing of ingenious, and at times extravagant, devices was one of the favorite pastimes of intellectuals in sixteenth- and seventeenth-century Europe, and influenced the iconography of publisher's imprints, medallions, and the decorative elements in art and architecture. Many notables of the period were eager to add luster to their names by associating them with a device of elevated moral content.

Although Andrea Alciato's *Emblematum liber* of 1531 was the progenitor of the emblem books to follow (which were legion), the celebrated Italian scholar Paolo Giovio (1483–1552) was the first to set down formal rules for the correct composition of a device, in *Dialogo dell'imprese militari e amorose*, which, unillustrated, was published in Rome in 1555. The French publisher Guillaume Rouillé produced the first illustrated edition in 1559. Since one of Alciato's five criteria was that a device, to be considered 'perfect,' must not contain the human figure, images of plants and animals were often employed instead. Classical texts (in particular Aristotle's *Historia animalium*, Pliny the Elder's *Naturalis historia*, and the works of Theophrastus), the traditions of classical mythology and medieval Christianity, and the latest books on the natural sciences offered a rich vein of images and mottoes that were eagerly mined. Animals, having been invested with symbolic meaning since Antiquity (one need only recall the virtues and vices that they traditionally represented), lent themselves with particular ease to this learned *divertissement*, and even Giovio sometimes used plants and flowers in the 'corpo' of his emblems.

The authors of the most important scientific texts of the late Renaissance, from Conrad Gesner to Pietro Andrea Mattioli (see No. 7) and Ulisse Aldrovandi, all sought to draw connections between 'physica' and 'philologia'—that is, between the data gathered by direct observation and the cultivated symbolic tradition that had always been associated with the world of nature. Therefore, as Aldrovandi wrote in the preface to *Ornithologiae* (1599), no author should neglect to discuss the 'Moralia, Mystica, Historica, Symbola, Emblemata, Fabulae et Apologi' (Morals, Mysticism, History, Symbolism, Emblems, Fables, and Allegories) of the natural sciences, an approach that would make their works not only 'useful' but also 'pleasing,' thus enlarging the circle of their readers to embrace the entire scholarly community.

With his combined humanistic and scientific background, Camerarius was eminently qualified to produce a work such as *Symbolorum et Emblematum*—a catalogue of naturalistic emblems accompanied by an erudite Latin text. On the recto of each page appears an image in an elegant circular frame echoing the typology of the classical medallion, with a brief motto above it and a poetic couplet below to explain its significance. An extended comment on the emblem, written in Latin prose and full of learned references, appears on the verso of the page.

Each *centuria* opens with an ornate title-page decorated with symbolic and mythological images, which announces the theme of the section. The first *centuria*—*Symbolorum et*

Emblematum ex Re Herbaria Desumtorum—was published by Johannes Hofmann and Hubert Caymox and presents one hundred emblems based on botanical subjects; the second and third, which were published by Paul Kaufman in 1595 and 1596, respectively, are devoted to quadrupeds (*ex Animalibus et Quadrupedibus*) and birds and insects (*ex Volatilibus et Insectis*); while the last, edited posthumously by the author's son Ludwig and published by Vogelin in 1604, has as its theme serpents and aquatic animals (*ex Aquatilibus et Reptilibus*). In the final *centuria* the commentary appears on the page facing the device rather than on its verso. In 1605, the works were issued collectively in a single volume.

Camerarius declares at the outset, in the title-page to the first *centuria*, that his work will link the outward appearance and hidden properties of plants to memorable stories and citations, while in his dedication to Jacopus Curtius he repeats that his intention is to draw ties between the science of botany (*res botanica*) and 'philologicas and ethicas meditationes partim a meo exiguo ingenio perfectas, partim ex aliis in primis Italicis scriptis collectas' (reflections on philology and ethics, in part the fruit of my modest ingenuity, in part drawn from the earliest Italian works on emblems). In this way Camerarius acknowledges his profound debt to the many Italian authors of emblems.

In compiling his work, the German physician drew his principal inspiration from *Dialogo dell'imprese militari e amorose* and *Ragionamento di Luca Contile sopra la proprietà delle imprese*, which was published by Contile (1505–1574) in Pavia in 1574. However, there are also citations from non-Italian treatises such as Claude Paradin's *Devises Héroïques* (Lyon, 1551) and *Hemblemata* by Hadrianus Junius (Antwerp, 1565). Therefore *Symbolorum et Emblematum* was based on a fully consolidated tradition, but as the originality of its theme set it apart, it was an immediate success all over Europe. Indeed Giovanni Ferro cites it in the bibliography of useful references works which he compiled for *Teatro d'Imprese*, published in Venice in 1623.

Camerarius was convinced that the emblem represented the ideal instrument for capturing and expressing the 'visibilia et invisibilia' of nature—not only the scientific aspects that fell within his domain as a naturalist, but also the hidden 'philological' meanings that he was able to discover through his classical learning. Furthermore, his work had the merit of presenting material that combined 'physica' and 'philologia' in the diverting form recommended by Gesner, Mattioli, and Aldrovandi. Camerarius claimed that he sought to adhere to scientific fact as much as possible in *Symbolorum et Emblematum*, but the humanistic influence of the tradition of emblems—as reflected in the moral and religious connotations that pervade the work—remained predominant. This orientation was already clearly present in the preliminary draft of the work (now conserved in the Mainz Stadtbibliothek), which Wolfgang Harms and Ulla-Britta Keuchen analyzed and published in a facsimile edition with detailed commentary in 1986. It is probable that Camerarius

XXXVI. 38

VULNERE VULNERA SANO.

Dic age cùm proprio tua vulnere vulnera sanem,
Stipite cur hominum durior hostis homo es?

K 2 Vitis

JOACHIM CAMERARIUS, *Symbolorum et Emblematum*, 1590. Man harvesting balsam resin (*Commiphora opobalsamum*), volume 1, page 38

JOACHIM CAMERARIUS, *Symbolorum et Emblematum*, 1590. Cluster of clover (*Trifolium* sp.) with two serpents, volume 1, page 76

LXXIV.

76

PROCUL ESTE PROFANI.

Sic tua majestas Deus une & trine profanos,
Vt reprobos angues alma triphylla fugant.

Theophra-

received further exposure to the erudite culture of symbology at the humanist academy in Altdorf near Nuremberg of which he was a member; indeed, the jurisconsult Conrad Rittershausen, who was also affiliated with the celebrated academy, made important contributions to the text of *Symbolorum et Emblematum*.

The Latin couplets, as we learn from the dedication addressed to the 'Candido Lectori' which opens the second *centuria*, were literary efforts by Joachim Jungermann, the author's nephew who was also an artist who had already helped to illustrate *Hortus medicus et philosophicus*. In his dedication, Camerarius expresses profound grief at the untimely death of his relative, to whom he was much attached, and declares that if anyone drew pleasure from these writings, the credit should go not to the author but to his friend Rittershausen and to his gifted nephew Jungermann. If we examine the sources punctiliously listed by Camerarius at the end of each *centuria* we discover that, in contrast to the large number of classical writers and authors of treatises on emblems, the number of naturalists mentioned is relatively small. The names of the botanists Luigi Anguillara, Bartolomeo Maranta, Carolus Clusius, Pietro Andrea Mattioli, Prospero Alpini, Garcia da Orta, and Nicolas Monardes appear more than once, but only a handful of zoologists can be found—Conrad Gesner, Olao Magno, Ippolito Salviani, Pierre Belon, and Pierre Gilles, who are just briefly cited in the commentaries on the emblems.

We know that Camerarius originally planned to include about two hundred emblems in *Symbolorum et Emblematum*, many depicting historical, religious, and mythological figures in the tradition of Giovio, but these were subsequently eliminated to preserve the homogeneity of the work. References in the draft manuscript to historical persons or events have been omitted from the final version as well. An unusual and pleasing feature of these emblems is that the subjects have not been isolated and decontextualized, but rather are portrayed against landscape settings that in many cases have been suggested by the plant or animal's natural habitat.

The emblems in the first *centuria*, which are distinguished by their refined composition and naturalistic style, may perhaps be attributed to the talented Joachim Jungermann, while the remainder were the work of other artists to whom Camerarius turned after the death of his nephew and whose names have not come down to us. The images were engraved by Hans Sibmacher of Nuremberg (?–1611), a well-known illustrator of military, allegorical, and landscape subjects. He generally signed his works with the initials 'HS' or 'JSI', as they appear on the title-page to the first *centuria* and in the emblem portraying a camel, the symbol of tolerance, in the second *centuria* (II.xvi). At a time when most scientific texts still relied on woodcut illustrations, Sibmacher employed the technique of etching, a noteworthy innovation, for he could achieve subtle modulations in tone and a sharpness of detail that were not possible by traditional methods.

We shall limit our discussion here to a handful of emblems from *Symbolorum et Emblematum ex re herbaria desumtorum*, which may be seen as a fascinating and exceptional herbal in the context of the period. We shall focus on images that are of particular significance from a botanical-medical standpoint, an area in which Camerarius possessed considerable expertise, as he had clearly demonstrated in *Hortus medicus et philosophicus*. If we study these emblems and the accompanying text in detail, we find ample evidence of the author's classical learning and his ability to collate knowledge from the most disparate of sources, in keeping with the encyclopaedic interests of the period. Relatively little of the text conveys knowledge of a purely 'scientific' nature, and most of this information is of limited originality. Nevertheless, the commentary does contain many pertinent observations that testify to Camerarius' extensive knowledge of botany, based in particular on his study of the texts of Antiquity. In contrast, the zoological information in the subsequent *centuriae* is much more limited.

In his comment on the crocus (I.lxxv), for example, Camerarius notes that if trod upon, the plant will not suffer but will grow back even more luxuriantly, citing Theophrastus as his source for this affirmation. Hence the motto 'Pulchrior attrita resurgo' (If trampled underfoot, I shall rise again more beautiful still), a reference to those men whose innate qualities of virtue and honesty are merely strengthened by adversity. The emblem consists of three crocus plants that are drawn with great realism, complete with their bulbs and roots, reflecting the iconography of the contemporaneous florilegia. The specimen on the left bears three flowers in various stages of inflorescence, while the other two plants are shown with their leaves bending earthward beneath the weight of a foot that emerges from a cloud (according to Giovio, the entire human figure could not be shown in a device, but individual parts of the body were allowed).

The illustration of the balsam (I.xxxvi) shows a man in elegant oriental dress cutting an incision in the stem in order to draw the resinous sap from the plant; included is the motto 'Vulnere vulnera sano,' refering to the 'wound that will heal other wounds.' In his text Camerarius takes the opportunity to cite the Italian physician Prospero Alpini, 'qui octo annis in Aegypto medicinam feliciter exercuit, & nunc magna cum laude Venetjis artem eandem factitat' (who exercised the profession of medicine for eight years in Egypt and exercised the same art with great success in Venice), and who had sent Camerarius a 'portiuncula' containing this precious substance. He notes that Alpini deserves credit for having furnished in his treatise *De Balsamo dialogus* (published in 1592 together with his book on Egyptian plants; see No. 15) a description of the 'true balsam.' The emblem itself is modelled after the woodcut in Alpini's work.

In *Hortus medicus* Camerarius described many plants recently arrived from the East and the New World that were being cultivated in the most prestigious gardens in Europe, and

LXXV. 77

PVLCHRIOR AT-
TRITA RESVRGO.

Pulchrior ut creſcat, teritur Crocus: ô furor ingens,
Lætior ut poßis ſurgere, nolle premi!

V *Valerius*

JOACHIM CAMERARIUS, *Symbolorum et Emblematum*, 1590. *Crocus* sp., volume 1, page 77

Symbolorum et Emblematum contains emblems inspired by some of these, including the fritillary (*Fritillaria imperialis*, I.lxxi) and the sunflower (*Helianthus annuus*, I.xlix).

In the emblem of the fritillary, two magnificent specimens are shown growing in an oval flower bed. The plant on the left is in full bloom with an erect stem bearing a corolla in the form of a crown. The flower on the right is beginning to wilt, evoking the themes of *vanitas* and the transient nature of life, which were recurrent motifs in the floral paintings of the period. The motto 'Modesta iuventus, honesta senectus' (Modest youth, honest old age) exhorts the reader to cultivate humility and eschew all forms of arrogance or self-conceit, thus earning the right to a venerable old age full of honors and dignity. In the text Camerarius informs us that this rare flower originally came from Constantinople, where it was called 'Tusai,' and that his great friend ('summus meus amicus') Carolus Clusius had identified it some years earlier as 'Hemerocallis vetus.' The German physician furnishes an accurate description of the plant, which suggests that the fritillary was cultivated in his garden in Nuremberg.

The emblem of the sunflower and its motto 'Non inferiora secutus' (I do not follow base things) were inspired by the heliotropic behavior of *Helianthus annuus*, whose majestic corolla follows the sun—considered a divinity by the ancients—across the horizon during the course of the day. This plant, which Camerarius referred to as 'Flos maximus' and 'Crysanthemum peruvianum,' was perhaps be found in his garden as well, for the text contains observations on its remarkable size, and on how to grow it from a seed. The author goes on to suggest that the sunflower and the peony were related species, an unfounded notion which, it may be hypothesized, originated with him. Joachim Jungermann executed a magnificent drawing of *Helianthus annuus* for the florilegium compiled by his uncle that is today conserved in the library of the University of Erlangen (see No. 42).

Camerarius' collection of floral emblems would not have been complete without the tulip, whose 'pulcherrimus flos' (most beautiful flower—I.lxxxviii) with its unusual form and vast array of colors captivated European gardeners. Since this species was 'new' to the European scene, no references to it could be found in the classical authorities, and Camerarius devised a succinct motto based on the plant's need for a sunny exposure: 'Languesco sole latente' (Without the sun I will languish). The emblem depicts three tulips, two with their flowers growing on tall, straight stems and turned towards the divine light of the sun, while the third, which faces in the opposite direction, is smaller and less vigorous. The naturalistic portrayal of the blossoms with their rippled leaves reflects the style of Jungermann, who painted a magnificent series of tulips for *The Camerarius Florilegium*.

The emblems in the following *centuriae* are more conventional, particularly those depicting animals connected with the symbolic tradition of Antiquity, such as the eagle, the lion, the chameleon, the bee, and the unicorn. Of greater interest are the images of exotic fauna (such as the armadillo) that had been brought to Europe from distant continents, especially the New World, by explorers.

Symbolorum et Emblematum was so popular that it was republished in Frankfurt in 1654 and in 1661, in Heidelberg in 1664, and in Mainz in 1668. Another German edition appeared in Mainz in 1697. The Oak Spring Garden Library owns a second copy in one volume printed in Nürnberg (colophon: Noribergæ impensis Johannis Hofmanni, & Huberti Comoxij).

53. Claude Duret (d. 1611)

Histoire Admirable des Plantes et Herbes esmerveillables & miraculeuses en nature: mesmes d'aucunes qui sont vrays Zoophytes, ou Plant'-animales, Plantes & Animaux tout ensemble, pour avoir vie vegetatiue, sensitive & animale: Avec leurs Portraicts au naturel, selon les histoires, descriptions, voyages, & navigations des anciens & modernes Hebrieux, Chaldees, Egyptiens, Assyriens, Armeniens, Grecs, Latins, Africains, Arabes, Nubiens, Ethyopiens, Sarrasins, Turcs, Mores, Persans, Tartares, Chinois, Indiens, Portugays, Espagnols, Fra[n]çois, Flaments, Anglois, Polonois, Moschouites, Allemans, & autres. Par M. Claude Duret, President à Moulins en Bourbonnois. [*Printer's device*]. A Paris, Chez Nicolas Buon, demeurant au mont S. Hylaire, à l'Image S. Claude. M.DCV. Avec Privilege du Roy.

8° 15.7 × 10.5 cm. a^4 e^8 A–X^8 Y^4 *i–xxiv* 1–295 *296*–302 *303*–308 *309*–315 *316*–318 *319*–341 *342–344* (255 as 155) [368 pp.].

PLATES: 28 woodcuts of plants.

BINDING: Modern full green morocco.

PROVENANCE: Library stamp on title-page: 'J^s A^c L[?] abaut P^r [?]

REFERENCES: *Age of Marvelous*, pp. 360–361; Nissen 571, Pritzel 2,553; Tavoni, Tongiorgi Tomasi and Tongiorgi, pp. 173–176.

PERHAPS the first book on the flora of the New World to be published in France was Claude Duret's *Histoire Admirables des Plantes et Herbes Emerveillables*. Printed in Paris by Nicolas Buon, the book was dedicated to the chevalier Maximilien de Béthune, Duke of Sully, Surintendant des Finances de Frances and Surintendant des Fortifications. The moralistic underpinnings of this work are set out clearly in the *Advertissement aux lecture*. The author begins with an Aristotelian definition of nature, but his central thesis was that the natural world represented the crowning achievement of God, offering an endless font of marvels that were all too often taken for granted.

Duret, a lawyer and civil servant of Moulins, was an omnivorous reader, devouring everything from biblical texts and the classics of Antiquity to the accounts of seafarers and cosmographers. He wrote several books on various topics, including a rather ambitious history of languages, *Thresor de l'histoire des langues de cet univers* (published by his widow in 1613). Since his eclectic interests ranged from religious history to contemporary events, his herbal, despite its novel subject matter, is replete with obscure citations that make for somewhat difficult reading.

Duret's aim in *Histoire des Plantes* was to describe some of the wonders of nature that were coming to the attention of naturalists during this period. Several explorers sent back to Europe a stream of plants and animals from newly discovered lands, some so bizarre that they might have been the fruits of some traveller's fevered imagination and others apparently identifiable with the monsters of popular mythology. It is unlikely that Duret ever visited the New World or had the opportunity to study at first hand the East and West Indian species presented in his herbal. Indeed, judging from the text, he had little direct knowledge of the plants in his book and drew most of his information from the handful of indispensable sources then available—the published works of Gonzalo Fernández de Oviedo y Valdéz (1478–1557), Cristobal Acosta, Nicolas Monardes (see No. 16), and Garcia da Orta (see No. 17). In every one of his descriptions, however, we find evidence of Duret's incessant search for confirmation of the existence of these 'marvels' in the texts of the classical authorities.

Histoire des Plantes reflects the myopia that characterized many of the scholars of the period, who seem to have had difficulty at first in distinguishing between fact and fiction. Thus, next to relatively straightforward descriptions of the Musa (banana tree), Indian fig, coconut palm, sunflower, and various 'plantes pudique,' we find uncritical accounts of the existence of 'animal-trees' such as the 'Arbre qui porte des feuilles, lesquelles tombées sur terre se tournent en oiseaux volants, et celles qui tombent dans les eaux se muent en poissons' and the 'Arbre de l'isle de Cimbubon, qui porte des feuilles qui vivent et cheminent' (Pl., p. 319).

Duret describes other strange natural phenomena that were widely accepted as fact, such as the notion that the barnacle goose (*Branta leucopsis*) hatched not from an egg but from a seashell growing on a tree. This legend probably sprang from sailors' accounts of trees that bore fruit in the form of shells—actually the goose barnacle (*Conca anatifera*), which grows on the sides of ships and on floating tree trunks, and whose heart-shaped white and black shell was thought to resemble the barnacle goose's head. This shell is mentioned in Gerard's famous *Herball*, published in 1597, just a few years before *Histoires Admirables* (see No. 12). Another bizarre creature cited by Duret was the 'Boramets' or *Agnus sciticus*, that was sometimes classified as an animal and sometimes as a vegetable. This fantastic hybrid, which had been described half a century earlier by the physician and philologist Julius Caesar Scaliger (1484–1558) in *De Subtilitate* (1557), was said to grow in the distant, myth-shrouded land of Scythia or Tartar and was portrayed as a down-covered animal hanging by its umbilical cord from the stem of the mother plant. The origin of this legend can be traced back to one of the practices of unscrupulous merchants, who would take the hairy root of the fern *Frutex tartareus*, trim it into the form of a quadruped, and sell it as a rare curiosity to collectors in Europe. Thus, reality and fantasy were often

CLAUDE DURET,
Histoire Admirable des Plantes et Herbes, 1605.
Barnacle Goose (*Branta leucopsis*) hatching from a seashell and growing on a tree (*Lepas anatifera*, Pelagic gooseneck barnacle, a Crustacean), page 303

CLAUDE DURET,
Histoire Admirable des Plantes et Herbes, 1605.
'Boramet' (*Agnus sciticus*) growing on a stem from the mother plant, page 330

intermingled in the study of the natural sciences. Duret defined the pineapple, for example, as 'une plante qui mange le fer'; after describing with reasonable accuracy the exotic fruit, which he notes came from Brazil, he repeats Acosta's assertion that if one left an iron knife in its pulp '[pour] l'espace d'un iour & une nuict, toute la partie dudit cousteau, qui aura esté dendans ladite etame, sera mangé & consommee par ce fruict' (p. 257).

Histoire Admirables des Plantes is accompanied by a series of twenty-eight beautifully executed woodcut engravings, many of which were not new to botanists, having already appeared in the Spanish sources consulted by Duret. Other plants, such as the 'Boramets' (p. 330), are illustrated here for the first time.

54. Pierre Pomet (1658–1699)

[*Printed in red and black with a double rule border*]: A Compleat History of Druggs, Written in French by Monsieur Pomet, Chief Druggist to the present French King; to which is added what is further observable on the same Subject, From Mess[rs]. Lemery, and Tournefort, Divided into Three Classes, Vegetable, Animal and Mineral; With their Use In Physick, Chymistry, Pharmacy, And several other Arts: Illustrated With above Four Hundred Copper Cutts curiously done from the Life; and an Explanation of their different Names, Places of Growth, and Countries from whence they are brought; the Way to know the True from the False, their Virtues, &c. A Work of very great Use and Curiosity. Done into English from the Originals. [*rule*] Vol. I. [–Vol. II.] [*rule*] London: Printed for R. Bonwicke, William Freeman, Timothy Goodwin, John Walthoe, Matthew Wotton, S. Manship, John Nicholson, Benjamin Tooke, Rich. Parker, and Ralph Smith. 1712.

4° 22.8 × 17.2 cm. Two volumes in one. a–b^4 B–2F^4 a–b^4 c^2 B–2B^4 2C–2F^2 *i–xvi* 1–224 [*20*] 225–419 [*13*] [468 pp.].

PLATES: 86 engraved and etched plates, 1–68 in Volume I and 69–86 in Volume II.

BINDING: Contemporary blind panel-stamped calf; rebacked in modern brown leather.

REFERENCES: Cleveland 295 and 328; Hunt 428; Plesch, p. 367; Pritzel 7,258.

A *Compleat History of Druggs*, the English translation of an herbal by the French pharmacist Pierre Pomet, appeared in London in 1712. Pomet, who had an apothecary shop in Paris at the sign of *La barbe d'or* in Rue des Lombards, was an enthusiastic collector with a great knowledge of the *materia medica*. Since his youth he had engaged in the buying and selling of medicines, travelling to various countries to discover new herbs and other agents. As ambitious as he was studious, Pomet eventually became chief druggist at the court of Louis XIV. His expertise was such that he was invited by eminent physicians to give demonstrations on the preparation of herbal remedies at the Jardin Royal des Plantes. At the apex of his career, in reward for his services Pomet was granted a pension by the king, but unfortunately he died on the same day that he received this gratifying news, at the age of forty-one. Another work by Pomet entitled *Droguier curieux* was published posthumously.

Pomet decided to compile the results of his many years of study and the knowledge that had brought him fame and fortune in a voluminous herbal that was a remarkably up-to-date work for its time—*Histoire générale des drogues* (Paris, 1694). Its history was marked by an unfortunate beginning; just before the work went to press, Pomet discovered that many of his notes and drawings had been stolen. He sought to have the persons suspected of pirating this material punished, but the court dismissed his case as 'a mere bagatelle,' as he notes bitterly in his Préface.

The English edition of Pomet's work was dedicated to Hans Sloane, secretary of the Royal Society, and boasted an impressive list of subscribers, including professors, pharmacists, physicians, surgeons, and booksellers. A German edition was published in Leipzig in 1717. In 1735 the son of the author, Joseph Pomet, who was a pharmacist at the Hôpitaux de Paris, published an expanded edition of his father's herbal entitled *Histoire générale des drogues simples et composées*. In the original French edition, which includes a portrait of Pierre Pomet, the reader is informed in the *Avis au lecture* that all of the drugs described in the herbal could be acquired, either in bulk or in small doses for one's personal use, at the sign of *La barbe d'or*. Copies of the herbal could also be purchased at the price of fourteen livres for the leather-bound edition and twelve livres for the edition bound *en blanc*.

Pomet's herbal was an extremely erudite and thoroughly documented work that reflected a very broad interpretation of the term 'drug.' According to the author this could include any substance—animal, vegetable, or mineral—with healing virtues, as may be seen from a perusal of the closing index. The success of Pomet's herbal can be attributed in part to the text; the author possessed the true writer's gift and his botanical descriptions and pharmacological observations are clear, succinct, and to the point. In the 1712 London edition the names of the drugs have been translated into English and supplementary information added by the botanist Joseph Pitton de Tournefort (1656–1708) and the pharmacist Nicolas Lémery (1645–1715), whose contributions are duly acknowledged. Lémery was the first scientist to make the distinction between the branches of vegetable (organic) and mineral (inorganic) chemistry.

The text is accompanied by a large number of etchings exactly copied from the original French edition. These include drawings of plants (sometimes four or five to a page), animals and minerals, and genre scenes depicting the harvesting or processing of different natural products. The section devoted to plants is the longest in the book and has been divided by the author into chapters on the various parts of the plant from which medicines could be extracted: the seeds, roots, woody parts, barks, leaves, flowers, fruits, resinous substances or gums (known since Antiquity to be effective in the treatment of illnesses), and juices. The botanical drawings are lucid and elegantly composed, while the genre scenes are full of interesting detail aimed at a public thirsty for information about the peoples,

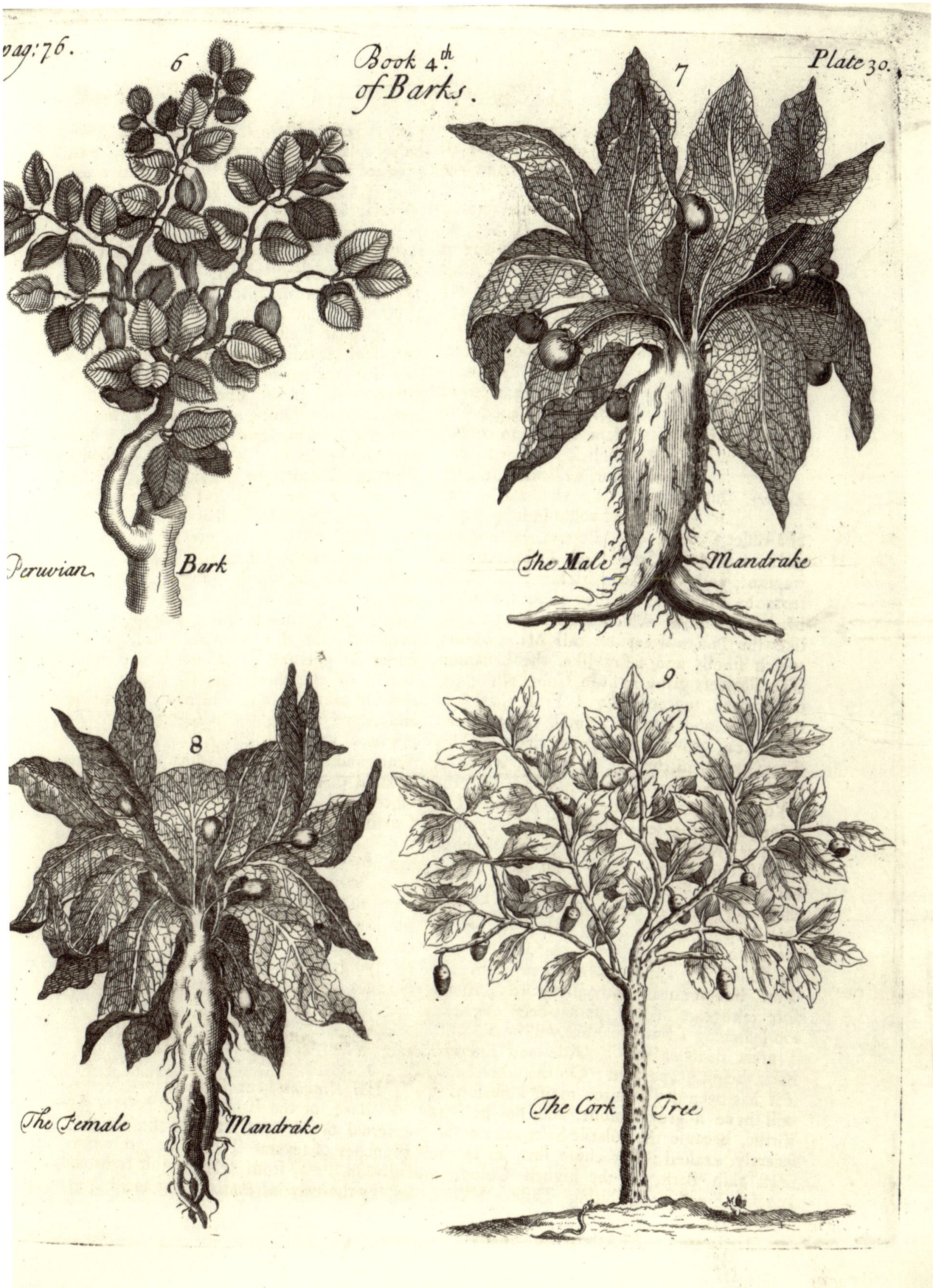

PIERRE POMET,
A Compleat History of Druggs, 1712. 'Peruvian Bark' (*Cinchona* sp.); 'The Male Mandrake' (*Mandragora officinalis*); 'The Female Mandrake' (*Mandragora autumnalis*); 'The Cork Tree' (*Quercus suber*), from 'Book the Fourth. Of Barks,' plate 30

PIERRE POMET,
A Compleat History of Druggs, 1712
Natives 'stringing & rolling tobacco,' from 'Book the Fifth. Of Leaves,' Plate 37

flora, and fauna of distant continents. For example, the artist has drawn a turkey in the foreground of the plate illustrating the drying of tobacco leaves, a scene typical of the New World (Pl. 37, opposite p. 94). Other engravings include the extraction of indigo, which depicts the plant 'Anil or Indigo' as well as 'Negroes' with their 'overseer' cutting, drying, infusing, and mechanically extracting the dye (Pl. 35, opposite p. 90); women preparing silk, depicting a cow feeding upon mulberry leaves, a butchered calf 'from whence ye silk worms are bred,' frames showing the worms feeding and the progression of their development (Pl. 79, opposite p. 281); and fishermen laying their catch on racks to dry in the sun (Pl. 84; opposite p. 293).

The second volume describes animals, metals, minerals, bitumens, stones, and earths. Only the first book, on animals, is illustrated. The zoological section contains many illustrations of land and marine animals. Some of the drawings are recognizable from earlier texts, such as the rhinoceros copied from the famous engraving by Dürer (Pl. 73). The second and third books of Volume II cover metals and minerals and it is interesting to note the establishment within the *materia medica* of mercury, lead, antimony, and arsenic.

55. Jean Pierre Rambosson (1827–1886)

Histoire et Légendes des Plantes Utiles et Curieuses Par J. Rambosson Ancien Rédacteur en Chef de la Science pour tous Ancien Président de la Classe des Sciences de la Société des Arts, Sciences et Belles-Lettres de Paris [*rule*] Ouvrage Illustré de vingt planches dessinées par Foulquier, Freeman, Gerlier et Lancelot gravées par Huyot et de cent vignettes insérées dans le texte [*rule*] Deuxième Édition [*rule*] Paris Librairie de Firmin Didot Frères, Fils et Cie Imprimeurs de L'Institut, Rue Jacob, 56 [*rule*] 1869 Droits de reproduction et de traduction réservés.

23.2 × 15.5 cm. *i–vi I* II–V *VI 1*–379 *380* [392 pp.].

PLATES: Engraved and etched portrait between pages 100–101 of 'Brillat-Savarin,' signed 'd'Après un dessin de Stanley,' and 'Passot sc.'; 20 full-page wood engravings and 104 wood-engraved illustrations within text.

BINDING: Red publisher's cloth embossed with gilt title on front board; re-backed in modern red calf; all edges gilt.

REFERENCES: *An Oak Spring Flora*, pp. 349–374; Pritzel 7,408; Staffleu & Cowan 8,605.

Histoire et Légendes des Plantes Utiles et Curieuses was just one of the many books on plants that were produced during the eighteenth century for a public made up of *amateurs* and *amatrices* of botany. The author, Jean Pierre Rambosson, was born in Saint Julien in the Haute-Savoie. President of the Classe des Sciences of the Société des Arts, Sciences et Belles-Lettres in Paris, he earned a modicum of fame as a writer specializing in works of popular science. One of his best known books was *Histoire des astres: astronomie pour tous*, published in Paris in 1874 to coincide with that year's historic Transit of Venus. *Histoire et Légendes des Plantes*, while written for the general public as well as specialists, was a serious and thoroughly researched botanical text. In it Rambosson presents forty-six different plants recounting the history, legends, superstitions, and uses of each in a pleasing, unpretentious style, embellished with literary references and poetic passages, in a manner typical of the period.

In his *Avis au Lecteurs* Rambosson describes the aims of his work. He begins by observing that anyone who cast even a brief glance over a typical botanical text would find only dry, arid descriptions and nomenclature sufficient to discourage the most ardent nature lover. He believed the reader would find *Histoire et Légendes des Plantes* clear, concise, and a pleasure to read, a rose without thorns (p. III). While he hoped the scientist would find nothing to quarrel with, he presented instead a wealth of information that might otherwise have to be searched for in thousands of different texts. Rambosson then explained that as he was writing the book he felt the necessity to embellish the text with a rich selection of illustrations. These were chosen with great care and inserted, generally at the end of each chapter, to fill what might otherwise have been a displeasingly empty space on the page. In fact, the author explains that he avoided encumbering his text with scientific descriptions, because he believed that accurate and realistic illustrations would be quite sufficient: 'Si peu que vaillent les dessins, ils donnent souvent du végétal une idée plus exacte que les phrases les plus claires et les plus nettes' (However poor the drawings, they often give a more exact idea of the plant than the clearest and plainest language) (p. v). Rambosson concludes by affirming that his name was frequently cited by a celebrated botanist, without however specifying the scientist's name.

The first chapter of *Histoire et Légendes des Plantes* contains a general discussion of plants and their 'influence, nostalgie, enseignements.' Rambosson describes the beauties of the forest and field in lyrical prose: '... orchestre divin où la brise murmure en gammes infinies l'hymne d'amour qui révèle le Créateur à la creature!' (divine orchestra where the breeze murmurs in an infinite cascade of notes the hymn of love that reveals the Creator to his creature!) (p.7). He then embarks upon a discussion of various plants, devoting one chapter to each, arranged in alphabetical order beginning with L'Acanthe and concluding with La Viorne. The author seems to have been interested above all in plants with applications in cooking (the lemon, citron, pear, and potato) and medicine (the aconite, agaric, and aloe). He also includes chapters on exotic species, such as the pineapple, banana, and the 'arbre du voyageur,' the name derived from the fact that the stem can be cut and the water inside consumed (Fig. 26). In the chapter 'Arbre du voyageur,' Rambosson's eclectic style is reflected in his description of the French Colonies located in the Reunion Islands near Madagascar, interwoven with odds and ends like the reference to *Paul et Virginie* (Paris, 1788) a romantic novel, by Bernardin de Saint Pierre (1737–1814), that takes place in the islands.

In his chapter on the Cigue (poison hemlock, *Conium maculatum*), Rambosson discusses the poisonous effects of the plant, its antidotes, and its uses in medicine. He recounts the story of the sublime death of Socrates (469–399 B.C.), as told by Plato in his dialogue *Phaedo*. Alphonse de Lamartine (1790–1869) also found inspiration in this historical episode for his poem, *La mort de Socrate*. In addition to a small wood engraving of the plant, Rambosson has included a full-page illustration depicting the Greek philosopher in his final hour, calmly accepting the cup of deadly poison (Fig. 46).

A long chapter is dedicated to the tea plant in which Rambosson describes its history and the arrival of the precious tea leaf in Europe. He furnishes a list of its chemical components 'tannin, volatile oil, wax, resin, gum, extracts, nitrogenous substances analogous to albumin, various salts, and an alkaloid called theine that is identical to caffeine' (p. 307), as

Fig. 26. — Arbre du voyageur.

JEAN PIERRE RAMBOSSON, *Histoire et Légendes des Plantes*, 1869. Man drinking water from the cut stem of the 'Arbre du voyageur' (*Ravenala madagascariensis*), plate 26

JEAN PIERRE RAMBOSSON, *Histoire et Légendes des Plantes,* 1869. Cityse commun ou false ébénier (*Laburnum anagyroides*), plate 42

Fig. 42. — Cityse commun ou faux-ébénier.

well as instructions on how to prepare the perfect infusion, and on how to store the leaves (in an airtight tin) so that they do not lose their delicate flavor. Regarding its therapeutic effects, Rambosson stated that tea could aid the digestion and was effective against stomach complaints, kidney stones, and sleepiness. He notes, however, that drinking too much tea could result in vertigo, tremors or tooth loss, and quotes Honoré de Balzac (1799–1850) on the Englishman's (already) excessive predilection for the beverage (p. 318). This chapter as well is accompanied by a botanical illustration and two large engravings depicting the tea harvest in China (Figs. 102, 103).

A chapter embellished with a fine illustration of *Tabac rustique* (Fig. 96) is devoted to the tobacco plant, its origins and history, its uses, and its abuse, particularly by the young. The latter topic was of special concern to the author, who felt strongly that the plant should be 'regardé purement comme substance pharmaceutique, et employé comme re-

Fig. 102. — La récolte du thé.

JEAN PIERRE RAMBOSSON, *Histoire et Légendes des Plantes*, 1869. Harvesting of tea, plate 102

mède seulement' (regarded purely as a pharmaceutical substance and utilized only as a medicine, p. 363).

The pleasantly bourgeois connotation of the work is reflected in the chapter dedicated to 'les Plantes à la Maison,' which offers practical advice on how to cultivate and care for houseplants. Here he describes different types of 'jardinières,' elegant containers for flower pots, planters with wooden frames to which climbing plants could attach themselves, acquariums for aquatic plants, shelving systems for succulent plants and cacti, and 'miniature greenhouses' that could be set up in a lady's sitting room, 'dont le prix est modique et dont l'élégance ne deparerait pas le plus riche mobilier' (whose price is modest and whose elegance would not detract from the most beautiful piece of furniture, p. 363).

Rambosson closes his work with some miscellaneous remarks on various curiosities. He provides a list of the flowers contained in Carolus Linnaeus' famous *horologium florae* or floral clock, which was divided into twelve sections and planted with flowers that would open in turn at the appropriate hour (see No. 46). Linnaeus claimed that his timekeeper was accurate to within thirty minutes, but Rambosson noted that the clock might run fast or slow depending on the climate at the reader's latitude. He also lists the flowers in Linnaeus' *Calendar of Flora*, a work in which the famous Swedish botanist described the plants that could be found blooming in France during each of the twelve months of the year. Rambosson concludes with a reference to the 'language of flowers,' a genteel diversion that was immensely popular during the nineteenth century, which consisted of assigning 'à la plupart des fleurs un attribut particulier, auquel elles servent d'emblème et d'expression' (to almost every flower a specific attribute, of which it would serve as the emblem and expression, p. 370) (see *An Oak Spring Flora*, pp. 349–374).

Based on the signature, the preparatory drawings for *Histoire et Légendes des Plantes* were the work of a certain 'Hayon,' perhaps Leon Albert Hayon (1840?–1885). The engravings were made using a new technique developed by the English artist Thomas Bewick (1753–1828), who employed very hard woods and the end of the block rather than the side, working the wood's end-grain with a graver or burin. This procedure resulted in highly detailed images that—unlike copperplate engravings—could be printed thousands of times along with the text on a conventional printing press, and therefore was widely adopted for book illustration in the nineteenth century. The blocks for Rambosson's work were engraved by several more or less well-known artists specializing in the technique, among them Jean Antoine Valentin Foulquier (1822–1898), Dieudonné Auguste Lancelot (1822–1894), and Freeman, Garlier, and Brenton, about which last three no information can be found.

VII

DRIED SPECIMENS AND NATURE PRINTING

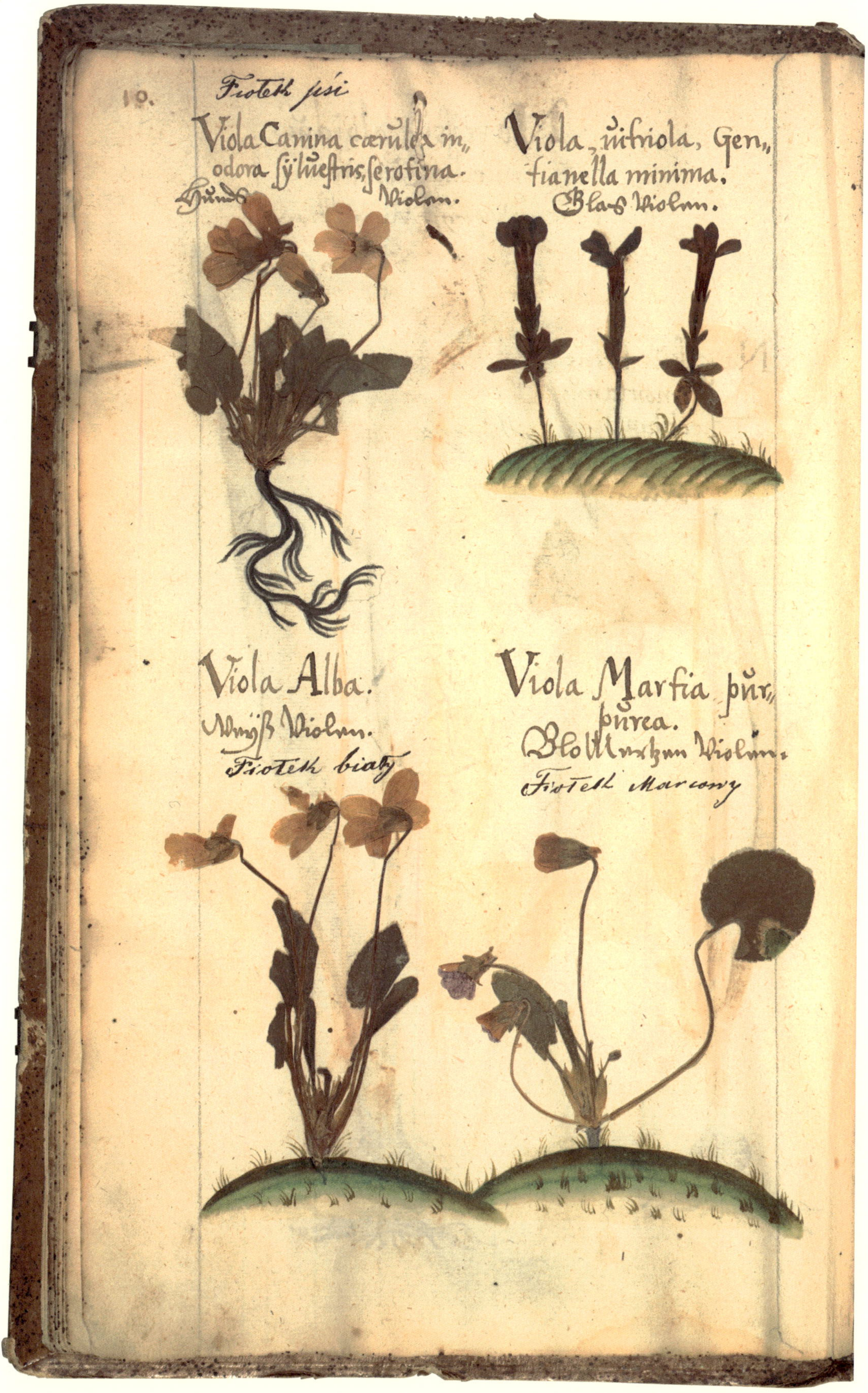

JOHANNES HARDER, *Historia Stirpium*, *c.* 1595. Three species of violets (*Viola canina, V. alba, V. odorata*) and a species of felwort (*Gentianella* sp.), folio 20

56. Johannes Harder (*c.* 1563–1606)

Historia Stirpium, Collecta, per Doctore[m] Ioanne[m] Harderum Medicu[m] Geislingensem [*circa* 1595].

31.3 × 19.7 cm. Title-page, six index leaves, six register leaves, 197 leaves numbered in a contemporary hand 1–201, 203–221, 223–224, 233–322, 324–406, containing 588 dried plant specimens, most with watercolor additions showing bulbs, roots, fruits, vegetables, grassy mounds, and other details, and ten blank leaves with 15 specimens loosely inserted. The plants have been identified in Latin and German by Harder.

BINDING: South German blind-stamped pigskin over wooden boards; remnants of metal and leather clasps and hasps.

PROVENANCE: Label on front paste-down endpaper: 'N° 1022 [*stamped numbers*] D XIV [*stamped roman numerals*] Bibljoteka Julińska.' Stamped on title-page and rear endpaper: 'Biblioteka Julińska.' 'Jeremias Forstner' inscribed on rear paste-down.

One of the oldest dried herbaria or *horti sicci* extant, *Historia Stirpium*, was compiled by Johannes Harder, son of the pharmacist Hieronymus Harder (1523–1607). Hieronymus Harder was one of the first naturalists to begin systematically collecting dried plants for scientific and teaching purposes in Germany. This practice spread quite early from Italy, forming a natural complement to the already well-established activity of botanical illustration. Approximately nine herbals compiled by Hieronymus have survived in different European collections (in Munich, Vienna, Salzburg, Linz, Uberlinger, Ulm, Heidelberg, and Rome). Johannes Harder studied medicine in Tübingen and then Strasbourg, where he received his degree in 1593. He practiced medicine in Geislingen an der Steige (where, based on the title, we may deduce that the herbal was compiled) and then in Ulm.

As we have already observed (see No. 13), the technique of preparing dried plant specimens can be traced back to the Italian botanist Luca Ghini (1490–1556), who assembled an *orto secco* during the early part of the sixteenth century. The practice was passed onto his students, among them Gherardo Cibo and Ulisse Aldrovandi. By the middle of the century, illustrious botanists in other parts of Europe had begun compiling similar collections: Konrad Gesner (1516–1565) in Switzerland, Caspar Ratzenberger (d. 1603) in Germany, and Jean Girault (*c.* 1558) in France. In his commentary on Dioscorides (1553), Amato Lusitano (1511–1562) mentions having admired the dried herbal of the English botanist John Falconer (*c.* 1553), while Michel de Montaigne (1533–1592) alludes to a remarkable collection of dried plants belonging to the celebrated professor Felix Platter (1536–1614), which he saw during his stay in the city of Basel (Arber, 1986, pp. 140–141). Beginning in the second half of the sixteenth century, the dried herbal became the most common system for the collection and scientific study of plants. Hieronymus and Johannes Harder were among the first to adopt the practice of conserving specimens by drying, marking the beginning of the modern study of botany and, indeed, the practice continues today. The

JOHANNES HARDER,
Historia Stirpium,
c. 1595. *Arum maculatum,* (Cuckoo-pint),
folio 21

herbarium constitutes a precious source of information and an invaluable instrument for studying the variability of specific plant species and their distribution around the world.

Historia Stirpium is preceded by a list of the 359 plants contained in the collection, arranged in alphabetical order by their Latin names and followed by their German equivalents. On every sheet of the herbal one or more dried plants has been carefully arranged. This work is unusual because of the hand-drawn details added by the author to clarify the structure and appearance of the plants. Nearly every page contains such additions; for example, the opening sheet presents a dried narcissus accompanied by a watercolor drawing of the bulb, which could not have been conserved in the herbal, given its shape. This may explain why so many of the dried plants are accompanied by a drawing of their roots. In some cases Harder went even further, adding naturalistic backgrounds to the lichen (fol. 4) and the violet (fol. 10) to illustrate their habitats.

Other interesting pages from this dried and painted herbal include the *Arum* (fol. 21), where the author has glued two dried leaves to the page and added drawings of the flower and seeds, and the red current (*Ribes rubrum*), which is accompanied by the sketch of a twig with some crimson berries (fol. 381). On occasion, flowers whose colors had faded during the drying process have been retouched by hand, as, for example, the botanist has added strokes of bright red watercolor to the petals of *Trifolium pratense* (fol. 324).

57. Nature prints of plants and trees from England (*early* 1700s)

20 × 15.8 cm. Two volumes. Volume I: Eight leaves of index, 125 leaves containing nature prints of approximately 383 plants, numbers 1, 60, 63 are painted green. Volume II: One leaf of index, one blank leaf, 122 leaves containing nature prints of approximately 153 plants. All are numbered and correspond to the indexes and some leaves have been inscribed with the plant's common names. One nature print loosely inserted.

BINDING: Contemporary Cambridge calf.

PROVENANCE: Bookplates: 'The Honble Tho. Trevor. Esqr' and Paul Mellon. Inscribed on front paste-downs: 'Phillipps MS 7373 2 Vol.'

REFERENCES: Cave & Wakeman, p. 7; Conihout, p. 12; Tognoni, pp. 347–370.

'NATURE PRINTING,' also referred to by the Latin term *typographia naturalis* and the German word *Naturselbstdruck*, is a very ancient technique that allows one to make direct impressions of plants on paper. One of the earliest surviving examples is a stem of sage (*Salvia officinalis*) in an Italian herbal that has been dated to the end of the fifteenth century, now conserved in the Musée d'Histoire Naturelle in Paris (MS. 326, fol. 19). The inclusion of this print in the herbal reflected the growing need of botanists to have more accurate and realistic illustrations to which to refer, rather than the stylized images contained in medieval herbals. Even the 'scientist-artist' Leonardo da Vinci (1452–1519) experimented with this technique. As is well known, many of his notebooks contain botanical, zoological, and anatomical drawings. In the imposing *Codice Atlantico* (Biblioteca Ambrosiana, Milan) we find the print of a sage leaf (fol. 72^{v}), whose veins and other raised parts can be seen in perfect detail against the paper. A manuscript note by the artist on the same sheet explains the procedure he used, which consisted of coating the leaf with lampblack and then applying brushstrokes of 'colla dolce' or glue. Leonardo did not repeat this experiment, which must be considered just one of the many promising ideas that he had but decided not to pursue, while other naturalists took it up with more interest. The practice of making impressions of leaves or other parts of plants on paper using different materials and techniques gradually won acceptance, particularly from the second half of the sixteenth century onwards, and indeed it was still being used until quite recently. The nature print thus constitutes a curious but engaging chapter in the history of botanical illustration, and in particular of the herbal.

An important Italian manuscript of nature prints compiled by the *aromatarius* or perfumer Zenobio Pacini of Florence entitled *Plantarum Icones* (*c.* 1520) was brought to France by Catherine de' Medici (1519–1589) (Bibliothèque Nationale, Paris, Est. Rés.j.d. 50). This work is composed of more than two hundred sheets printed on both the recto and verso sides with images of plants in brown ink that were then colored with watercolors and coated with a protective layer of varnish. A note dating from the eighteenth century explains that the impressions were made by placing each specimen between two sheets of paper over which a heavy cylinder was passed several times (see Conihout, p. 12). A similar herbal of nature prints by Pacini is conserved at the Biblioteca Nazionale in Florence (N.A. 90).

Although this technique of nature printing was at first kept secret, as its use spread amongst physicians and naturalists, manuals explaining the procedure began to appear, such as the *Liber de secretis naturæ* by the polymath and man of letters Girolamo Ruscelli (*c.* 1500–1566). Ruscelli published the work under his pseudonym, Alexis Pedemontanus, in Milan in 1556 and in Paris in 1573. Among the collection of scientific drawings produced for the Botanical Garden of Pisa are two sheets dating from the end of the sixteenth century containing the images of various leaves printed using lampblack (Pisa, Biblioteca Universitaria, MS. 513, fols. 213, 216). Botanists continued to avail themselves of the technique during the seventeenth century and in his work *Isagoges*, first published in 1606, the Belgian naturalist and physician Adriaan van de Spiegel (see No. 13) provides a detailed description of nature printing. A later edition of *De Subtilitate libri XXI* (Lugduni: 1663, p. 578) by Gerolamo Cardano of Milan (1501–1576) also contains a section on nature printing.

A monumental herbal of about five hundred nature prints prepared by Fabio Colonna (see No. 11), member of the prestigious Accademia dei Lincei in Rome, has recently been discovered in the library of Blickling Hall, Norwich (Tognoni, pp. 347–370). The title, *Icones ipsis plantis ad vivum expressae quaod fieri potuit nova quaedam arte excogitata ab ipso auctore,* shows that Colonna, like many authors, believed that he was the first to describe the procedure, whereas he was merely introducing variations and improvements to what was in fact a very elementary principle. After Fabio Colonna, the next important contribution to nature printing was made by Paolo Boccone of Sicily (1633–1723), a Cistercian monk and well-known botanist who, before he died in Oxford, seems to have explained his method to William Sherard (1659–1728). Boccone left behind a manuscript of nature prints now conserved in the Bodleian Library at Oxford, and another filled with prints of plants from the Botanical Garden of Florence, today conserved in the Austrian National Library in Vienna (Cave & Wakeman, p. 7). The technique was gradually perfected: the specimen was prepared by removing excess moisture or drying the plant entirely, and the impression was made by laying the plant between two sheets of paper, which were then pressed between two metal plates, one of soft lead and the other of hard steel. This procedure resulted in a remarkably detailed image and with further improvements during the nineteenth century it became possible to obtain a series of prints from a single inked plate.

During the course of the eighteenth and nineteenth centuries, nature printing became more popular, especially in the German-speaking regions of Europe where botanists began to publish *Naturselbstdruck*. They were convinced that the technique could replace the botanical illustrations produced by woodblock and copperplate engraving, although it never actually succeeded in doing so. This could have been in part been because the technique, which was well adapted to the reproduction of single leaves and parts of plants, could not be used to make impressions of an entire plant or those parts that were rounded rather than flat, such as the roots. In 1733 Johann Hieronymus Kniphof (1704–1763), professor of anatomy and botany at the University of Erfurt, published *Botanica in originali seu Herbarium vivum in quo tam indigenae quam exoticae plantae*, with the close collaboration of the printer J. M. Funcke. This work of two hundred nature prints was so well received that expanded editions were printed in 1747 and in 1757–1764, the latter containing no less than 1,200 prints. The Silesian physician and botanist Christian Gottlieb Ludwig also published a work illustrated with nature prints, *Ectypa vegetabilium* (see No. 59), in 1760. In colonial America in 1737, Benjamin Franklin even incorporated a variation of the process into the first issue of paper currency in the Western world, exploiting the unique vein structure of a leaf as a foil to counterfeiting.

The anonymous two-volume English herbal conserved at the Oak Spring Garden Library, which dates to the first half of the eighteenth century, constitutes one of the first

2
4
5

ENGLISH SCHOOL,
Nature prints of plants and trees from England, early 1700s. Double-page spread of nature prints: 1. 'Asrabacca,' or European wild ginger (*Asarum europaeum*); 2. 'Silver Weed' (*Potenilla anserine*); 3. 'Indian Cresses,' or Indian Cress (*Tropaeolum majus*); 4. 'Sweet Brier' (*Rosa rubiginosa*); 5. 'Fool's Parsley' (*Aethusa cynapium*). Volume I

examples of nature printing to be produced in England following the improvements introduced by Boccone. The herbal contains nearly five hundred botanical impressions made using the most ancient technique of printing with lampblack. The nature prints, which depict for the most part leaves, small stems, and a few flowers with simple corollas, are admirably precise and detailed. Each print is numbered, and the author provides an index of the plants (listed by their English names) together with their corresponding numbers at the beginning of the manuscript. Short descriptions and notes on the first thirty-three plants follow, providing information on the native habitat of each plant and including, where appropriate, its therapeutic properties and chemical components. The consideration given to the latter aspect reflects the growing importance of the 'modern' science of chemistry in the traditional study of botany. The author notes that most of the plants collected for the nature prints were from the north of England, such as Bramham in Cumberland, and Newborough in Yorkshire.

An attempt was made to color by hand the first print, an *Asarum europeaum* leaf, but unfortunately the green watercolor completely obscured the veins and the author only repeated this experiment once or twice more in his herbal. Of the *Asarum*, he noted that 'Chymically analyz'd it yeilds [*sic*] a copious quantity of oil, Acid Phlegm, Earth with a small Portion of an brinous spirit … Cathartick, Diuretick, Diaphoritick, Hepatick … reduc'd to powder … it opens obstruction' and it was in addition effective for many other illnesses.

58. Carlo Sembertini (*c.* 1720) [*attributed*]

[*Within landscape and floral border*]: [*erasure*] ⋆Pharmaco⁑
[*erasure*]
Diversarum ⋆ Plantarum ⋆ Studiosum ⁂
⋆ Congestum ⋆
Opus labore, ac maiori, qua potuit
⋆diligentia⋆
⋆ proprÿs manibus peractu[m] ⋆

37.1 × 16 cm. The front leaves consist of: (1) Title-page; (2) frontispiece, 'In Nomine Domini [*within top cartouche*] Deus. Misit. Medicinam de. Coelo' [*within lower carthouche*]; (3) two leaves of dedication and encomiastic poems with text and drawings; (4) leaf of flowers and landscape with putti holding a ribbon inscribed 'In herbis, In verbis, Et In Lapidibus;' (5) four index leaves in red and black; (6) leaf with allegorical drawing of the sorceress Circe, inscribed 'Colligit ex dumis ingens medicamina Circe.' 186 leaves of dried plant specimens, mounted with decorative cloth strips and silk ribbons, numbered 1–187 with the plant names inscribed in red and black below each specimen and each leaf bordered with red and brown rules. Leaf seven has been cut out; 84 leaves are blank for additional specimens to complete collection, as listed in index.

BINDING: Original red morocco tooled in blind; all edges painted in flower and vine motif.

CARLO SEMBERTINI, *Pharmaco*, *c.* 1720. Title-page

THE AUTHORSHIP of this singular *orto secco* or dried herbal presents a genuine enigma, beginning with the fact the name of the author has for some unknown reason been erased from the title-page. The antiquarian who sold the herbal claimed to have identified the author as a certain 'Carlo Sembertini,' while the provenance (Verona) and approximate date of the manuscript (1720s) can be deduced from the watermarks. Research into the history of the botanical and pharmaceutical sciences in the city of Verona suggests the author of the herbal could have been a member of the Sembenini family, one of whom, Giovanni Battista, was active as a pharmacist and author of popular scientific works between 1806 and 1870.

Whoever the author may have been, *Pharmaco* represents an admirable achievement, compiled with loving attentiveness by a plant collector of modest artistic talent. The work opens with an impressive title-page decorated with two allegorical landscapes painted in vivid watercolors and surrounded by a floral border. In the center is the title, which alludes to the effort and diligence required of the author to create this work 'proprijs manibus' (with his own hands). A decorated frontispiece follows, consisting of a group of winged putti who hold up two cartouches bearing the inscription: 'In Nomine Domini Deus Misit Medicinam de Coelo' (In His name the Lord sent down medicine from the heavens), a reference to the divine origin of the healing arts. A coat of arms with the Holy Lamb and three stars set against a landscape is depicted beneath.

The next page contains a dedication in Latin addressed to an unknown 'Eccellentissimo ac praestantissimo Viro Angelo Barberio Medico Fisico' (Most excellent and important physician Angelo Barberio). It is signed 'Humilis^{us} Seruus' (most humble servant), the author apologizing for his lack of skill and trusting that his exalted patron might find this gift nonetheless pleasing. There follows an epigram addressed to the reader, and two short poems in praise of the author, one written by Camillo Guidotti and the other by Federico Dendronico. Nothing is known of these three personages, and no specific information can be found regarding the works, which are trite examples from the literary tradition of celebratory verse-making. On the next leaf is the drawing of a large flower set against a landscape, with a putto holding up a cartouche: 'In herbis in verbis et in lapidibus' (In herbs, in words, and in stones).

This is followed by an index of the plants in the herbal, most of them well-known indigenous medicinal species, arranged in alphabetical order by their Latin names from 'Anemone' to 'Vincitoxicum.' The herbal itself opens with a painting of Circe, shown gathering plants in a landscape filled with human figures and animals. A cartouche reads: 'Colligit ex dumis ingens medicamina Circe' (Circe gathers from the thickets powerful medicines) and the entire scene hovers over the panorama of a harbor town. According to ancient tradition, Circe was the daughter of the Sun and a sorceress who possessed an

CARLO SEMBERTINI, *Pharmaco*, *c.* 1720. Leaves of two unidentified monocots and *Ipomoea quamoclit* at the center, folio 29

CARLO SEMBERTINI, *Pharmaco*, *c.* 1720. Two specimens of *Leontopodium alpinum*, *Geranium argenteum* at the center, two specimens on each side of *Antennaria dioica* s.l. and *Micropus erectus* (*Bombycilaena erecta*) at the center, folio 53

intimate knowledge of herbs and potions; in Book x of the *Odyssey*, Homer recounts how she transformed Odyseus' sailors into a herd of swine.

The one hundred and fifty pages that follow present a series of dried plants, charmingly mounted with silk ribbon and strips of decorated cloth, and surrounded by red and brown ink borders. Beneath most of the specimens the author has written the Latin name of the species. While not a gifted artist, the author possessed a genuine expertise in the drying of plants, and even today his specimens retain their most ephemeral parts, such as petals, and the most delicate leaves. This work is endowed with an immense charm and elegance that is only enhanced by the mystery surrounding its origins.

59. Christian Gottlieb Ludwig (1709–1773)

Ectypa Vegetabilium usibus medicis praecipue destinatorum et in Pharmacopoliis obviorum variisque modis praeparatorum ad Naturae similitudinem expressa accedit eorumdem culturae proprietatum viriumque brevis descriptio moderante D. Christiano Gottlieb Ludwig Ordinis Medici in Academia Lipsiensi Decano. [*rule*] Halae Magdeburgicae 1760. Impressit Joannes Godofredus Trampius Prostat Lipsiae In Officina Breitkopfiana. [*double rule*] Nach der Natur verfertigte Abdrûcke der Gewâchse, welche besonders zu dem medicinischen Gebrauche bestimmt sind, und in Apotheken aufbehalten auch auf verschiedene Art zubereitet werden. Nebst einer Kurzen Beschreibung deren Wartung, Gigenschaften und Wirkungen, unter der Aufsicht D. Christian Gottlieb Ludwigs Dechants der Medicinischen Facultât zu Leipzig, herausgegeben. [*rule*] Halle im Magdeburgischen 1760. Berfertiget von Johann Gottfried Trampe, und zu bekommen bey Bernh. Christ. Breitkopf in Leipzig.

2° 26.8 × 22.6 cm. ⋆² A–2A² *i–iv 1* 2–96 (74 as 47) [100 pp.].

PLATES: 199 hand-colored nature-printed leaves of plants, numbered 1–200, lacking number 67, with letterpress of plant's name in Latin and German, and most with accompanying annotations.

BINDING: Contemporary half calf and gray paste paper boards.

PROVENANCE: Previous owner's manuscript entry on front free endpaper.

REFERENCES: Brunet III.1230; Dunthorne 188; Hunt 569; Nissen 1,252; *An Oak Spring Flora*, No. 83; *An Oak Spring Sylva*, No. 7; Plesch, pp. 319–20; Stafleu & Cowan 5,068.

NATURE PRINTING reached remarkable levels of quality and precision in eighteenth-century Germany and books of images produced using this technique became extremely popular. The first to attempt such a work was Johanns Hieronymus Kniphof (1704–1763), professor of botany and anatomy at the University of Erfurt, whose *Botanica in originali* containing two hundred images was printed by Funcke in Erfurt in 1733. A second edition containing no less than 1,200 nature prints was produced between 1757 and 1764 in Halle by the publisher Trampe.

Another celebrated work that represented 'along with J. H. Kniphof's *Botanica in originali*, one of the eighteenth century's most valiant attempts at nature printing' in the words of Hunt, was Christian Gottlieb Ludwig's *Ectypa Vegetabilium Usibus Medicis*. A physician, botanist, and mineralogist from Silesia, professor of medicine at the University of Leipzig and director of Leipzig's Botanical Garden, Christian Gottlieb Ludwig made significant contributions to the study of botany in eighteenth-century Germany. He undertook an important scientific expedition to North Africa from 1731 to 1733, bringing back many specimens that unfortunately were later lost. He was a friend of the naturalist Christoph Jacob Trew (1695–1769), and supervised the last part of Elizabeth Blackwell's *Herbarium Blackwellianum* (*An Oak Spring Flora*, No. 83). Ludwig himself was the author of many books, perhaps the most important being *Definitiones generum plantarum in usum auditorum*, which appeared in Leipzig in 1737.

Ectypa Vegetabilium was printed in Latin and German in Halle by Johann Gottfried Trampe, who also produced the second edition of Kniphof's *opus magnum*. In his preface Ludwig explains that he had compiled this condensed book of medicinal plants for those readers who were unable to afford Kniphof's twelve-volume work, and *Ectypa Vegetabilium* was distributed in modest, affordable issues consisting of twenty-five plates each between 1760 and 1764.

In contrast to *Botanica in originali*, Ludwig's herbal was printed in color using a new, somewhat laborious and, as it turned out, not entirely satisfactory technique that is briefly described in the introduction. The botanical specimen was coated with printer's ink, placed on a sheet of paper, and sprinkled with black powder. After the impression was made the print was colored by hand, and the process of printing the text could then be completed. In his introduction Ludwig underlined the importance of color in the study of plants and also expressed his view, understandable in a physician, that the primary aim of botany was to discover useful treatments for illnesses. *Ectypa Vegetabilium* represented a laudable attempt to overcome one of the great challenges of book production—color printing, and indeed the solution would only be found at the end of eighteenth century with the invention of lithography. Concerning Ludwig's innovation, Brunet wrote in his *Manuel du Libraire*: 'Par ce moyen on s'est épargné les frais de la gravure, mais il n'en est résulté qu'un ouvrage informe, qui est à très bas prix' (By this technique one was saved the expense of engraving, but the result was merely a shapeless work at a low price—III.1230), and the fact was that, on the contrary, the cost of this process must have been anything but modest.

The plants in Ludwig's herbal are almost all native species. Each illustration is accompanied by a brief description in German and Latin that covers the plant's medicinal properties and includes recipes for simple remedies. As many scholars have commented, from an aesthetic point of view these prints are disappointing, even if some of them do possess a certain naïve charm. The naturalism that resulted from the technique of printing from live specimens, as can be seen in the illustrations of *Dictamnus albus* (Pl. X) and the *Alliaria*

CHRISTIAN GOTTLIEB LUDWIG, *Ectypa Vegetabilium*, 1760–1764. Black salsify or Spanish salsify (*Scorzonera hispanica*), plate 96

CHRISTIAN GOTTLIEB LUDWIG, *Ectypa Vegetabilium*, 1760–1764. Common thorn apple (*Datura stramonium*), plate 108

(Pl. LXXVII), was all too often marred by inept hand-coloring that masked important details such as the veins of the leaves. Each illustration bears the name of the plant printed in Latin and in German, and this copy contains further hand-written notes regarding the plant's nomenclature.

Other works of nature printing can be found in the Oak Spring Garden Library, including a German manuscript compiled in 1824 that contains prints in black and white of deciduous leaves, bound together with a treatise entitled *Sammlung von 50 in Kupfer gestochenen Abdruecken* (perhaps written by the Breslau bookseller, Johann Friederich Korn the Elder) that was printed in Breslau in 1797 (see *An Oak Spring Sylva*, No. 7, pp. 29–32).

VIII

AMERICAN HERBALS

SAMUEL HENRY,
A New and Complete American Medical Family Herbal, 1814.
Title-page

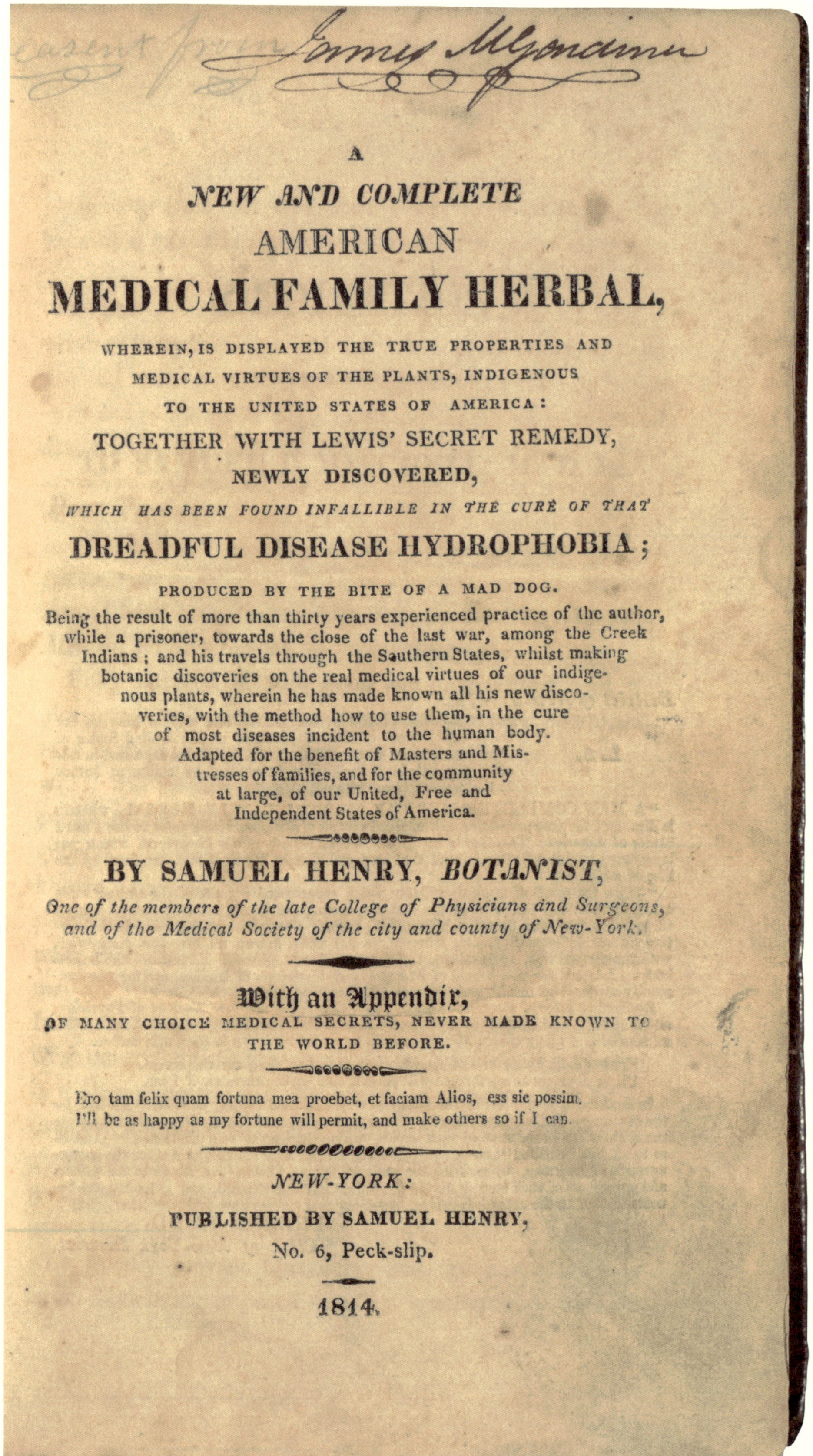

A

NEW AND COMPLETE

AMERICAN

MEDICAL FAMILY HERBAL,

WHEREIN, IS DISPLAYED THE TRUE PROPERTIES AND MEDICAL VIRTUES OF THE PLANTS, INDIGENOUS TO THE UNITED STATES OF AMERICA:

TOGETHER WITH LEWIS' SECRET REMEDY,

NEWLY DISCOVERED,

WHICH HAS BEEN FOUND INFALLIBLE IN THE CURE OF THAT

DREADFUL DISEASE HYDROPHOBIA;

PRODUCED BY THE BITE OF A MAD DOG.

Being the result of more than thirty years experienced practice of the author, while a prisoner, towards the close of the last war, among the Creek Indians; and his travels through the Southern States, whilst making botanic discoveries on the real medical virtues of our indigenous plants, wherein he has made known all his new discoveries, with the method how to use them, in the cure of most diseases incident to the human body. Adapted for the benefit of Masters and Mistresses of families, and for the community at large, of our United, Free and Independent States of America.

BY SAMUEL HENRY, *BOTANIST*,

One of the members of the late College of Physicians and Surgeons, and of the Medical Society of the city and county of New-York.

With an Appendix,

OF MANY CHOICE MEDICAL SECRETS, NEVER MADE KNOWN TO THE WORLD BEFORE.

Ero tam felix quam fortuna mea proebet, et faciam Alios, ess sic possim.
I'll be as happy as my fortune will permit, and make others so if I can.

NEW-YORK:

PUBLISHED BY SAMUEL HENRY,

No. 6, Peck-slip.

1814.

60. Samuel Henry (*fl.* 1800s)

A New and Complete American Medical Family Herbal, Wherein, is displayed the True Properties and Medical Virtues of the Plants, Indigenous to the United States of America: Together with Lewis' Secret Remedy, Newly Discovered, Which has been found Infallible in the Cure of that Dreadful Disease Hydrophobia; Produced by the Bite of a Mad Dog. Being the result of more than thirty years experienced practice of the author, while a prisoner, towards the close of the last war, among the Creek Indians; and his travels through the Southern States, whilst making botanic discoveries on the real medicinal virtues of our indigenous plants, wherein he has made known all his new discoveries, with the method how to use them, in the cure of most diseases incident to the human body. Adapted for the benefit of Masters and Mistresses of families, and for the community at large, of our United, Free and Independent States of America. [*decorative rule*] By Samuel Henry, Botanist, One of the members of the late College of Physicians and Surgeons, and of the Medical Society of the city and county of New-York. [*swelled rule*] With an Appendix, of Many Choice Medical Secrets, Never Made Known to the World Before. [*decorative rule*] Ero tam felix quam fortuna mea proebet, et faciam Alios, ess sic possim [*sic, i.e., ... praebet, et faciam Alios esse, si possim*]. I'll be as happy as my fortune will permit, and make others so if I can. [*decorative rule*] New-York: Published by Samuel Henry, No. 6, Peck-slip. [*swelled rule*] 1814.

8° 22.1 × 13.8 cm. 1–49^4 χ^2 *i–iii* v (i.e., iv) *v* vi *7–9* 10–336 *337–339* 340–384 *385* 386–390 *391* 392–393 *394* [*2*] [396 pp.].

PLATES: 157 wood engravings of plants within text, all hand-colored.

BINDING: Contemporary calf; re-backed with original gilt leather.

REFERENCES: *The American Register*; *Analectic Magazine*; Goodman; Rosenberg; Shafer; Wickes; Wolfe; Young.

In America, as in Europe, the healing arts diverged into many streams. Yet from the traditional remedies of the home nurse to the established therapies of the professional physician, medicine in the early 1800s, as in many other aspects of life in the new nation, was dependent upon old-world antecedents. Indeed, the first medical book to be published in America was *The English Physician: Containing, Admirable and Approved Remedies, For Several of the Most Usual Diseases, Fitted to the Meanest Capacity, by N. Culpepper* [sic] (Boston, 1708), an adaptation and compilation of Culpeper's herbal of 1652 (see No. 41). The frontiersman and his family and the homesteader far afield from urban centers, as well as those of scanty means within towns, were all by necessity self-reliant. For these persons, the published manual remained the source of medical lore, remedies, and comfort.

A rich heritage of printed materials, supplied with advice and medicinal recipes, preceded Henry's *American Medical Family Herbal*. The colonial physician John Tennent of Virginia published *Every Man his Own Doctor, or, The Poor Planter's Physician* (1734), which was thrice reprinted by Benjamin Franklin. British imports into Colonial America were also widely disseminated, the most popular being the Edinburgh-trained physician William Buchan's *Domestic Medicine, or, The Family Physician* (Edinburgh, 1769; first printed in Philadelphia in 1771). Even the founder of Methodism, John Wesley, produced such a book advocating the use of common herbs, *Primitive Physick, or, An Easy and Natural Method of Curing Most Diseases* (1747), which was reprinted in Philadelphia in 1764 and went through multiple editions. Anticipating Henry's *Herbal* by several years was Samuel Stearns' *The American Herbal, or Materia Medica*, printed in Walpole, New Hampshire, in 1801.

Native medical schools began to organize themselves, meanwhile, and by 1800 four important faculties had been established at the College of Philadelphia, Pennsylvania (1765), King's College (later Columbia), New York (1768), Harvard College, Cambridge, Massachusetts (1783), and Dartmouth College, Hanover, New Hampshire (1798). The work upon the New-World *materia medica* was initiated by the botanical studies of Mark Catesby (1683–1749), John Bartram (1699–1777), Peter Kalm (1716-1779), Johann David Schoepf (1752–1800), and Manasseh Cutler (1742–1823). At Philadelphia, the center of medical science in the new republic, Benjamin Smith Barton (1766–1815), Professor of Natural History and Botany as well as *Materia Medica* at the College of Philadelphia, published *Collections for an Essay Towards a Materia Medica of the United States* (1798–1804).

In order to place Henry's *Herbal* within its context, it is important to understand the prevailing medical doctrines of the traditional physician and his therapeutics, which were the product of the new American medical schools. This profession may be represented by one of Barton's colleagues at the College of Philadelphia, Doctor Benjamin Rush (1745–1813), whom Barton succeeded as Professor of the Theory and Practice of Physic. Rush, one of the fathers of the Republic and signatory of the Declaration of Independence, was trained at the University of Edinburgh, then one of the leading medical faculties of Europe. He practiced in his home country according to the received doctrines of certain eighteenth-century physiological theorists—the most directly influential upon Rush being William Cullen (1710–1790) and John Brown (1735–1788), both of Edinburgh. Ascribing one aetiology to disease, Brown saw sickness as a systemic imbalance of the natural order wherein spasms of the extreme arteries (*sthenia*)—or a diminution thereof (*asthenia*)—resulted in disease of the body. An overanimated state called for treatment and drugs to lower bodily excitement, while the opposite state called for stimulants. In practice, however, in antebellum America, nearly all illness was viewed as resulting from a morbid superabundance of excitement, as manifested, for instance, in the rapid heartbeat concomitant with a fevered condition. The prescribed therapy was therefore calculated to deplete or lower the perceived overactive state. Such depletive therapeutics involved an array of mechanical or chemical assaults meant to wrest the offending superfluity from the digestive tract and even from the very pores and veins: by the letting of blood in truly vampiric amounts, either by means of the lancet (venesection or arteriotomy), by the suction of heated cups applied to an incision (wet cupping), by the direct laceration of skin (scarification), or by leeches; by the deliberate inflammation of tissue through the placement of a plaster of mustard, cantharides (blister beetles), or antimony over expanses of skin (counterirritant blistering); by the violent purging of the bowels through the oral administration of calomel (mercurous chloride) and of the stomach by tartar emetic (potassium antimony tartrate) or arsenic; by the induction of profuse sweating and by the administration of oral diaphoretics (such as antimony, mercury, arsenic, and turpentine); by the promotion of salivation by the direct action of, it is now understood, iatrogenic mercurial and arsenical poisoning; and, finally, by a low, debilitating diet.

This regimen of therapeutics was aptly named the Heroic Therapy, and while meant to portray the active, bold intervention of the physician, the hero in question was undoubtedly the patient. Poisonous metallic and semi-metallic mineral elements—derived from mercury, lead, antimony, and arsenic—were the products of Arabic alchemical processes and introduced into the western *materia medica* in the sixteenth and seventeenth centuries by the alchemists and iatrochemists, Paracelsus at their head (see No. 19). Chemical processes were found to purify gold and the dramatic action of chemicals on a living organism was fraught with this metaphor as a cathartic or cleanser. Moreover, the older Galenic theory of humoral imbalance was by no means a dead letter and 'ill humors' were still the object of purgation as well. In addition, stimulants, among which were counted alcohol and strychnine, along with the ubiquitous opium as a depressant, were prescribed heavily. The heroic therapy—with calomel and antimony as the panaceas of the age—was visited upon all classes of society capable of paying the fee or on those who were the beneficiaries of governmental solicitude, from the drawing rooms of the wealthy and parlors of the middle class to soldiers in the field, sailors at sea, and paupers in almshouse and hospital.

In addition, patent or proprietary medicines were a regular item of import from the mother country and were soon a prolific native industry as well, so much so that a newspaper editor of 1800 could write 'Perhaps no past age in the history of this country has teemed with such a multitude of medical mountebanks as the present. The venders of patent medicines in almost every capital town in the United States are fattening on the weakness and folly of a deluded public.' A polypharmacy of herbs, opiates, and active poisons of every description, most often offered to the palate in a potent draft of alcohol, were hidden in bottles adorned with artful labels emblazoned with evocative names.

In contradistinction to traditional practice, Henry's *American Medical Family Herbal* represents a variation upon the 'botanic doctor' in the tradition of Culpeper, with the added benefit that the medicinal offerings were augmented with the 'secrets of the Indians.' This mode of practice was exceedingly popular and long-lived, with titles such as *The Indian Doctor's Receipt Book* or *The Indian Doctor's Practice of Medicine* being published well into the 1850s and with the Indian theme a mainstay in patent medicines as well. The tradition was scorned and resisted by physicians generally; Benjamin Rush stated that 'it would be a reproach to our schools of physic, if modern physicians were not more successful than the Indians.' This collegial disagreement was possible inasmuch as medical education and licensure were by no means standardized and were as yet unregulated. Medical schools, especially the proprietary or private institutions, outstripped one another in shortening their curricula, cutting fees, and diminishing requirements for degrees in order to fill their halls. Outside of formal education, apprenticeship was the other route to the profession, which process, again, was afforded little or no regulation. Those who took the title of doctor, therefore, ranged from the barely literate operative, who had little or no primary education, to the classically trained scholar, who took a degree from an eminent European university. Henry was decidedly closer to the former type.

The *Family Herbal* describes 157 plants and continues with an appendix 'Containing many choice medical secrets, Never made known to the world before,' which fact Henry never tires of proclaiming throughout the publication. The appendix is filled with recipe-cures for a variety of illnesses, starting with 'An infallible cure for the Chronic Rheumatism' and ending with 'A cure for a withered limb, &c.' The herbal's remedies provide a rare glimpse into the plant and animal products used to deal with health issues that perennially assail the human frame. They are, in effect, a published formulary of what must have lay hidden in many patent remedies. Despite the prevailing mineral and chemical therapeutics, however, Henry's remedies to his credit are largely botanic. Some of his recipes do indeed call for calomel, such as 'The author's famous Anti-bilious Pills (XXXII),' but Henry also exhibited an awareness of his colleagues' excesses, as seen, for instance, in his 'Cure for violent Pain in the Bones, proceeding from the imprudent use of mercury' (XXXVII), which consists in a relatively soothing draft of 'matheglin' (metheglin or mead) and purgative Epsom salts (magnesium sulfate). Bilious pills were the first nostrum to be patented in America (in 1796) and competitors vied with various brands and formulas, most of which contained calomel. One brand entered the market in New York City by 1800 as a direct competitor to 'the author's famous Anti-bilious Pills.' Patent nostrums were themselves a battleground in the war between chemical (mineral) and botanical medicaments, but with the adjective 'vegetable' often belying their contents.

In Henry's publication of Lewis' secret remedy for Hydrophobia we may see a demonstration of the altruism expressed in the title-page's Latin tag, translated as 'I'll be as happy as my fortune will permit, and make others so if I can.' A certain physician of New Jersey, Lawrence Vanderveer, claimed to have discovered in the herb *Scutellaria lateriflora* (skullcap, of the mint family) a preventative and cure for hydrophobia. One of Vanderveer's patients, Daniel Lewis of Westchester, New York, who suffered the bite of a mad dog in 1783, survived his ordeal and, capitalizing on his good fortune, advertised a cure for rabies. Lewis

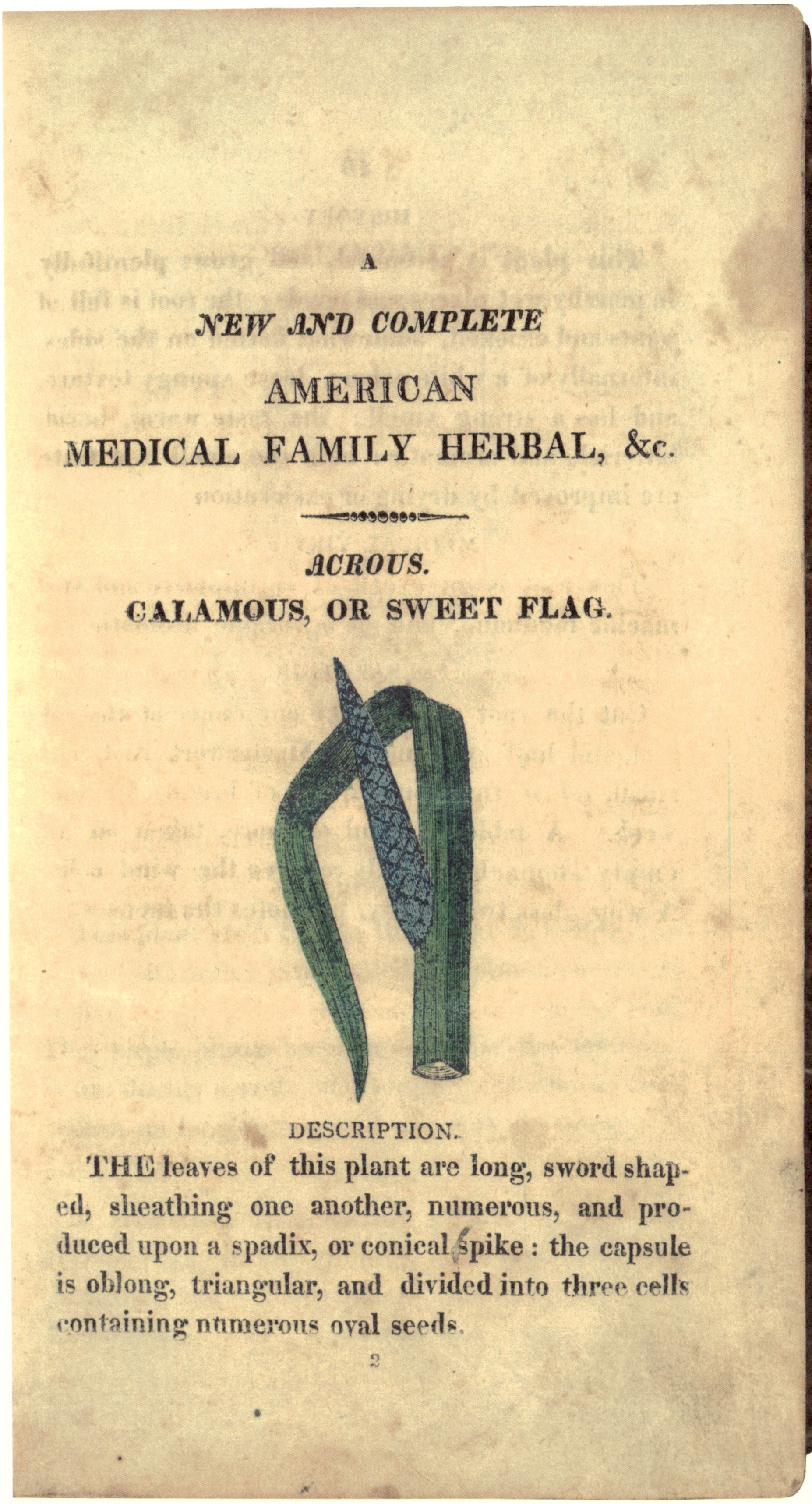

A

NEW AND COMPLETE

AMERICAN

MEDICAL FAMILY HERBAL, &c.

ACROUS.

CALAMOUS, OR SWEET FLAG.

DESCRIPTION.

THE leaves of this plant are long, sword shaped, sheathing one another, numerous, and produced upon a spadix, or conical spike: the capsule is oblong, triangular, and divided into three cells containing numerous oval seeds.

2

SAMUEL HENRY,
A New and Complete American Medical Family Herbal, 1814 Common Sweet Flag (*Acorus calamus*), page 9

received much acclaim for and interest in his cure, as is evidenced by an unsuccessful effort that was made to obtain from Congress an appropriation for the purchase of his secret remedy. This 'Newly Discovered' cure was spread abroad by Henry, who procured the 'secret preparation ... from a physician in New-Jersey' (the cure is revealed under Henry's entry for 'Scull-Cap. Scutellaria Galericulata'). It should be noted that Dr. Lyman Spaulding—who would later work with Jacob Bigelow on the preparation of the first *Pharmacopoeia of the United States* (see No. 61)—read a paper on the plant before the Historical Society of New York in 1819, but his researches were pronounced to be 'empirical' by William P. C. Barton, among others.

The volume also includes two short essays: 'The Virtues of the Different Essences,' which discusses hemlock, penny-royal, sweet fennel, cinnamon, and wintergreen, and the author's insight and advice on 'Collections and Preservation of Simples.' An index of plant names in English and Latin completes the work.

Not much is known about Samuel Henry's life, other than what is found in his preface. Henry claimed to have been a prisoner among the Creek Indians during the War of 1812; his knowledge of plant cures is derived from time spent among this tribe and from travels 'through all the Southern States.' He practiced in lower Manhattan and his book advertises his remedies sold at 'No. 6, Peck-slip, New-York.' Never loath to advertise, in the headnote to his list of remedies Henry states:

> The following has been the means, in my hands, for several years, of restoring many to the use of their limbs, who had been given up as incurable, as may be seen by their several attested certificates, by calling at my shop.

Here was an opportunity not only to obtain remedies, but to peruse testimonial certificates as to their efficacy, and to purchase a hand-colored practical herbal into the bargain. As is attested by the index of subscribers at the end of the volume, Henry's book points to trends in the publication of such handbooks as well, since the subscribers were offered the option of either black-and-white or hand-colored illustrations, and of specially bound and gilt copies. The author describes himself, on his title-page, as 'One of the members of the late College of Physicians and Surgeons and of the Medical Society of the city and county of New-York.' The College of Physicians and Surgeons within Columbia University was re-established in 1807 from King's College medical school founded in 1768 (while suffering several subsequent tumultuous reorganizations). King's College itself did not survive the Revolution, but its original founding was second only to the establishment of the medical school of Philadelphia in 1765.

Henry's *Herbal* was immediately subject to the biting ridicule of the *American Medical and Philosophical Register* (1814). The reviewer lampoons Henry's 'infallible cures,' as for

102

ELDER, BLACK.

SAMBUCUS NIGRA.

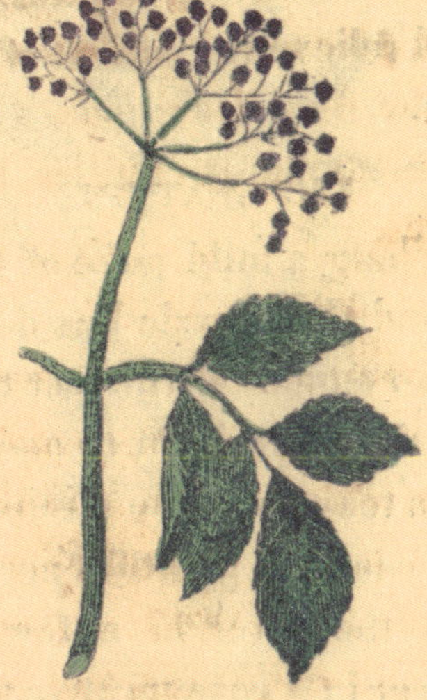

DESCRIPTION.

It rises to the height of a small tree, and is much branched towards the top: the young shoots are full of pith, the old ones empty: leaves are pinnated consisting of two or three pair, with an odd one at the end: flowers are sweet smelling, white and produced on large flat umbels or clusters: the fruit is a round succulent berry, of a blackish purple colour, and contains three seeds.

HISTORY.

This tree grows in hedges and along the borders of meadows in every part of the United States: flowers in July, and the berries are ripe in September.

SAMUEL HENRY, *A New and Complete American Medical Family Herbal*, 1814 Elderberry (*Sambucus nigra*), page 102

instance that for deafness (XIV), which involves dripping the fat of a roasted beaver's tail, or the heated urine of a polecat, into the patient's ear. The reviewer wonders whether some might 'prefer the disease to the remedy,' and confesses that he 'cannot discover what is the rule by which Beavers and Pole-cats are classed under the head of American plants.' The reviewer is also amused to note that, under 'Recommendations,' the author has collected the praise of four learned gentlemen, not for the volume as published, but only for its planned outline. Finally, with mock charity, the reviewer attributes Henry's grammatical lapses to the fact that he spent some time 'among a savage nation.'

Another notice in *The Analectic Magazine* of 1814 characterizes Henry's *Herbal* as falling outside of orthodox medical practice, as many of the plants described 'are altogether unknown to the pharmacopia of the regular physician.' 'The author is evidently an unlettered man,' the reviewer continues, since 'the scientific names which he professes to give are often grossly misspelt, or erroneous…,' as C. S. Rafinesque would also point out in his herbal (see No. 62). *The Analectic Magazine and Naval Chronicle* of 1816 published a fully fledged review by William Baldwin (1779–1819), a Philadelphia physician and botanist who then resided, for reasons of health, in Savannah, Georgia. The reviewer asserts that, of the plants described, more than half are European. A great many of the woodcuts as well as the descriptions are in fact copied from William Woodville's *Medical Botany* (3 vols., plus supplement, London, 1790–1794[5]). Henry himself mentions Woodville's book in his preface: 'The only book on Botany, worthy of notice and public usefulness, is the celebrated Doctor Woodvill's Medical Botany.' Baldwin examines in detail thirty-six plants that, despite Henry's inclusion, are exotic to North America, as well as the same number of indigenous plants. Of the exotics, twenty-nine are copied directly from Woodville; one has been naturalized to the United States; three are exotics that must have been depicted from a source other than Woodville; and three actual native species are wrongly described and illustrated as foreign plants. Of the true native North-American species, most are so ambiguously named and illustrated that Baldwin fears 'masters and mistresses of families,' who mistake a poisonous for a benign plant, may 'repent of their folly when too late.'

Henry's critics urge caution in the use of his herbal, and Henry himself counsels caution in regards to the works of his competitors, which is expressed in the opening paragraph of his preface:

> The author of the work here presented to the public, having read many books on the subject of Botany, as well as several of such as have been styled Herbals, on the virtues of plants, has had occasion to discover not only an almost immemorable quantity of errors and defects, but also a multiplicity of directions for their uses, which on examination and trial, he has proven to be altogether ineffectual in the cure of the complaints to which these treatises have directed their application.

Certainty in nineteenth-century medicine was a scarce commodity. And yet little would Henry's smug and condescending reviewer of 1814 have believed that both the heroic practitioners and the 'herb doctors' were empiricists alike, or that Henry's herbal remedies likely did less harm—and perhaps effected more cures—than the heroic therapy of his more learned colleagues.

61. Jacob Bigelow (1787–1879)

American Medical Botany, Being a Collection of the Native Medicinal Plants of the United States, Containing their Botanical History and Chemical Analysis, and Properties and Uses in Medicine, Diet and the Arts, With Colored Engravings. [*double rule*] By Jacob Bigelow, M.D. Rumford Professor and Lecturer on Materia Medica and Botany in Harvard University. [*double rule*] Vol. I [–Vol. III]. [*row of four asterisks*] Boston: Published By Cummings and Hilliard, At the Boston Bookstore, No. 1, Cornhill. University Press.... Hilliard and Metcalf. 1817 [–1818, –1820].

8° 27 × 19 cm. Three volumes in six parts. Volume I, Part I: 1^4 2^2 3–14^4 *i–v* vi–xi *xii* *17–32* 33–38 *39–51* 52–59 *60–65* 66–74 75–83 *84–89* *90–95* *96–110* *111–112* [108 pp.]; Volume I, Part II: 15–25^4 *i–ii* 113–124 *125–132* 133–141 *142–148* *149–154* *155–160* *161–168* *169–176* 177–186 *187–191* *192–197* *198* [88 pp.]; Volume II, Part I: *1*–13^4 *i–v* vi–vii *viii–ix* x–xvi (xiv as xvi) *15–26* 27–33 *34–40* *41–50* *51–58* *59–66* *67–72* *73–81* *82–96* *97–104* [104 pp.]; Volume II, Part II: *14*–25^4 *i–ii* 107–114 *115–120* 121–136 *137–141* *142–147* *148–153* *154–159* *160–165* *166–170* *171–187* *188–199* *200* [96 pp.]; Volume III, Part I: *1*–12^4 13^2 *i–v* vi–x *11–98* *19–31* *32–42* *43–48* *49–54* *55–60* *61–75* *76–81* *82–91* *92–98* *99–100* [100 pp.]; Volume III, Part II: *14*–25^4 26^2 *i–ii* 101–106 *107–118* *119–128* *129–133* *134–140* *141–146* *147–150* *151–155* *156–162* *163–173* *174–175* *176–177* *178–179* 180–187 *188–189* 190 *191* 192–193 *194–195* 196–197 *198* [100 pp.].

PLATES: 60 aquatint and hand-colored plates, ten in each part.

BINDING: Original printed green paper over boards.

PROVENANCE: Inscribed on front pastedown of first part: 'From Dr. Augustus Torrey, Beverly, Mass.' 'Sarah Howe Beverly 1828' inscribed on fly-leaf of each part. Bookplates: Library of the University of Vermont and State Agricultural College and Paul Mellon. All title-pages are stamped: 'Library of the University of Vermont.'

REFERENCES: Berman; Bigelow; Cleveland 804 and 771 (biography); Ellis; Holmes; Pritzel 773; Stafleu & Cowan 514; Warner; Weimerskirch; Wolfe.

JACOB BIGELOW was born in Sudbury, Massachusetts, in 1787. His father was a Congregational minister and he spent his childhood on the family farm. In 1802, at the age of sixteen, he attended Harvard College and was introduced to anatomy by the lectures of John C. Warren, son of the founder of the medical school. He graduated with honors in 1806 and while teaching Latin and Greek, first at Worcester and then at the Boston Latin School, he attended the medical lectures of Warren on anatomy, Aaron Dexter on chemistry, and Benjamin Waterhouse on botany. In 1808 he became a private pupil of Dr. John Gorham, successor to Dexter as Professor of Chemistry and *Materia Medica*. Having found his vocation, he matriculated at the prestigious Medical School of the College of Pennsylvania in 1809, where he attended the lectures of Benjamin Rush, Professor of the

Theory and Practice of Physic, Caspar Wistar (1761–1818), Professor of Chemistry and Anatomy, Philip Syng Physick (1768–1837), Professor of Surgery, and John Redman Coxe (1773–1864), an early American advocate for vaccination. But the greatest influence upon Bigelow's future was to be Benjamin Smith Barton (1766–1815), Professor of Natural History and Botany, *Materia Medica*, and—after the death of Rush—Theory and Practice of Physic.

Benjamin Smith Barton was long engaged in botanical studies, but because of ill health, he financed the expeditions of Federick Pursh (1774–1820) and Thomas Nuttall (1786–1859) to supplement his own researches. He wrote *Elements of Botany* in 1803, the first botanical textbook published in the United States (with illustration by William Bartram). Barton's outline for his proposed flora of North America took the form of *Collections for an Essay Toward a Materia Medica of the United States* (in two volumes, 1798 and 1804). Thomas Jefferson employed Barton as an advisor to the Lewis and Clark Expedition, Meriwether Lewis being a former student, and at the conclusion of that endeavor the botanic specimens came to Barton for analysis. His untimely death at the age of forty-nine left his researches incomplete.

Bigelow undertook private studies with Barton and received his medical degree in 1810. The following year he embarked on private practice in Boston. In 1812 Bigelow's fortunes were enhanced enormously when he joined the practice of Dr. James Jackson (1777–1867). Jackson, one of the most successful physicians in Boston, was the second professor of Theory and Practice of Physic at Harvard and later Dean of the faculty of medicine. In the same year, Bigelow began to lecture on botany at Harvard College. He edited an American edition of James Edward Smith's *An Introduction to Physiological and Systematic Botany* in 1814 and published *Florula Bostoniensis*, a census of the local flora. An enlarged edition of *Florula* was published in 1824, which encompassed the flora of New England, including the mountains of New Hampshire and Vermont, and a third edition appeared in 1840. Bigelow was appointed Lecturer of *Materia Medica* in Harvard Medical School in 1815 and was made Professor the following year, while continuing to lecture on botany at Harvard College.

In 1816, at the age of twenty-nine, he received the unique honor of dual professorships when he was appointed Rumford Professor of the Application of Science to the Useful Arts. Keenly interested in the applied arts, Bigelow assiduously studied processes of art and manufacture and it was he who coined the modern usage of the word 'technology.' With his influential writings concerning educational reform, he laid the groundwork for the establishment of Massachusetts Institute of Technology. He published *Elements of Technology* in 1829, the elaboration of his class lectures, with a second edition issued in 1831 and a third in 1840, the last under the title *The Useful Arts*.

In 1820 the United States Pharmacopoeia (USP) was formed and Bigelow was one of five committee members to compile the first *Pharmacopoeia of the United States*—including Dr. Spaulding of New York, the proponent of an herbal cure for rabies (see No. 60). His system simplified the British drug nomenclature, substituting a single name, where practicable, for the customary double or triple Latin names. In 1822 he published *A Treatise on the Materia Medica, Intended as a Sequel to the Pharmacopoeia...*, which expatiated upon the character, qualities, and uses of the listed drugs.

Bigelow's *American Medical Botany* is a pioneering survey of the pharmacological uses of plants in North America. It is also the first color-printed book produced in the United States. The work was originally issued in six parts—the state of the publication conserved by the Oak Spring Garden Library—that were later bound in three volumes, dated 1817, 1818, and 1820. Bigelow explains in his Introduction to the first volume that 'The Figures of the present volume have been engraved and coloured from original drawings, made principally by myself. Dissections of the flower and fruit have been added to each for the use of botanical students. The subsequent portions of the work will be issued as rapidly as is consistent with their faithful execution.' One thousand copies were published and each volume contains twenty full-page color plates (with sixty plates, total) accompanied by descriptive text. The aquatint process, which originated in France in the 1760s, was used, but with multiple colors applied to each plate. He explains the process in more detail in an unpublished autobiography:

> After many trials and experiments a tolerably successful mode was discovered, which consisted in engraving the plates in *aqua tinta*, thus producing a continuous surface, to the parts of which separate colours could be applied, and the surplus wiped off in different directions, so as not to interfere with each other. In this way the simple plates, or those with few colors, could be delivered from the press complete, without requiring to be retouched. But those which had small or insulated spots were obliged to be finished with the pencil [i.e., brush—for European productions of color printing using multiple colors on a single plate, see Nos. 27 and 36 and *An Oak Spring Sylva*, No. 1].

Bigelow's original intention was to produce the plates in the conventional manner with engravings colored by hand, but issues of speed, quality, and expense intervened. About 200 copies of the first part, containing ten plates, were produced and hand-colored by a quorum of young Boston ladies until the prospect of producing 60,000 hand-colored plates proved impracticable. Bigelow's conversance with the applied arts stood him in good stead, however, and he devised with his engravers the method described above; the remainder of Part 1, and all subsequent parts, were issued as color-printed aquatints. There are, then, two states of the first number, one with hand-colored, and the other with color-printed, plates.

The exemplar in the Oak Spring Garden Library is the color-printed second state. In addition, some copies combine hand-colored and color-printed plates in Volume I, Part I. In all the parts, the color-printed plates are, as he describes, occasionally touched with hand coloring, especially where it would have been impractical to add etched color (see for example Pl. XXVII, Magnolia Glauca, where there are large brushed areas of blue color and delicate touches of yellow). Some plants, which do not depict flower or fruit, are printed in green and black inks alone.

An advertisement for *American Medical Botany* in the 1816 issue of *The New-England Journal of Medicine and Surgery* states that 'A work of similar character to this was announced sometime before his death by the late Professor Barton of Philadelphia [namely, Barton's projected flora of North America], but has never been carried into execution.' In fact, another alumnus of Barton's course in *Materia Medica*, Barton's nephew, William P. C. Barton, was simultaneously at work to bring to fruition their teacher's plan. With each unaware of the other's work, William P. C. Barton published in the years 1817 and 1818 his two-volume work *Vegetable Materia Medica of the United States*. Barton's work contained about fifty hand-colored plates, which, in some cases, the reviewers found to be more pleasing than Bigelow's aquatints. C. S. Rafinesque in his *Medical Flora* (see No. 62) considered that, of the two, 'Bigelow's is the most learned, accurate and useful, while Barton's has often the best figures.' Bigelow wrote in his Preface:

> The Materia Medica, comprising the great body of medicinal agents now in use in the hands of physicians, cannot be said to need an increase in the number of its articles. It is already incumbered with many superfluous drugs; even its active substances are more numerous than can be of use to any one physician, so that it seems quite as susceptible of benefit from reduction as from augmentation in the number of its materials. [a principle held by Bigelow's teacher, Benjamin Rush] … It is the policy of every country to convert as far as possible its own productions to use, as a mean of multiplying its resources, and diminishing its tribute to foreigners. The plants of the United States are various in their character in proportion to the extent of latitudes and climates, which our country embraces.
>
> Among those which have been medicinally investigated, are many of useful properties and decided efficacy. Several departments of the Materia Medica may be amply supplied from our own forests and meadows, although there are others, for which we must as yet depend on foreign countries. … Under the title of American Medical Botany, it is my intention to offer to the public a series of coloured engravings of those native plants, which possess properties deserving the attention of medical practitioners. The plan will likewise include vegetables of particular utility in diet and the arts; also poisonous plants which must be known, that they may be avoided. In making the selection, I have endeavoured to be guided by positive evidence of important qualities, and not by the insufficient testimony of popular report. In treating of each plant, its botanical history will be given; the result of

JACOB BIGELOW,
American Medical Botany,
1817–1821. Thorn apple
(*Datura stramonium*),
volume 1, part 1,
plate 1

> such chemical examinations as I have been able to make of its constituent parts, and lastly its medical history. The botanical account will be found more diffuse than is necessary for exclusive botanists. The chemical inquiries are made chiefly with a view to the pharmaceutical preparations of each plant, or to interesting principles it may contain. Its medical history will contain such facts, relative to its operation on the human system, as are known to me from my own observation, or the evidence of those, who are qualified to form correct opinions on the subject....

In Volume 1, Plate 1, Bigelow begins with the Thorn Apple, *Datura stramonium*. He discusses in detail its use in treating epilepsy, pulmonary diseases, and as a pain reliever: 'Like opium and like other powerful medicines, this plant, when taken in small quantity, and under suitable regulations, proves a remedy of importance, and a useful agent in the hands of physicians.' Bigelow's descriptions of all the plants are in essay format, relating various personal experiences that touch upon their medicinal qualities. Each entry concludes with botanical and medical references.

This work of Bigelow's early career must be seen, however, in its broader context. The mid-nineteenth century was a turbulent period indeed for American medicine. In the course of Bigelow's successful Boston practice and academic duties a revolution was taking place in American therapeutics and, as the century progressed, Bigelow both witnessed a tempering of heroic therapeutics—of which tempering he himself was a moving force—and, more dramatically, saw the establishment of rival medical movements, including several botanic schools and, posing a more professional threat, homeopathy.

While in the medical epicenters of Boston, Philadelphia, and New York practitioners had already begun to moderate somewhat the excesses of the heroic therapy, in the cities of the West and in the rural areas patients continued to be bled white, starved down, sickened with horrific doses of calomel, and stupefied with opiates and alcohol. A growing mistrust of the professional physician in general and of his therapeutics in particular—already seen in Henry's *Herbal*—was manifested in an egalitarian movement put in motion by an unlettered farmer. Samuel Thomson (1769–1843) of Alstead, New Hampshire, after witnessing the ministrations of the heroic therapy upon various family members, sought the good offices of an herbalist ('the widow Benton'). Thomson began his own alternative practice locally and, under the name of the 'Friendly Botanic Society,' formed an empire of sub-practitioners and proprietary patent medicines in direct contravention to the orthodox physicians. This sectarian movement took on the force of a popular revolt. In 1822 Thomson broadened his influence by the publication of two works that were to transform the medical landscape: *A Narrative of the Life and Medical Discoveries of Samuel Thomson* and *The New Guide to Health, or, Botanic Family Physician* (the latter of which, subsequently published along with the former, would be issued in twenty-six editions through

JACOB BIGELOW,
American Medical Botany,
1817–1821. Skunk cabbage (*Ictodes foetidus*),
volume II, part I,
plate 4

the span of twenty years). The Thomsonians set themselves against the highly toxic mineral compounds of the regular practitioners, and—reverting to a Galenic and quasi-vitalistic theory of disease—endeavored to accomplish the same end of purgation through the more benign power of herbs and steam baths. The movement was even given impetus by Bigelow's predecessor as Harvard Lecturer in Botany and Professor of Theory and Practice of Medicine, Benjamin Waterhouse (1754–1846). Waterhouse published essays in praise of Thomson, which were reprinted in Thomsonian tracts, and Thomson dedicated the thirteenth edition of his book to him.

Soon after early successes, however, the Thomsonians split into rival factions. One such group was formed by Thomson's early disciple Alva Curtis (1797–1880), who founded the Physio-Medicalists, which group, to the disgust of Thomson (the inveterate populist and anti-intellectual), founded medical schools, state medical societies, and boards of censors in competition with the 'regulars.' In addition, Wooster Beach (1794–1868), trained as a physician, published *The American Practice of Medicine* (1833) and thereby founded the Reformed Medical Society, an independent botanic school. Later called the Eclectics, Beach's movement became well organized and established, eventually overtaking the Physio-Medicalists in membership numbers.

Concurrently in Europe, a rival theory of therapy was put forth by Samuel Hahnemann (1755–1843), whose *Organon der Heilkunst* (Leipzig, 1810), steadily gained converts. Unlike the allopaths (so-called by Hahnemann), or regular physicians, who violently set against a disease, the homeopaths sought to rouse the body's own defence systems with botanic and mineral drugs, but in infinitesimal dosages deemed to elicit symptoms similar to the disease—thus, like cures like (*similia similibus curantur*). Homeopathy was introduced into the United States around 1825 and, after a series of devastating cholera epidemics through 1848 to 1852, its ranks swelled. In 1849, for instance, 1,000 physicians joined the Homoeopathic Society in Cincinnati, Ohio, in the aftermath of the epidemic. In 1844 the American Institute of Homeopathy was organized as the first national medical society and in 1848 the Homeopathic Medical College of Pennsylvania was established. In addition, homeopaths sometimes joined forced with the Eclectics within their medical schools, most notably at the Eclectic Medical College in Cincinnati, beginning at its founding in 1845. The practice attracted a fashionable clientele, which gladly swallowed the harmless infinitesimal doses offered in palatable sugar pills or alcoholic tinctures. Bigelow, however, as the voice of regular practitioners, decried homeopathy as 'a specious mode of doing nothing ... by the formal and regular administration of nominal medicine.' The Homeopaths, the Physio-Medicalists, and the Eclectics established sectarian medical colleges and societies that, in prolific and protean forms, flourished as rival medical schools that lasted into the early twentieth century.

JACOB BIGELOW,
American Medical Botany,
1817–1821. Elliotts' gentian, or American gentian, or Bottle gentian (*Gentiana catesbaei*), volume II, part II, plate 34

Moreover, within the ranks of the regular physicians themselves, the clinical practices developed in Parisian and London hospitals undertook to observe and quantify the supposed efficacy of traditional therapeutics. Pierre-Charles-Alexandre Louis (1787–1872) of l'Hôpital Charité, Paris, was able to prove by statistical analysis that bloodletting was a positive harm to patients. Bigelow—one of the more prominent members of the orthodox medical profession—assimilated and articulated the growing scepticism regarding the efficacy of therapeutic systems. Late in his life, Bigelow wrote 'In 1818 [*sic*], I began to publish a work on American Medical Botany.... My attachment to botany and exaggerated estimate of therapeutics led me at that time to attach greater value to such an enterprise than I have since done.' In 1835 Bigelow delivered an epoch-making address to the Massachusetts Medical Society 'On Self-limited Diseases.' Indeed, Dr. Oliver Wendell Holmes asserts that 'this remarkable essay has probably had more influence on medical practice in America than any similar brief treatise, we might say than any work ever published in this country. Its suggestions were scattered abroad at the exact fertilizing moment when public opinion [that is, that of orthodox physicians] was mature enough for their reception.'

A self-limited disease, Bigelow wrote, '[is] one which receives limits from its own nature, and not from foreign influences: one which, after it has attained foothold in the system, cannot, in the present state of our knowledge, be eradicated or abridged by art....' When it was observed that sufferers of diseases who refused treatment often recovered without medical intervention, or that creditable physicians often endorsed opposite modes of treatment for the same disease, or that cures occurred at the hands of quacks, then one could only but be sceptical of the efficacy of medical art. It was a dramatic demonstration of this point when, in the midst of an outbreak of epidemic disease, for instance, patients of homeopathic or botanic practitioners were seen to survive, whereas patients of the heroic therapy did not. Bigelow came to trust more in a restorative therapy, holding that patients fared better when given over to the healing power of nature (*vis medicatrix naturæ*), an ancient doctrine whose impeccable pedigree may be traced back to Hippocrates. This was a long way removed from Benjamin Rush, whom Bigelow recalled as having admonished his preceptors 'We can have no reliance on Nature, gentlemen—we must turn her out of doors in our practice, and substitute for her efficient art.' Notwithstanding, it should be noted that Bigelow by no means abandoned the heroic armamentarium: 'Cathartics,' he writes, 'laxatives, emetics, leeches, counter-irritants, cupping, &c., are of benefit in various local and general maladies.' Bigelow's colleague, Dr. Holmes—who walked the corridors of l'Hôpital Charité with Pierre Louis—went too far, it was thought, when he stated in an 1860 address that 'I firmly believe that if the whole materia medica, *as now used*, could be sunk to the bottom of the sea, it would be all the better for mankind, and all the worse for the fishes.'

In 'On Self-limited Diseases,' Bigelow expressed to his fellow colleagues: 'It [the loss of a patient] brings with it the humiliating conclusion, that while other sciences have been carried forward, within our own time and almost under our own eyes, to a degree of unprecedented advancement, Medicine, in regard to some of its professed and most important objects, is still ineffectual speculation.' Before the advent of bacteriology, immunology, and synthetic organic and antibiotic drugs, this stated the case plainly and honestly. To a profession under assault on many fronts, Bigelow did no more than state the obvious. And yet, by the commencement of the twentieth century, the ascendency and supremacy that Bigelow's American Medical Association (AMA) would achieve brought to pass, in fact, the very medical monopoly that was denounced and feared by its botanic and homeopathic detractors.

Bigelow led a long and astonishingly productive life, combining the skills of an artist, technician, teacher, botanist, pharmacist, physician, and scholar. He attempted to winnow the superfluity of existing drugs and discover new ones that could replace foreign imports. His influential writings affected the state of education in America, advocating a curriculum of modern science and literature in parallel to the traditional classical and theological, and he was one of the early vice presidents of Massachusetts Institute of Technology. Along with John C. Warren and James Jackson, he founded the *New England Journal of Medicine*. As a public health issue, he proposed the establishment of burial within a garden landscape instead of church crypts and founded in 1831 the first of the great municipal garden or botanic cemeteries, Mount Auburn. He served as Mount Auburn's landscape architect as well as designing its fence, gateway, tower, and chapel. He published anonymously a book of witty poem parodies and, in his old age, amused himself by translating Mother Goose rhymes into Greek and Latin verse, which were published privately in 1871 as *Chenodia, or, The Classical Mother Goose*. Bigelow's son, Henry Jacob (1816–1890), was an outstanding surgeon at Massachusetts General Hospital and professor of Surgery at Harvard Medical School. Jacob died in 1879 at the age of ninety-two and was buried in Mount Auburn Cemetery.

62. Constantine Samuel Rafinesque (1783–1840)

Medical Flora; Or, Manual of the Medical Botany of the United States of North America. Containing a Selection of above 100 Figures and Descriptions of Medical Plants, with their Names, Qualities, Properties, History, &c.: And Notes or Remarks on Nearly 500 Equivalent Substitutes. In Two Volumes. [*double rule*] Volume The First, A=[*double rule as dash*] H With 52 Plates [–Volume The Second, With 48 Plates.] [*rule*] Medical Plants are compound Medicines prepared by the hands of Nature, &c.—Med. Princ. 31. [*double rule*] By C. S. Rafinesque, A.M...PH.D. Ex-Prof. of Botany, Natural History, &c. in Transylv. University of Lexington, the Franklin Institute of Philadelphia, &c. Member of the Medical Societies of Cincinnati and Lexington—the Philos. Soc.

and Lyceum of New York—the Acad. of Nat. Sc. Of Philadelphia—the Amer. Antiq. Society—the Kentucky Institute—the Linnean Soc. of Paris—the Imp. Nat. Cur. Soc. of Bonn.—the Imp. Economical Soc. of Vienna—the R. Italian Inst.—the R. Inst. of Nat. Sc. of Naples, &c. &c. [*double rule*] Philadelphia: Printed and Published by Atkinson & Alexander, No. 112 Chestnut Street. [*swelled rule*] 1828 [–1830].

12° 19.7 × 11.8 cm. Two volumes. Volume I: π^2 A–C^6 2C^6 D–X^6 Y–Z^4 [*4*] *i* ii–xii *1* 2–268 [284 pp.]. Volume II: A–Z^6 *1–3* 4–6 *7* 8–276 [276 pp.].

PLATES: 100 wood engravings printed in green: 1–52, Vol. I; 53–100, Vol. II.

BINDING: Original printed boards; untrimmed.

PROVENANCE: Inscribed on front board in ink: 'John Millington / Williamsburg Va./27938.' Bookplate of Paul Mellon.

REFERENCES: Call; Cleveland 946 and 748 (biography); Gray; Nissen 1,579; Pritzel 7,401; Rafinesque; Stafleu & Cowan 8,579; Warren.

CONSTANTINE SAMUEL RAFINESQUE was—as his letter of introduction to John James Audubon characterized him—an 'odd fish.' He had thoroughly imbibed the *philosophe*'s ideal of the French Enlightenment and with great exuberance made the American shores the object of his researches. He was brilliant and ever sanguine, if not manic. Eccentric in dress and manner, he retained his French accent and gentlemanly European deportment throughout his long years in his adopted country. His work ranged over the whole of the natural kingdom, with a speciality for botany and ichthyology, running into natural history, archaeology, anthropology, ethnology, geology, linguistics, and even to poetry and a scheme for equitable banking. Harmless as a dove in his social and financial dealings, he lacked the wisdom of a serpent. Entirely innocent of diplomatic skills, trusting by nature, but ruthless when critiquing his American colleagues' work in print, he was continually surprised by the resultant animus towards him by these brother scientists, whose concern, as he supposed, should be for 'truth' alone. He was an autodidact, whose unorthodox insights exhibited flashes of undoubted brilliance well in advance of his time, but whose unguided theoretical flights sometimes overran prudence and, in his later isolation, bordered on the quixotic or, after long years of neglect, the paranoiac. His prolific writings were often ignored, greeted with scepticism, or even subject to thoroughgoing hostility. His prolificacy extended to a positive superabundance of botanical naming and his new genera and species overwhelmed and alarmed his colleagues. While realizing several commercial successes in early life, suffering unusual vicissitudes of fortune, he was increasingly subject to material privations as he devoted himself single mindedly to scientific pursuits. Unlike his fellow French expatriate, John James Audubon (who cast Rafinesque in one of his overblown and exaggerated tales in *Ornithological Biography*), Rafinesque became increasingly alienated from his adopted colleagues and died destitute in Philadelphia. Without regard to the dictates of his will, his amassed botanic and natural collections, his books, and his papers were discarded or dispersed after his death, the proceeds from which did not raise enough capital to cover his burial.

Towards the end of his life, Rafinesque wrote an autobiographical summary of his career, *A Life of Travels* (Philadelphia, 1836). He was born in 1783 in Constantinople (ancient Byzantium, modern-day Istanbul—hence his given name, Constantine). His mother was born in Greece but of German descent—née Schmaltz—and his father was a French merchant of Marseilles. During his boyhood in Marseilles he interested himself in botany and zoology early on and, as he expressed it, 'became a Naturalist.' Because of the excesses of the French Revolution, the family quit France for Livorno, staying in exile from 1792 to 1796. While on a trading venture in Philadelphia in 1793, his father succumbed to an epidemic of yellow fever. In 1797 the diminished family returned to Marseilles where Rafinesque continued a desultory education. By his own account, before the age of twelve he had read over 1,000 books and spoke French, Italian, and English and had taught himself 'Latin, Greek, Hebrew, Sanskrit, Chinese, and fifty other languages.' In 1802, Rafinesque and his younger brother, taking up their father's profession, sailed in a ship laden with goods to Philadelphia.

There the young naturalist was offered an apprenticeship to Benjamin Rush—what strange bedfellows the two would have made!—but was forced to flee an outbreak of yellow fever, thus avoiding his father's fate. He reached Germantown and the hospitality of the horticulturist Colonel Forrest. From there Rafinesque began the first of his numerous botanical expeditions into the pristine American landscape, studying the flora as well as fishes, reptiles, and birds. Along the way he saw John Bartram's and Humphrey Marshal's botanic gardens (see *An Oak Spring Sylva*, No. 16), visited G. H. E. Muhlenberg (1753–1815), Benjamin Smith Barton (1766–1815), Frederick Pursh (1774–1820), and André Michaux (1746–1802), whose *Flora of North America* [*Flora Boreali-Americana*] (1803) he sought to supplement by his own researches.

The entire year of 1804 was spent botanizing, during which time he travelled nearly 1,200 miles—mostly afoot with little more equipment than a collecting bag, as he maintained that only in this mode is the botanist able to adequately and intimately study his subjects. He considered joining the Lewis and Clark Expedition, but declined when not appointed as the expedition's botanist. Although he met Thomas Jefferson, unable to obtain a university post through the president's offices, Rafinesque sailed to Sicily on a business venture that would keep him in those climes for the next ten years. He brought with him his American herbarium of nearly 10,000 specimens and continued his naturalist studies in Sicily, where he made his fortune in trading such medicinal materials as sea squill, rosemary, and wormwood. He married during this time, but not happily, and decided to return to America in 1815, succinctly epitomizing Sicily as offering 'a fruitful soil, delightful climate, excellent productions, perfidious men, [and] deceitful women.'

Tragedy befell, however, for his ship foundered off Connecticut. Lost were Rafinesque's manuscripts, 2,000 maps and drawings, 300 copperplates, 6,000 specimens of shells, and a herbarium so large that part of it had been left behind in Sicily—'I had lost everything, my fortune, my share of the cargo, my collections and labors of 20 years past, my manuscripts, my drawings, even my clothes.'

Rafinesque recovered himself by serving as a tutor for several years until he had regained the resources to continue his studies of nature. In 1818 he undertook a tour of 2,000 miles through Ohio, Kentucky, and Illinois. He spent some time with Audubon and, most importantly, met John D. Clifford (d. 1820), the Kentucky paleontologist who proposed a professorship at Transylvania University (founded 1780), in Lexington, Kentucky. With Clifford as his protector and sponsor, Rafinesque began his professorship in 1820 as lecturer on Natural History, refusing a more lucrative post as chair of chemistry. Rafinesque thus became the first professor of the natural sciences west of the Appalachians. He supplemented his income by teaching modern languages as well. Unfortunately, Clifford died the same year and Rafinesque's seven-year tenure at the university was tentative and tumultuous. The university curriculum, based as it was in classics and divinity, was ill-suited to accommodate the new studies of the natural sciences. Although the university contained a school of medicine, Rafinesque was refused the chair of *Materia Medica* because he was not a physician. A sense of the autodidact's frustration, caught in the bureaucratic web of the new university, can be felt in the words of his *Life of Travels*: 'I was refused for awhile in Lexington the Diploma of M.A. because I had not studied Greek in a college! altho' I knew more languages than all the American Colleges united; but it was granted me at last; altho' that of M.D. was never granted, because I would not assist to anatomical dissection for which I entertain a dislike. . . .' Throughout his tenure, Rafinesque never ceased his naturalist expeditions and the fruit of this work was seen in the publications of *Florula Ludoviciana, or, A Flora of the State of Louisiana* (New York, 1817) and *Ichthyologia Ohiensis, or, Natural History of the Fishes Inhabiting the Ohio River* (Lexington, 1820).

Rafinesque spent a great amount of energy attempting to establish a botanic garden in Lexington, for which he solicited support from the state legislature in Frankfort, as well as from private subscriptions. In the end, the university offered the post of librarian and keeper of the museum in lieu of his scheme for a garden. Despite being refused the chair of *Materia Medica*, beginning in 1823, Rafinesque taught within the medical college a course on medical botany, which he continued until the end of his tenure in 1826. His focus on medical botany is witnessed by the fact the he visited Jacob Bigelow in Boston.

In the introduction to the first volume of his *Medical Flora* (1828), Rafinesque summarizes the state of medical practice in the United States and his views are of particular interest because of his perspective as an outside observer. He classes practitioners under three heads: the Rationalists, that is, the true scientists; the Theorists, the various sectar-

CONSTANTINE SAMUEL RAFINESQUE, *Medical Flora*, 1828–1830. Blue Cohosh (*Caulophyllum thalictroides*), volume 1, plate 19

ians within the regular medical practitioners; and the Empirics, that is, the quacks. 'The Rational medical men,' Rafinesque explains, are 'liberal and modest, learned or well informed, neither intolerant nor deceitful, and ready to learn or impart information.' One would suspect that Rafinesque would place Bigelow into this category (although Bigelow never acknowledged Rafinesque's work). The Rationalists are further sub-divided into the Improvers, Eclectics, and Experimentalists. The definition of the Eclectics was to be important subsequently, for the label was taken up by Wooster Beach's Reformed Medical movement and the Eclectics counted Rafinesque as one of their own. According to Rafinesque, 'Eclectics are those who select and adopt in practice, whatever is found most beneficial, and who change their prescriptions according to emergencies, circumstances and acquired knowledge.'

The status quo of regular practitioners—those dispensers of the heroic therapies against whom the popular botanic movement had arisen—Rafinesque characterizes as 'often illiberal, intolerant, proud, and conceited; they follow a peculiar theory and mode of practice, with little deviation, employing but few vegetable remedies, and enlisting under the banner of a teacher or sect,' among which sects are mentioned the Brownists, Galenists, Chemicalists, Calomelists, et cetera.

The Empirics are afforded especial opprobrium by Rafinesque as being 'commonly illiterate, ignorant, deceitful and reserved [that is, secretive]: they follow a secret or absurd mode of practice, or deal in patent remedies.' These include '*Herbalists*, vulgarly called *Indian* or *Root Doctors*, and the *Steam Doctors* [i.e., the Thomsonians], who follow the old practice of the natives, the *Quacks* or dealers in Nostrums, the *Patent Doctors*, the *Prescribers* of receipts ...&c.'

Of previously published works, Rafinesque recommends Johann David Schoepf's *Materia medica Americana* ... (Erlangen, 1787), which soundly followed Linnaean principles but which was of necessity cursory and limited. The recent books of Bigelow (see No. 61) and William P. C. Barton (*Vegetable Materia Medica of the United* States, 1817–1818) 'have each their merit, and although not free from errors and omissions, are useful assistants to those who can afford them.' Samuel Henry (see No. 60), a 'spurious Botanist,' is mentioned only to warn against his work, which is a 'worthless book, with incorrect names, wrong descriptions, erroneous indications, and figures mostly fictitious or misapplied.' Rafinesque does allow, however, that Henry's book 'contains some of the Empirical concealed knowledge, available in a few instances,' that is, some of Henry's remedies are of value.

Finally, Rafinesque urges a return to 'pristine Linnaean simplicity' in a work graced with 'cheap but correct' illustrations. Such is the aim, in fact, of Rafinesque's work, 'which is intended as a portable manual of Medical Botany, for the daily use of medical Students,

CONSTANTINE SAMUEL RAFINESQUE, *Medical Flora*, 1828–1830 Twinleaf (*Jeffersonia diphylla*), volume II, plate 55

Physicians, Druggists, Pharmacians, Chemists, Botanists, Florists, Herbalists, Collectors of herbs, heads of families, Infirmaries, &c.'

It is unfortunate that, by and large, this list of persons did not make good use of Rafinesque's accurate and far-sighted manual. The rigors of systematic medical botany were not embraced generally and Rafinesque's characterization of regular practitioners was not likely to engender the warm embrace of that group. Indeed, during the time of the publication of the first and second volumes of *Medical Flora*, Rafinesque had aroused the positive wrath of physicians against him by setting himself up as a rival, alternative practitioner. During the period when Rafinesque quit Transylvania University for Philadelphia, he contracted phthisis (pulmonary tuberculosis) and applied his own scientific powers towards effecting a cure. Rather than subjecting himself to the depletive heroic therapies, he employed a restorative therapy, with a balanced diet of milk, fruit, and green vegetables, as well as an herbal remedy of his own invention. This approach was practiced as well by Rafinesque during his time in Lexington, where, he reports in his *Life of Travels*, '[I] took the measles then prevailing, and was very sick on my return to Lexington; but I recovered in spite of the Physicians, by taking none of their poisons, antimony and opium, while many died in their hands.' Beginning in November of 1827, Rafinesque, under the pseudonym 'Medicus,' ran a series of advertisements in *The Saturday Evening Post* of Philadelphia announcing 'Pulmel,' a cure for consumption. In 1829, under his own name, Rafinesque produced a pamphlet *The Pulmist, or, Introduction to the Art of Curing and Preventing the Consumption or Chronic Phthisis* (Philadelphia; also published in French in Paris, 1833). He introduced the idea of medical specialization: 'I became a Pulmist, who attended only to disease of the lungs, as a Dentist attends only to the teeth.' His consultations were all by correspondence, however, which he attended to with great solicitude. Pulmel was described by Rafinesque as:

> a peculiar compound substance, formed by the chemical combination of several powerful vegetable principles, acting on the lungs and whole system. It contains no pernicious nor poisonous substance. The taste and smell are fragrant and balsamic.

As the century progressed the heroic therapy was diminished by degrees, while restorative therapy came to be adopted generally. This is yet another instance in which Rafinesque was ahead of his time.

Medical Flora was enthusiastically embraced by the less anti-intellectual factions of the botanic movements, especially by the Eclectics. After his death, Rafinesque's plates were acquired and first used by Thomas Cooke in the journal *Botanico-Medical Reformer* and then by Wooster Beach in later editions of *American Practice of Medicine*. The great Eclectic

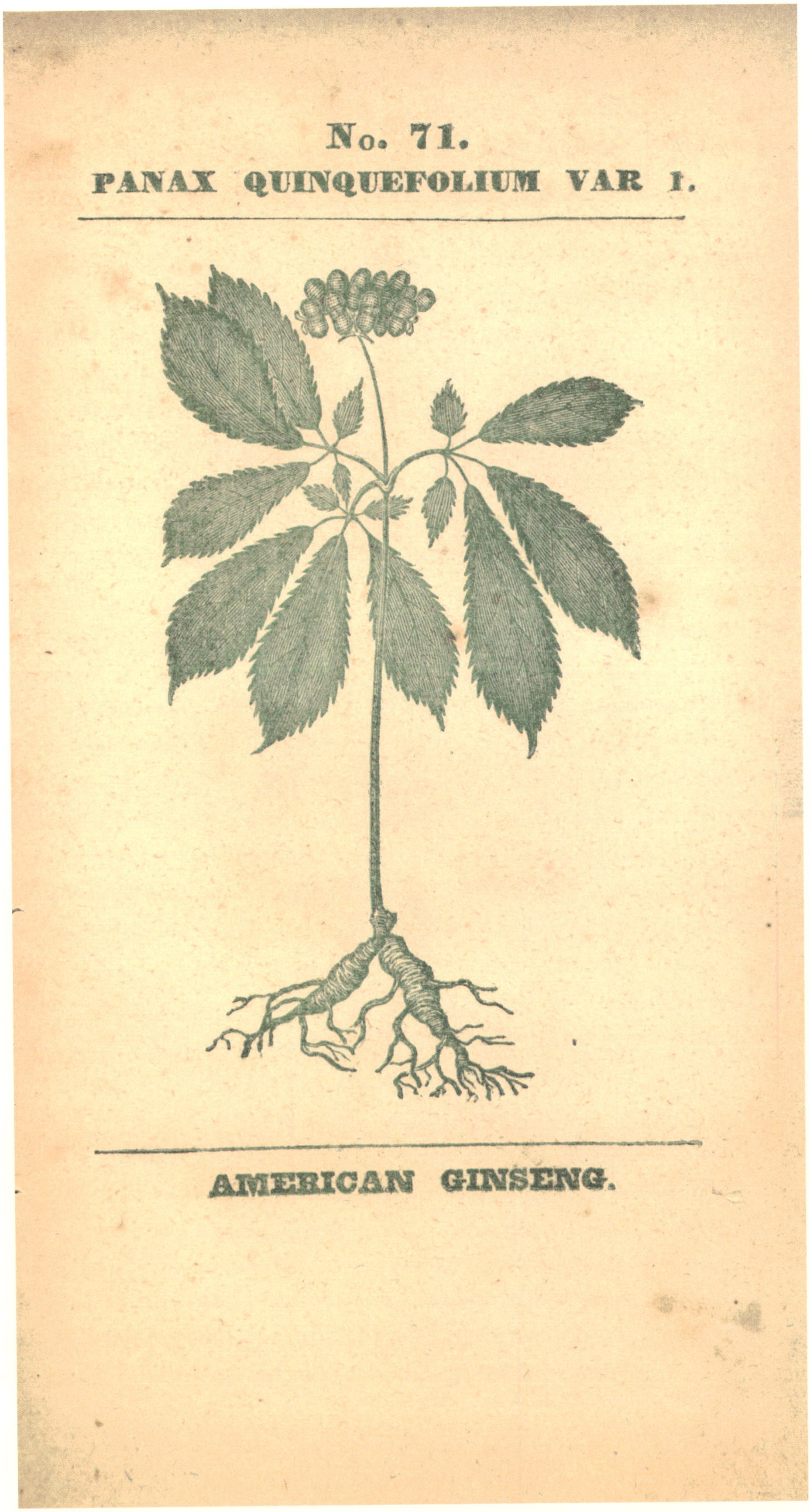

CONSTANTINE SAMUEL RAFINESQUE, *Medical Flora*, 1828–1830. American ginseng (*Panax quinquefolius*), volume 1, plate 711

pharmacist John Uri Lloyd took up the call in *Medical Flora* to investigate, along the lines of the Society of Pharmacists of Paris, the concentrated active principles of medicinal plants. The chemical production of *materia medica* was to be the future.

Rafinesque's *Medical Flora* illustrates 100 plants and covers an additional twenty-five species that were not included in either Barton or Bigelow. Following his introduction, an essay entitled 'General Principles of Medical Botany' is included. In his concluding remarks, he adds that the best medicinal gardens for purchasing herbs are those of the Shakers or modern 'Essenians' (which group he contracted to prepare his Pulmel remedy). He warns against the evils that exist in purchasing imported medicinal plants because they are often adulterated. Rafinesque ends with a list of principal reference books and a list of botanical definitions. Volume I consists in an alphabetized description of plants A–H. Each plant is labelled with its binomial, English, French, German, officinal, and vulgar names, a general description followed by a brief history, description of locality, qualities, properties, substitutes, and remarks. Each entry is accompanied by a wood engraving printed in green. The plant drawings were accomplished by Rafinesque, since he mentions in his introduction that when possible he copied the illustrations from Barton and Bigelow.

In the second volume the illustrations are bound at the end of the book, and there is a second alphabetic sequence to cover plants not included in the original schema. In the introduction to Volume II, Rafinesque explains the changes he has made and makes suggestions for future scholars. He adds that John Bartram's (1699–1777) Botanical Garden near Philadelphia, at that time owned by Colonel Carr, was extremely helpful, as being the oldest and best garden of its kind in the United States, including a living collection of many native plants.

The last plant covered is the grape or *Vitis*. Rafinesque takes a fascinating tangent with the grape, including its history, cultivation, details of making wine, and a monograph on the numerous species. In the Oak Spring Garden Library there is a copy of this small section of the book, published separately and entitled, *American Manual of the Grape Vines and the Art of Making Wine* (Philadelphia, 1830). Rafinesque travelled in 1825 through Virginia along his way to Washington where he met President Adams, Major M'Kinney (head of the Indian department), and Major John Adlum, who owned a vineyard. It was his studies at Adlum's vineyard that were the source of his knowledge of American grapes.

Although for the purposes of medical botany the *Medical Flora* stresses 'pristine Linnaean simplicity,' Rafinesque was long an advocate of the Natural System of taxonomy, as against Linnaeus's Artificial or Sexual System. The former—expounded by, among others, Antoine-Laurent de Jussieu (*Genera plantarum*, Paris, 1789)—expanded the plant characteristics beyond the sexual system alone. Rafinesque writes in *Life of Travels*:

> But all those learned men [American botanists] were mostly strict Linneists, deeming the principles of Linneus, the *nec plus ultra* of natural science, as some do yet to this day. They blamed me for adopting the views of the French School of Science. My guides for Botany were already [Antoine-Laurent de] Jussieu [(1748–1836)], [Etienne Pierre] Ventenant [(1757–1808)], [Michel] Adanson [(1727–1906)], [Noël Joseph de] Necker [(1727–1793)], &c.

This progressive view was first given credence in the United States by the publication of John Torrey's American edition of John Lindley's *Introduction to the Natural System of Botany* (1831) (see No. 50).

During his lifetime, and well beyond, Rafinesque was considered a creator of taxonomic chaos. His exuberant tendency to publish new genera and species—in *Life of Travels* he even relates his discovery of a new species of plant as he disembarks and first sets foot upon the American shores, as well as his discovery of a new fish descried as his ship sinks around him upon his return in 1815—caused American botanists to turn reactionary at Rafinesque's onslaughts. One of his many critics, however, the esteemed botanist Asa Gray (1810–1888), acknowledged in 1841 (the year of Rafinesque's death):

> It is indeed a subject of regret, that the courtesy which prevails among botanists of the present day, (who are careful to adopt the names proposed by those who even suggest a new genus,) was not more usual with us as some twenty years ago. Many of Rafinesque's names should have been adopted; some as a matter of courtesy, and others in accordance with strict rules.

Beyond Rafinesque's unbridled introduction of new genera and species lay a more radical overview of the natural order, however. Before the full development of the concept of organic evolution in Charles Darwin's *Origin of Species* in 1859, it was believed that genera and species were immutable and fixed. The work of the botanist was to uncover this order imposed by the Creator, which view was cherished by Linnaeus himself. The fact that Rafinesque suggested otherwise—as early as 1832—goes some way to explain his taxonomic free-play. He wrote:

> The truth is that Species and perhaps Genera also, are forming in organized beings by gradual deviations of shapes, forms and organs, taking place in the lapse of time. There is a tendency to deviations and mutations through plants and animals by gradual steps at remote irregular periods. This is part of the great universal law of perpetual mutability in every thing.

This perspicacious observation could only come from one who had studied nature so closely throughout his life. The same point is made, in twenty parts, by his epic poem, *The*

World: Or, Instability (1836), whose Preface states: 'The great aim of this poem is to prove that *Instability* is as much a law of nature, as attraction or gravitation.'

The fact that much of Rafinesque's work was ignored in his adopted country ends his *Life of Travels* with the melancholic notice of the 'neglect of my discoveries by Nuttall, Bigelow, and others in Botany…' He asks 'Why should I not find protectors or enlightened patrons, as were found by Audubon and so many others?'

63. Peter Peyto Good (1789–1875)

[*Volume I*]: The Family Flora and Materia Medica Botanica, Containing the Botanical Analysis, Natural History, and Chemical and Medical Properties of Plants: Illustrated by Colored Engravings of Original Drawings, Copied from Nature. By Peter P. Good, A.M. Editor of an Improved Edition of the "Memoirs of the Late John M. Good, M.D., F.R.S., F.R.S.L., Mem. Am. Phil. Soc., and F.L.S. of Philadelphia, &c. &c.["] [*rule*] Volume I. [*rule*] Elizabethtown, N. J.: Published by the Author. [*rule*] [1845].

[*Volume II*]: The Family Flora and Materia Medica Botanica, Containing the Botanical Analysis, Natural History, and Chemical and Medical Properties and Uses of Plants. Illustrated by Colored Engravings of Original Drawings, Copied from Nature. By Peter P. Good, Editor of the Materia Medica Animalia. *Κατανοήσατε τὰ κρίνα, πῶς αὐξάνει· οὐ κοπιᾷ, οὐδὲ νήθει. Λέγω δὲ ὑμῖν, οὐδὲ Σολομῶν ἐν πάσῃ τῇ δόξῃ αὐτοῦ περιεβάλετο ὡς ἓν τούτων.* [Matt. 6:28–29]—Jesus the Christ. Volume II. A New Edition, Revised and Enlarged. Cambridge, Mass.: Published by Peter P. Good, Jr. [1854].

8° 23.4 × 14.5 cm. Originally a periodical published quarterly, 96 numbers bound in two volumes.

PLATES: Volume I: Hand-colored lithographs of plants numbered 1–48. Volume II: added lithographed title before title-page (signed 'Lith of E. C. Kellogg—Hartford, Conn.'). Hand-colored lithographs of plants numbered 49–96.

BINDING: Original blind-stamped black cloth; 'Good's Family Flora' and volume number stamped in gilt on spines. The following advertisements are bound in back of Volume I: *Good's Family Flora, and Materia Medica Botanica* (on yellow paper); *A New Translation of the New Covenant Or Testament of Our Lord and Saviour Jesus Christ; The Memoirs of the Late John M. Good,* (on yellow paper), with lithograph portrait of John M. Good, signed 'Lith. of Kelloggs & Comstock.'; *Principles of Physiology*. The following advertisement is bound in the back of Volume II: *Specimen of Good's material medica animalia*, with one colored lithograph: 'No. 16 Cervus Elaphus. The Stag.,' signed 'Lith. of E. C. Kellogg, Hartford, Conn.'

PROVENANCE: Volume I: Bookplate: Arpad Plesch. Inscribed on title-page: 'E. J. Richardson.' Volume II: Inscribed on front free endpaper: 'O. M. Winagen[?] NH/ Willie Blaisdell Goffstown N.H. July 9, 1898.'

REFERENCES: Arpad Plesch, p. 239; Good.

THE PREFACE to Good's *Family Flora* states that 'The Writer cheerfully acknowledges the assistance he has derived from several large and important works written exclusively for professional readers,' but the author's 'favorite source has been a collection of books and manuscripts of his Uncle … who contemplated a work of this nature more than thirty years ago, and from whose loose and desultory mass of scraps collected for this purpose, a very considerable part of this publication has been culled.' Peter Peyto Good's uncle was John Mason Good, M.D. (1764–1827), a multi-faceted scholar, whom his nephew

PETER PEYTO GOOD,
The Family Flora and Materia Medica Botanica, 1845–1854. Common dandelion (*Taraxacum officinale*), volume 1, plate 29

described in detail in the appended *Memoirs of the Late John M. Good.* John Mason Good, M.D., the son of the Reverend Peter Good, was born in 1764 and named after his maternal uncle, John Mason, a nonconformist minister who wrote the then well-known *A Treatise on Self-Knowledge Showing the Nature and Benefit of that Important Science and the Way to Attain It* (1744). From a young age John Mason Good was taught by his father, who organized a seminary for young children, teaching Latin, Greek, and French. He went on to apprentice with an apothecary and then a surgeon. Working as a doctor in a London prison, he received an award from the city's Medical Society for a dissertation on diseases that most frequent prisons and poorhouses and how to treat and prevent them. The pamphlet *On the Mischief Incidental to the Tread-wheel as an Instrument of Prison-discipline* (1813), later broadened his advocacy for prisoners. He engaged in many literary projects as well, most noted among them a metrical translation of Lucretius' *De rerum natura*, as well as translations from Hebrew of the Song of Solomon and the Book of Job. He wrote a history of the apothecary profession in 1795, *The History of Medicine so far as it Relates to the Profession of the Apothecary*, a taxonomy of diseases in 1817, and *The Study of Medicine* (London, 1822; 4 vols.). His lectures at the Surrey Institution were published as the *Book of Nature* (London, 1853), a popular compendium of natural philosophy. In 1808 he delivered a lecture before the Medical Society of London 'On the General Structure and Physiology of Plants Compared with those of Animals.' This last may lend a clue to the material handed down to his nephew and used in the compilation of Peter Peyto Good's *The Family Flora*.

Peter Peyto Good was born in 1789 to Peter (the younger brother of John Mason Good) and Agnes Leckie Good, of Bocking, England. The young Peter Peyto Good immigrated to America and met his wife Susan, née Tappan, in New York City. They were married in 1830 and had two children. In the same year Good published *Exercises Designed to Assist Young Persons to Pronounce and Spell Correctly* (with an imprint of Woodstock, Vermont), which reveals Good's early vocation both as a publisher and lower-form teacher. Their first child, Agnes, died in 1833, the same year that their second child was born, whom they named Peter Peyto, Jr. The family of three moved to Elizabethtown, New Jersey, where Peter is described in the census as a book publisher. It was during this time that he brought out Volumes I and II of *The Family Flora* in 1845. Good published a sequel in 1853: *A Materia Medica Animalia: Containing the Scientific Analysis, Natural History, and Chemical and Medical Properties and Uses of the Substances that are the Product of Beasts, Birds, Fishes, or Insects*. This publication, and also 'A New Edition, Revised and Enlarged' of *The Family Flora* (1854), were issued from Cambridge, Massachusetts, the latter with Peter Peyto Good, Jr., as publisher, indicating his brief involvement with the family publishing ventures. Peter, Jr., graduated from Harvard Law School in 1856 and practiced law in New York. By 1860 Peter, Sr., had settled in Plainfield, New Jersey, where he taught school, and where Peter,

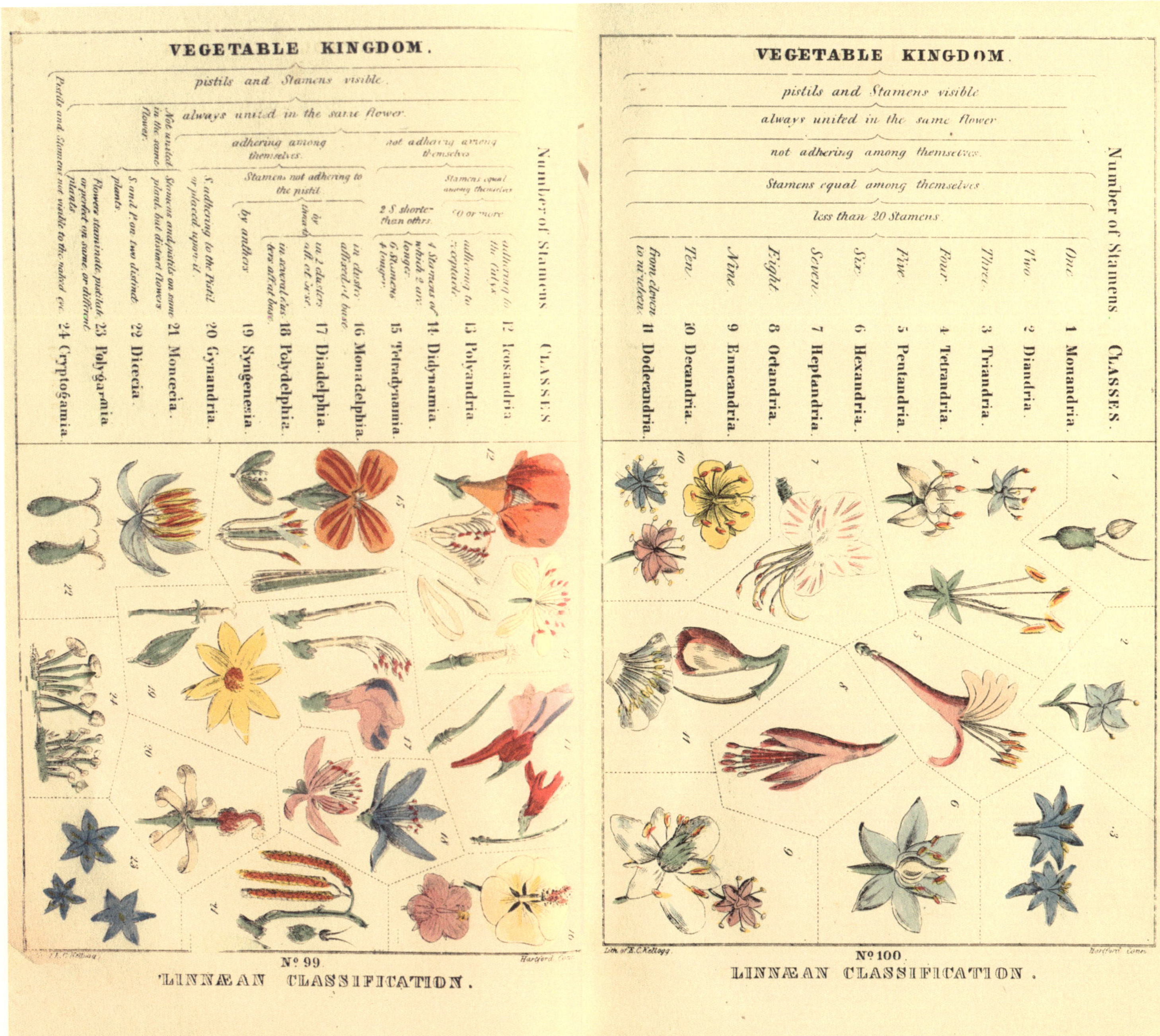

PETER PEYTO GOOD, *The Family Flora and Materia Medica Botanica*, 1845–1854. Chart of the vegetable kingdom, according to Linnaeus, volume II

Jr., presided as a city judge. From that town Peter, Sr., published in 1865 *Stenography: An Original System for Quick Writing, Eminently Eclectic and Useful.* The first volume of *A Family Flora* also advertises the titles *A New Translation of the New Covenant or Testament*, Peter P. Good, *Editor* ('Published by the Editor'), and *Principles of Physiology* by J.C. Comstock and B. N. Comings, published by Pratt, Woodford and Co., New York, and E.C. Kellogg, Hartford. Good was described by one reviewer (reprinted under 'Commendations' in Vol. II) as being 'his own engraver [*sic*], printer [i.e., publisher], agent, and editor.'

Susan Good died in 1865 and Peter, Sr., moved back to New York City, where he died in August 1875 at the age of eighty-six. Following the death of his parents, Peter, Jr., moved to Seattle, Washington, in 1885, and died there the following year, having succumbed to pleurisy after a brief imprisonment for his role in the instigation of anti-Chinese labor riots associated with the early settlement of the Puget Sound Co-operative Colony, his utopian vision of a community of shared ideals, labor, and land, inspired by Peter's visit to the 'Palais Social,' a co-operative community of ironworkers at Guise, France.

The complex publishing history of *A Family Flora*—brought out with different title-pages, imprints, cities of issue, dates, and even publishers—is due to the fact that the peripatetic Good published the work in quarterly parts of ninety-six numbers. There also exists a two-volume set with the title *A Materia Medica Botanica, Containing the Botanica Analysis, Natural History, and Chemical and Medical Properties of Plants*..., with the imprint: 'New York: Published by J. K. Wellman; New Milford, Conn.: Peter P. Good, 1845,' as well as an imprint from Elizabethtown in 1847, and one with the same title from Cambridge, 1853 (of Vol. I). In the new issue of *A Family Flora,* Volume II, published in 1854, there is an advertisement for *Animalia* with an inserted lithograph of a stag. *The Family Flora* is in turn advertised in *Animalia* (1853), from which can be gained a view of the publication process:

> *The Family Flora* ... is published, in pamphlet form, regularly, in quarterly parts, in January, March, June, September, and December of each year, and each year forms one volume alternately and consecutively. Each *quarterly* part contains twelve numbers and plates, with the letter-press matter attached, embracing the Botanical Analysis, the Natural History, and the Chemical and Medical Properties and Uses of each plant.
>
> With these four parts, containing No. 1 to No. 48 inclusive, completing the first volume, is published (*without additional charge*) an extra Part, in January, containing a handsome Frontispiece, with the Title, Preface, Index, and an extensive Glossary of Botanic Terms, together with an uncommonly striking likeness of the late John M. Good, M.D. F.R.S., &c., &c., with a notice of his life, writings, character, &c.
>
> Cash terms, $3 a year, in advance,
>
> Four quarterly parts, containing No. 49 to No. 96 inclusive, similar in all respects to the above, but embracing different plants, are published for the second year with these four parts, completing the second volume, is published (*without additional charge*) an extra Part, containing the Title, Index, and Introduction to the Study of Botany, with two beautiful plates representing the Linnæan Classification.
>
> Cash terms, $3 a year, in advance.
>
> ☞ The third year and volume is not yet published.

PETER PEYTO GOOD,
The Family Flora and Materia Medica Botanica, 1845–1854. 'Helonias dioica,' Blazing-star or Rattlesnake root (*Chamaelirium luteum*), volume II, plate 76

> The two volumes already published, perfect, interleaved, and handsomely bound, will be immediately forwarded (*free from postage*) to any part of the United States, on receipt of orders with money. Cash terms, $4 *each* volume, in advance.
>
> Address Peter P. Good, North Cambridge, Middlesex Co., Mass.

The volumes were at least two years in preparation and a third, projected volume was never issued. The Oak Spring Garden Library's first volume of *The Family Flora* is from the original edition, published in Elizabethtown, New Jersey, in 1845, while the second is from the new edition, published in Cambridge, Massachusetts, in 1854 by the author's son (although Volume II issued from Elizabethtown in 1845 is also designated 'A New Edition, Revised and Enlarged'). Each volume covers forty-eight plants for a total of ninety-six hand-colored lithographs. Commercial lithography was introduced to America in the 1830s, while chromolithography—which was poised to flood with color American books, prints, and advertisements—was not common until the 1850s. The plates are signed by the Kellogg family of lithographers, a prolific firm, second only to Currier & Ives. The company was founded around 1830 in Hartford, Connecticut, and later opened offices in New York City. Trading under the initials of various Kellogg brothers and appending names of various partners, the firm lasted to the beginning of the twentieth century. The laborious process of hand coloring was sped along by the use of stencils to block in colors, which method, never exact, can be seen clearly in Figure 1, Rose Centifolia, of the Oak Spring Garden Library's copy, where the stencil is dramatically mis-registered. The fact that Good advertises *Principles of Physiology*, published by E.C. Kellogg, indicates that he was engaged in several ventures with the Kellogg firm.

At the beginning of Volume I there is a detailed glossary of both botanical and medical terms. The second volume of *The Family Flora* begins with a two-page illustration of the vegetable kingdom according to the Linnaean classification. This sets the stage for a detailed introduction to the study of Botany. As an indication of the progress of botanical science in the new nation, we may notice that Good discusses the details of the Linnaean system as well as its faults, and compares it to the Natural System of classification, which by then had superseded the Artificial System. What for C. S. Rafinesque had been a lifelong campaign for the supplanting of Linnaeus' artificial with the French system was now patiently explained in Good's Introduction to the amateur botanist.

In Volume II, Good prints 'Commendations' of the *Family Flora*. Many are from the journals of various heterodox medical schools then in evidence. From the *Medical Journal* of Ohio comes the comment

> The author appears familiar with many of the plants not in use as remedial agents in the allopathic school of medicine, but which are in very common use amongst the Eclectic class

of physicians. This speaks well for his liberality, and clearly manifests a disposition to keep up with the improvements of the science.

Good cites as references under specific plants a range of botanic works, including Linnaeus, *Species plantarum* (1753), Frederick Pursh's *Flora Americae Septentrionalis* (London, 1814), John Lindley's *Flora Medica* (1838) (see No. 50), and the American herbals of Henry (see No. 60), William P. C. Barton, Jacob Bigelow (see No. 61), and Rafinesque (see No. 62). The U.S. and London dispensatories are also cited, and, like Henry, Good extensively employs William Woodville's *Medical Botany* (London, 1790–1794[5]). In addition to these are found works of the Thomsonians: John Thomson's (Samuel's son) compendious *The Thomsonian Materia Medica, or, Botanic Family Physician* (1841), and Horton Howard's Thomsonian work *An Improved System of Botanic Medicine* (1832). Most in evidence, however, are the various works of the Eclectics: namely, the later version of Wooster Beach's manual, *The American Practice Abridged, or, The Family Physician* (1846), John Kost's *Elements of Materia Medica and Therapeutics* (1849), and the *Eclectic Dispensatory of the United States* (1852). As the reviewer suggests, the selection of the plants themselves also reflects the prevailing Eclectic pharmacopeia. Alphonso Wood's *Class-Book of Botany: Being Outlines of the Structure, Physiology, and Classification of Plants* (1845) is cited as well in both the work's entries and in the 'Prospectus.' Good recommends to those who adopt the *Family Flora* as a textbook that they may use 'in connection with it' Wood's *Class-Book*.

By the 1840s, Good's book on the *materia medica* was an anomaly, if not an anachronism. It was marketed as a family guide to self-prepared and administered medication (in the tradition of Culpeper), but issued in the form of an *édition de luxe*. The price of $8 for the complete set is equivalent to over $200 in current values. It represents an odd transition from handbooks of necessity in a new nation, to an ornament meant for the parlor of a new middle-class with time for amusement in the pursuit of a botanic hobby. Good advocated its use not only for 'heads of families' and as a textbook, but also as a 'most acceptable and appropriate Parlor or Lady's Book.' Save for the unpublished researches of Good's illustrious uncle, *The Family Flora* is not a work of original scholarship. Indeed, even Good's *Memoir* of his uncle, which he calls 'an improved edition,' is but a reworking of Olinthus Gregory's *Memoirs of the Life … of the Late John Mason Good,* originally published in London in 1828. Lacking contemporary sales accounts, we have no way to discover whether Good's publishing venture paid dividends.

INDEX

INDEX

Abbey, John Roland, 45, 46
Abdachim, 99
Abdella, 99
Abguillara, Luigi, 182
Académie Royal des Sciences, Paris, xlii, 256–7
Academy of Science, Berlin, 226
Accademia dei Lincei, 74, 100, 174, 300, 331
Accademia dei Secreti della Natura, 298
Accademia di San Luca, Rome, xxvii
Acciaiuoli, Donato, 62
Acosta, Cristobal, 63, 312, 315
Adanson, Michel, 220, 375
Adet, Pierre, 161
Aegineta, Paulus, 67, 77
Agricola (Georg Bauer, *called*), 112
Aiton, Mr., 133
Albert, Duke of Prussia, 40
Albertus Magnus, xxxi, 4, 24
Alciato, Andrea, 303
Aldini, Tobia, 282
Aldrovandi, Ulisse, 50, 51, 64, 246, 300, 303, 304, 327
Alexander VII (Pope), 277–8
Allde, Edward, 99, 103
Alpini, Prospero, xxxviii, 95–99, 141, 245, 307, 308
Altieri, Francesco, 142
Amman, Jost, xl, 248
Amsterdam Physic Garden, 261
Anderson, Frank, 79
Anguillara, Luigi, 248, 307
Antoniis, Marco de, 285
Apelles, 32
Apollo, 32, 70, 268
Apuleius Platonicus, xxx, xxxi
Arber, Agnus, 15, 51
Aristotle, xxviii, 8, 297, 303
Artemisia, Queen of Caria, 70
Asclepius, 70, 216, 268
Ashmole, Elias, 204
Audubon, John James, 366, 368, 376
Augustus, Elector of Saxony, 46, 48
Austrian National Library, Vienna, 331
Avicenna, 8, 24, 140

B

Bacci, Andrea, 170, 282
Baciocchi, Elisa, Grand Duchess of Tuscany, 287
Baglione, Giovanni, 174
Baker, George, 80, 82
Baker, Henry, 149
Baldwin, William, 354
Balzac, Honoré de, 322
Bämler, Johann (Hans), 3
Banckes, Richard, 24
Banks, Sir Joseph, 133, 147, 292
Barbaro, Ermolao, 31, 42
Barberio, Angelo, 336
Barker, Nicolas, 78
Bartolomaeus Anglicus, 24
Barton, Benjamin Smith, 348, 356, 358, 367
Barton, William P. C., 352, 358, 370, 372, 374, 383
Bartram, John, 147, 348, 367, 374
Bartram, William, 356
Bascarini, Nicolò de, 48
Bassano, Leandro (*also known as* Da Ponte), 98
Batthynay, Balthasar de, 93
Bauer, Georg, *see* Agricola
Bauhin, Caspar, 229, 276
Bauhin, Jean, 82, 276
Beach, Wooster, 362, 370, 372, 383
Beauharnais, Alexandre, 158
Beauharnais, Eugéne de, 158
Beauharnais, Josephine, xxxviii
Belon, Pierre, 99, 307
Benalio, Bernardino, 14, 20
Benedetti, Ignazio, 285
Benedict XIV (Pope), 142, 278
Benvenuti, Baldassarre, 287, 288, 290
Berès, Pierre, 23
Bericchia, Giacomo, 170
Bernaud, Arsenne Thiébault de, 161
Bernis, Francesco Gioacchino de Pierre de (Cardinal), 285
Besler, Basil, 250
Bewick, Thomas, 324
Biblioteca Alessandrina, Rome, 51
Biblioteca Ambrosiana, Milan, 274, 330
Biblioteca Angelica, Rome, 51, 87
Biblioteca del Museo Civico di Storia Naturale, Milan, 274
Biblioteca Nazionale, Florence, 330
Biblioteca Universitaria, Bologna, 56
Bibliothèque de Drapiez, 160
Bibliothèque Nationale, Paris, 20, 23, 78
Bibliothèque Royale, Paris, 141
Bieler, Ambrossius Carolus, 212, 217
Bigelow, Jacob, xlv, 355–65, 368, 370, 372, 374, 376, 383
Birckman, Arnold, 56, 60
Blackwell, Elizabeth, 340
Blake, John Bradby, 132–137
Blake, John Bradby, Sr., 133
Blégny, Nicolas de, xxxviii, 115–20
Blunt, Wilfrid, 63, 180, 212, 260
Boccaccio, 177
Boccone, Paolo, 331, 334
Bock, Hieronymous (Tragus, *known as*), 31, 42, 57, 82
Bodleian Library, Oxford, 331
Boerhaave, Hermann, 276

Böhme, Jakob, 204
Bollifant, Edmund, 72, 80
Bombarde, M., 220
Bonafadini, Bartholomeo, 169, 170
Bonaparte, Josephine (Empress), 158, 162
Bonato, Giuseppe Antonio, 154
Bonelli, Giorgio, xlii, 277–85
Bonvicino, Alessandro (*called* 'il Moretto'), 52
Borch, Ole, 112
Borcht, Pieter van der, 64, 66, 70, 107
Bosse, Abraham, 256, 257, 260
Botanical Garden (Hortus Medicus), Amsterdam, 140, 261–6
Botanical Garden, Bologna, 73
Botanical Garden, Florence, 138, 331
Botanical Garden (Royal), Kew, 132, 290–2
Botanical Garden, Leiden, 63, 85, 88–9, 256
Botanical Garden, Leipzig, 340
Botanical Garden, Padua, 39, 64, 73, 86, 96, 98, 153–4, 158–9, 182, 200, 256
Botanical Garden (Jardin [Royal] des Plantes [Médicinales]), Paris, xl, xlii, 108–9, 220, 251–7, 315
Botanical Garden, Philadelphia (John Bartram), 374
Botanical Garden (Giardino dei Simplici), Pisa, xxxviii, 39, 63, 64, 73, 86, 140, 256, 286–7, 330
Botanical Garden (Jardin des Plantes), Poitiers, 182
Botanical Garden, Rome, 73, 277–87
Botanical Garden, Tübingen, 39
Botanical Garden, Turin, 270–4
Botanical Garden, Vienna, 63, 246
Botanical Library, University of Florence, 11
Botanical Museum, University of Florence, 87
Bouchard, Giovanni, 278, 282, 285
Bouchard, Maddalena (*also known as* Cleonice Aphrodisia), 282, 285
Bouchet, [A.], 116
Bourbon-Condé, Prince Henri de, 180, 182, 184
Boym, Michel, 133
Bradford Medical Institute, xlv–xlvi
Brançion, Jean de, 63
Braun, Georg, 167
Bresciani, Orazio, 200
British Museum, London, 80, 126, 256
Brochini, Marchioro, 200
Brooke, Thomas, 20, 23
Brosse, Guy de la, xl, 108, 251–57
Brosse, Isaïe de la, 251
Brown, John, 348
Brown, Patrick, 151
Brunacci, Elena Amici Mastiani, 287
Brunet, J. C., 340
Brunfels, Otto, xxxiv, 31–36, 37, 42, 58, 70, 82
Buchan, William, 347
Buffon, Georges Louis Leclercq de, 220
Buillon, Charles de, 251, 254
Buon, Nicolas, 311
Buoncompagni, Ignazio (Cardinal), 285
Busbecq, Ogier Ghiselin de, xxix, 51, 63, 64, 66

C

Caccia, Giovanni Bartolomeo, 270, 278
Caffi, Pietro, 199
Calceolari, Francesco, 199, 200
Camerarius, Joachim, the Elder, 39, 50, 245
Camerarius, Joachim, the Younger, xl, xlii, 50, 64, 95, 245–51, 302–311
Camerarius, Ludwig, 304
Campanella, Augusto, 285
Canacci, Giuseppe, 287, 288, 290
Cardano, Gerolamo, 330
Carlino, Giovanni Giacomo, 74
Casabona, Giuseppe (Goethuysen, Jodocus de, *known as*), 245, 248, 250
Casserio, Giulio, 85, 86
Castelli, Pietro, 282
Catesby, Mark, 348
Cathcart, William, Earl of, 121
Cattani, Isabella, 174
Cattrani, Baldassarre, xxxviii, 152–159
Cattrani, Tomaso, 154
Caymox, Hubert, 304, 311
Cecil, William, First Baron of Burghley (Lord Burghley), 80, 82
Ceres, 84
Cesalpino, Andrea, 87, 250, 287
Cesi, Prince Federico, 74, 174, 300
Charles I, King of England, 51
Charles V, Emperor, 70
Charles VIII, King of France, 19
Chatillon, Louis, 257
Chromis, 32
Cibele, Angela Nardo, 154
Cibo, Gherardo, 51, 87, 327
Circe, 336
Ciuti, Giuseppe, 287
Civinini, Giovanni Domenico, 138–41
Civinini, Lodovico, 138
Civinini, Marcantonio, 138
Clement XIV (Pope), 278
Clesio, Bernardo (Cardinal, Bishop of Trent), 46
Clifford, George, 266, 268
Clifford, John D., 368
Clusius, Carolus (Jules Charles de l'Écluse), xxxviii, 62–67, 68, 70, 82, 88, 93–5, 100, 107, 109, 182, 146, 248, 250, 264, 276, 307, 310
Cockerell, Sydney Carlyle, Sir, 45, 46
Codex Juliana Anica (*Codex Vindobonensis*), xxix, 51
Colbert, Jean-Baptiste, 258
Coles, William, xl, 202–5, 237
Collegio Alberoni, Piacenza, 272
Collegio Ghislieri, University of Pavia, 274
Collinson, Peter, 147
Colonna, Fabio, 72–79, 84, 331
Colonna, Girolamo, 72

Commelin, Jan and Caspar, xlii, 260–6
Contant, Paul, 111, 180–6, 192
Contant, Jacques, 180–6
Contile (publisher), 304
Cook, James, 147
Cooke, Thomas, 372
Copernicus, Nicolaus, 37
Cordo, Valerio, 42
Cornut, Jacques, 111
Corterius, Gabriel, 169
Coxe, John Redman, 356
Crateuas of Pontus, xxix
Crema, Liberale, 86
Cremyeu, Seigneur de, 167
Cullen, William, 348
Culpeper, Nicholas, xl, 202, 235–41, 347, 349, 383
Curtis, Alva, 362
Curtius, Jacopus, 304
Custodia Foundation, Paris, 64
Cutler, Manasseh, 348

D

d'Este, Luigi (Cardinal), 298
Darling, William, 226
Darwin, Charles, 152, 210, 375
Darwin, Erasmus, 210
Davy, Jacques (Archbishop), 178
De Bry, Theodore, 86
De Giraud, Bernardino (Cardinal), 285
De Rossi, Andrea, 282
Demoges, P., 180, 184
Dendronico, Federico, 336
Department of Printing and Graphic Arts, Harvard University, 56
Dewes, Gerard, 70
Diani, Tito, 169, 170
Diaz, Hernando, 100
Didot, Pierre François, *le jeune* (printer), 218
Dietrich, Jacob of Bergzabern, *see* Tabernaemontanus
Dietrich, Johann Georg, 212
Dietrich, Ludwig Michael, 212
Diocles or Carystus, xxviii
Dioscorides Pedanius of Anazarbus, xxix, xxx, xxxii, 31, 41, 42, 44, 46, 48, 51, 73, 77, 82, 84, 169, 170, 182, 184, 236, 237, 264, 287, 327
Dobbs, Arthur, 147
Dodart, Denis, xlii, 257–60
Dodoens, Denys, 67
Dodoens, Rembert, 58, 63, 66, 67–72, 80, 82, 198, 264
Dombey, Joseph, 162
Donati, Antonio, 198–202
Doria, Giovanni Andrea, 98
Drake, Sir Francis, 64, 93, 94
Drapiez, Pierre Auguste Joseph, 160
Du Pinet, Antoine, Sieur de Noroy, xl, 51, 167–9
Duflos, Pierre, 218
Duhamel du Monceau, Henri Louis, 52, 56, 149
Dumbarton Oaks, Washington, D.C., 52, 56, 158
Durante, Castore, xl, 169–77, 282
Dürer, Albrecht, 27, 32, 34, 245, 319
Duret, Claude, xlii, 311–15

E

Eco, Umberto, xxviii
Edwards, Edward, 223
Edwards, John, 223-26
Ehret, Geog Dionysius, xlii, 151, 212, 217, 268
Eliot, Lawrence, 93
Elizabeth I, Queen of England, 58, 60, 70, 80, 93, 95
Ellis, John, 136, 145–52
Elzevir, Louis, 85
Emo, Giorgio, 96
Empedocles, 4
Endem, Johannes de, 107
Ettmüller, Michael, 114
Evelyn, John, 256

F

Faber, Johannes, 282
Fabrici, Girolamo d'Acquapendente, 85, 86, 96, 245
Faciotti, Guglielmo, 74
Fagon, Guy Crescent, 256
Falconer, John, 58, 87, 327
Falloppia, Gabriele, 245
Fehrt, Carl de, 218
Felix Platter Herbarium Systematisch-Geobotanisches Institut, University of Berne, 34
Ferdinand I, Emperor, 11, 50, 51
Ferdinand, Archduke of Further Austria and Tyrol, 48
Ferro, Giovanni, 304
Feyerabend, Sigmund, 246
Fialetti, Odoardo, 86
Fiori, Antonio, 285
Fitzwilliam Museum, Cambridge, 167
Fleetwood, J. (Rev.), 236
Flora, 85, 216
Fothergill, John, 151
Foulquier, Antoine Valentin, 324
Fourgeron, Ignace, 224
Fragoso, Juan, 107
Frampton, John, 99, 103
Franchi, Sebastiano, 138
Franklin, Benjamin, 230, 331, 347
Fransiosi, Angiolo (Abbot), 154
Fuchs, Leonhart, xxxviii, 37–45, 57, 58, 60, 68, 70, 72, 82, 86, 110, 169, 172, 174
Fugger, Anton, 64
Fugger, Jacob, 64
Füllmaurer, Heinrich, 44
Funke, J. M., 331, 339

G

Gaglier, Gaetano, 290
Galen (Claudius Galenus), xxx, 4, 8, 20, 73, 77, 169, 170, 270
Galilei, Galileo, 74, 297, 300
Galland, Antoine, 141
Garden of Adonis, 32
Garden of the Hesperides, 32, 70
Garsault, François Alexandre Pierre de, 217–23
Garzoni, Giovanni, 51
Gassendi, Pierre, 251
Gastone, Gian, 140
Gaultieri, Niccolò, 138

Gavelli, Niccolò, xxxviii, 141–145
Gemmingen, Johann Conrad von (Prince-Bishop), 250
Gentius, King of the Illyrians, 70
Geoffroy, Etienne François, 218
Georg, Prince, Margrave of Brandenburg, 39
Gerard, John, xxxviii, 79–84, 177, 312
Gesner, Konrad, 58, 246, 248, 250, 276, 303, 304, 307, 327
Gheyn, Jacques de, the Younger, 64
Ghini, Luca, 40, 58, 86, 327
Gilles, Pierre, 307
Giovio, Paolo, 303, 307, 308
Girault, Jean, 87, 327
Good, John Mason, 376, 383
Good, Peter Peyto, Jr., 378
Good, Peter Peyto, Sr., 376–83
Gorham, John, 355
Gravier, Giovanni Giuseppe, 278, 285
Gray, Asa, 375
Griffin, Edward, 72
Grignion, Charles, 206, 208
Grolier Club, New York, 78
Guerin and Delatour, publishers, 52
Guidobaldo II della Rovere (Duke of Urbino), 51
Guidotti, Camillo, 336
Guilandino, Melchiorre (Melchior Wieland), 96, 98
Gutenberg, Johann, 15

H

Hahnemann, Samuel, 362
Haid, Johann Jacob, 212, 216, 217
Hainzelmann, Johann, 118
Hájek, Tadéas, 50
Handsch, Georg, 50
Happe, Andreas Friedrich, 226–29
Harder, Johannes, xlv, 327–29
Harms, Wolfgang, 304
Hayon, Leon Albert, 324
Henri II, King of France, 144
Henri II, Prince of Condé, 251
Henri IV, King of France, 108, 109
Henrietta Maria, Queen of England, 51
Henry VII, King of England, 19
Henry VIII, King of England, 58, 60
Henry, Nicolas, 167
Henry, Samuel, xlv, 347–55, 360, 370, 382
Hercules, 32, 70
Hermann, Paul, 89
Hernández, Francisco, 74, 100
Héroard, Jean, 254
Herolt, Johanna Helena, 266
Hill, John, 206–10
Hippocrates, 39, 73, 364
Hofmann, Johannes, 304, 311
Hogenberg, Franz, 167
Holmes, Oliver Wendell, Sr., 364
Homer, 141, 336
Hopper, Samuel, 224
Hortus Cliffortianus, 266–70
Howard, Horton, 383
Hunt Institute for Botanical Documentation, Carnegie Mellon University, Pittsburg, 56
Huydecoper, Joan van Maarseveen, 264
Huysum, Jacob van, 216

I

Ibn al-Baitar, Abdallah, xxx
Iconoteca Bassiana, University of Bologna, 277
Imperato, Ferrante, 73, 74
Isidore of Seville, 8
Isingrin, Michael, 40, 45, 68

J

Jackson, James, 356, 365
Jagiellonian Library, Cracow, 64
James I, King of England, 144
Janssen van Kampen, Gerard, 66
Jefferson, Thomas, 356, 367
Johann Georg III, Elector of Saxony, 111
Johnson, Thomas, 84, 204
Joppi Library, Udine, 11
Junch, Paolo, 278
Jungermann, Joachim, xl, 248, 250, 307, 310
Junius, Hadrianus, 304
Jussieu, Adrien-Laurent-Henri de, 161, 162
Jussieu, Antoine de, 151, 218, 256
Jussieu, Antoine-Laurent de, 218, 374
Jussieu, Bernard de, 218, 229
Jussieu, Joseph de, 162

K

Kaempfer, Engelbert, 133, 137
Kalm, Peter, 348
Kaufman, Paul, 304
Kellogg, E. C., 379, 382
Kelly, Thomas, 235, 236, 241
Kepler, Johannes, 297
Keuchen, Ulla-Britta, 304
Kiggelaer, Frans, 261
King Solomon, 84
Kirkall, Elisha, 216
Kniphof, Johann Hieronymus, 331, 339
Knowlton, Thomas, 95
Konrad von Megenberg, xxxi, xxxiv, 2–9
Korn, Johann Friederich the Elder, 342
Kost, John, 383

L

L'Écluse, Jules Charles de, *see* Clusius, Carolus
L'Obel, Mathias de, 66, 67, 80, 82, 103
La Pérouse, Jean François de, 160
Lamaritne, Alphonse de, 320
Lambertini, Prospero, *see* Benedict XIV
Lancelot, Dieudonné Auguste, 324
Langlois, N., (Paris printer), 162, 164
Le Coq, Pascal, 182
Le Moine, Wilhelmus, 99
Leclerc, Sebastien the Elder, 258
Leest, Antoine van, 66
Lémery, Nicolas, xxxiii, 316

Lennox, Charles, 2nd Duke of Richmond, 206
Leonardo da Vinci, 8, 330
Leopoldo, Grand-Duke of Tuscany, 287
Lestiboudois, François Joseph, 160
Lewis, Daniel, 350–52
Lewis, Meriwether, 356
Libavius, Andreas, 112
Liberale, Giorgio, 45, 50, 52, 56
Ligozzi, Jacopo, 95
Linck, I., 114
Lindley, John, 290–93, 375, 383
Linnaeus, Carolus (Carl von Linné, *called*), xlii, 66, 79, 99, 146, 146, 149-152, 206, 208, 224, 229, 266–70, 282, 285, 286, 287, 288, 290, 323, 324, 374, 375, 382, 383
Linocier, Geofroy, 110
Lipsius, Justus, 62
Lloyd, John Uri, 374
Loe, Henrick van der, 70, 72
Loe, Jan van der, 67, 68, 70, 72
Longhi, Pietro, 188
Löschenkohl (banker), 212
Louis XIII, King of France, xl, 251
Louis XIV, King of France, xxxiii, 116, 140, 151, 257, 258, 260, 315
Louis, Pierre-Charles-Alexandre, 362, 364
Louise of Lorraine, Queen of France, 167
Lownes, A. E., 204
Ludwig, Christian Gottlieb, 331, 339-43
Luigné, Sophie de, 162
Lusitano, Amato, 327
Luther, Martin, 31
Lysimachus, 70
Lyte, George, 79
Lyte, Henry, 70

M

Macer Floridus (Macer, Amelius), 172
Magietto and Pasquali (publishers), 86
Magno, Olao, 307
Malie, Thomas, 121–31
Manuel I, King of Portugal, 104
Maranta, Bartolomeo, 73, 307
Maratti, Francesco, 278, 282
Marchant, Nicolas, 257
Maria of Hungary, 70
Maria Theresa, Queen of France, 116
Marie de Brimeu, Princess of Chimay and Duchess of Aarschot, 62
Marshal, Humphrey, 367
Marsili, Giovanni, 153
Martelli, Niccolò, 278, 282, 285
Martyn, John, 216
Mary Tudor, Queen of England, 58
Mattioli, Ferdinando, 48
Mattioli, Pietro Andrea, xxxii, xxxviii, 45–56, 57, 58, 70, 82, 110, 167, 169, 172, 174, 246, 303, 304, 307
Mauk-Sow-U, 131–137
Maurocenus, Marcus Antonius, 88
Maximilian I, Emperor, 11
Maximilian II, Emperor, xxix, 48, 50, 63, 68
Maximilien de Béthune, Duke of Sully, 184, 311
McKinley, Daniel, 147
Medici, Catherine de', 144, 330
Medici, Cosimo I de', Grand-Duke of Tuscany, 40
Medici, Cosimo III de', Grand-Duke of Tuscany, 138, 140
Medici, Ferdinand I de', Grand-Duke of Tuscany, 86, 248
Medici, Marie de', 109
Melanchthon, Philipp, 39, 64, 245
Melantrich, Jiří, 48, 50
Mellon, Paul, 100, 235, 329, 355, 366
Mercati, Michele, 282
Mercuriale, Gerolamo, 96
Merian, Maria Sibylla, 126, 266
Mersenne, Marin, 251
Mesuë, Joannes, the Younger, 24
Metlinger, Pierre, 24
Meydenbach, Jacob, 14–15, 24
Meyer, Albrecht, 44
Meyer, Frederick, 44
Meyerpeck, Wolfgang, 45, 50, 56
Michaux, André, 367
Micheli, Pietro Antonio, 138
Mierdman, Steven, 58
Miller, John (Johann Müller), 151
Minich, Hans, 50
Mirbel, Charles François (*known as* Brisseau), 159, 162
Mithradates VI Eupator, King of Pontus xxix, 70
Mnasilus, 32
Monardes, Nicolas, xxxviii, 63, 84, 99–103, 104, 107, 180, 200, 307, 312
Moniglia, Gaetano, 138
Moninckx, Jan, 264
Moninckx, Maria, 264
Montaigne, Michel de, 327
Morandi, Giovanni Battista, xlii, 270–7
Moretus, Jean, the Elder, 66
Morgan, Hugh, 95
Morin, René and Pierre, 111
Muhlenberg, G. H. E., 367
Müller, Cornelis, 66
Muntzer, Thomas, 31
Musée d'Histoire Naturelle, Paris, 256, 329
Museo dell'Accademia, Venice, 188
Museo Provincale d'Arte, Trent, 11

N

Naironi, Anton Fausto, 140
Napoleon I, Bonaparte, Emperor, 287
National Library, Vienna, 40
Natural History Museum, London, 121, 126, 133, 226, 270
Necker, Noël Joseph de, 375
Nelson, Charles, 147
New York Public Library, 251
Newton, Niman, 72
Nicholas V (Pope), 277
Nicolai, Arnaud, 66, 70, 107

Nicolaus de Regio, 20
Nicot, Jean, 103, 144
Niebuhr, Casten, 160
Nobilis, Pietro di, 78
Norton, Bonham, 99, 103
Norton, John, 80
Norton, William, 100, 103
Nuttall, Thomas, 356, 376

O

Öllinger, Jörg, 246, 248
Orléans, Gaston d' (Duke), 116, 257, 260
Oroborus (serpent), 268
Orta, Garcia (Abraham) da, xxxviii, 58, 63, 93, 104–7, 307, 312
Ortelius, Abraham, 62, 251
Osborne, T., 206, 208
Ottoboni, Pietro (Cardinal), 272
Oviedo y Valdés, Gonzalo Fernández de, 130, 312

P

Pace, Antonio, 74
Pacini, Zenobio, 78, 330
Palisot de Beauvois, A.M.F.J. de, 159–164
Palissy, Bernard, 8
Pancras, Gerbrand, 264
Paracelsus (Theophrastus Bombastus von Hohenheim, *called*), xlii, 112, 251, 349
Paradin, Claude, 304
Paradiser family (of Trisoz), 9, 11
Paradiser, Helena, 9, 11
Parasole, Leonardo (*also called* Norsino), 174
Parkinson, John, 84, 204
Pasquati, Lorenzo, 86
Patin, Guy, 109
Payne, John, 84
Peale, Charles Willson, 161
Pedemontanus, Alexis, *see* Ruscelli, Girolamo
Peiresc, Nicolas Claude Fabri de, 64
Pena, Pierre, 103
Pennsylvania Academy of Fine Arts, Philadelphia, 161
Pentecchi, Felice, 290
Perrault, Claude, 257
Petit Trianon, Versailles, 220
Petit, Alfred, 14
Petrarch, 34
Petre, Robert James, 8th Baron of Writtle, 206
Peyroleri, Giovanni Ignazio Francesco, 172
Pezzana, Niccolò, 46
Philip II, King of Spain, 100
Physic Garden, Chelsea, 132, 292
Physick, Philip Syng, 356
Pigouchet, Philippe, 27
Pillone Library, 23
Pillone, Giorgio, 20, 23
Pillone, Odorico, xxxiv, 20, 23
Pinelli, Giovan Vincenzo, 73
Pinson (artist), 180
Piranesi, Giovanni Battista, 278
Plantin, Christophe, 56, 62, 66, 67, 68, 72, 84, 93, 103, 107
Plantin-Moretus Museum, Antwerp, 56, 63, 67, 70
Platerius, Matthaeus, 24
Plato, 320
Platter, Felix, 87, 327
Plesch, Arpad, 180, 277, 376
Pliny the Elder, xxix, xxx, 31, 34, 73, 77, 170, 303
Plumier, Charles, 177
Pomet, Joseph, 316
Pomet, Pierre, xlv, 315–19
Pomona, 84
Pona, Giovanni, 199
Porta, Giovanbattista della, xlii, 73, 74, 202, 237, 258, 297–302
Pradel, Abraham du (Nicolas de Blégny, *pseudonym of*), 115
Prêtre, Jean Gabriel, 159, 162, 164
Prevost, Benoit-Louis, 218
Priest, Dr., 80
Pritzel, G. A., 229, 256
Prüss, Johann, 19, 20
Prüssiche Staatsbibliothek, Berlin, 64
Puccini, Tommaso, 138
Pullein, Samuel, 146
Pursh, Federick, 356, 367, 383

Q

Quincy, Dr., 230

R

Rafinesque, Constantine Samuel, xlv, 358, 365–75, 383
Raliegh, Sir Walter, 80, 82
Rambosson, Jean Pierre, xlv, 319–324
Ratzenberger, Caspar, 87, 327
Ravelingen, Frans van (Franciscus Raphelengius), 66
Recchi, Nardo Antonio, 100
Redi, Francesco, 140
Regensburg (Ratisbon), 3
Reimarus, J. A. H., 210
Reneaulme, Paul de (Paulus Renealmus), 178–80
Richardson, Richard, 95
Richelieu, Armand Jean du Plessis de, (Cardinal), 251
Ridinger, Johann Elias, 212, 216, 217
Rittershausen, Conrad, 307
Robert, Nicolas, xlii, 257, 258, 260
Roberts, Henry, 208
Roberts, James, 147
Robin, Jean, the Elder, xxxviii, 108–11, 182, 254
Robin, Jean, the Younger (*known as* Vespasien), 109, 254
Rogers, William, 84
Rondelet, Guillaume, 62
Rouillé, Guillaume, 303
Royal Academy, London, 224
Royal Horticultural Society, London, 290, 292
Royal Oak Foundation, New York, 78
Royal Society, London, 141, 316
Ruaber, Christoph, 9, 11
Ruano, Dr. (Garcia da Orta), 107

Rudolph II, Emperor, 50, 68, 298
Ruelle, Jean, 42
Ruscelli, Girolamo (Alexis Pedemontanus, *pseudonym of*), 330
Rush, Benjamin, 348, 350, 355, 358, 364, 367
Rusticucci, Girolamo (Cardinal), 170
Ruysch, Frederik, 261

S

Sabatti, Costantino, 278
Sabatti, Liberato, 278
Saccardo, Pier Andrea, 154
Sachsische Landesbibliothek, Dresden, 46
Saint Pierre, Bernardin de, 320
Salerno, medical school at, xxxi, 8
Salis, Giovanni, 107
Salviani, Ippolito, 78, 307
Salviani, Orazio, 74
Santacroce, Prospero (Cardinal), 172
Santi, Giorgio, 286, 287, 288
Savi, Gaetano, 286–90
Scalberge, Frédéric, 254
Scaliger, Joseph Justus, 182
Scaliger, Julius Caesar, 312
Schoepf, Johann David, 348, 370
Schöffer, Peter, 15
Schönsperger, Johann, 3
Schott, Johann, 36
Scott, George, 151
Sembertini, Carlo, 334–39
Sembertini, Giovanni Battista, 336
Serantoni, Antonio, 287
Seuter, Bartholomaüs, 212, 216, 217
Seymour, Edward, Lord Protector, Duke of Somerset, 58
Sherard, William, 331
Shipton, J., 206, 208
Sibly, Ebenezer, 236
Sibmacher, Hans, 307
Sichem, George Van, 50
Silenus, 32
Sivry, Philippe de, Governor of Mons, 63
Sixtus V, Pope, 170
Sloane, Sir Hans, 80, 316
Smith, James Edward, 152, 356
Smithson, Hugh, Earl of Northumberland, 206, 208
Società Botanica Fiorentina, 138
Society of Artists, London, 224
Socrates, 320
Solander, Daniel Carl, 136, 147
Sorg, Anton, 3
Sousa, Martim Afonso de, 104
Spaulding, Lyman, 352, 357
Speckle, Veit Rudolph, 44
Spiegel, Adriaan van de (Spigelius), 85–9, 330
Staatsgalerie, Stuttgart, 98
Stafford, Simon, 72
Stearn, William, 268
Stearns, Samuel, 348
Süleyman I the Magnificent, Sultan, xxix, 51

T

Tabernaemontanus (Jacob Dietrich of Bergzabern, *called*), 82
Tacuino, Giovani, 14, 20
Tavernier, Jean Baptiste, 116
Taylor, Simon, 151
Tempesti, Giovan Battista, 288
Tennent, John, 347
Teyler, Johannes (Teijler), 216
Thal, Johann, 246
Theophrastus of Eresos, xxviii, xxix, xxx, 73, 84, 88, 206, 264, 287, 303
Thomas de Cantimpré, xxxi, 4, 8
Thompson, James, 230
Thomson, John, 383
Thomson, Samuel, 360–2
Thoreau, Julian, 184
Tilli, Michelangelo, 140
Titian (Tiziano Vecellio, *called*), 23
Tornabuoni, Alfonso, 144
Tornabuoni, Niccolò (Cardinal), 144
Tornieri, Giacomo, 170
Torrey, John, 375
Tournefort, Joseph Pitton de, 44, 79, 162, 224, 229, 276, 282, 287, 316
Tozzetti, Ottaviano Targioni, 286
Tragus, *see* Bock, Hieronymous
Trampe, Johann Gorrfried, 339, 340
Treveris, Peter, 23–4
Trew, Christoph Jacob, xlii, 340
Trionfetti, Giovanni Battista, 278, 282
Turner, William, xxxviii, 56–61, 84

U

Ubertini, Cesare, 278, 282, 285
Ungnad, David, Count of Weissenfeld, 63
Università la Sapienza, Rome, 170
Urban VIII (Pope), 144

V

Vaillant, Sébastien, 156
Valesio, Francesco, 86
Valgrisi, Vincenzo, 46, 48, 50, 52, 169
Valle, Domenico, 200
Vallet, Pierre, 109
Vasari, Giorgio, 50
Vecellio, Cesare, 23
Ventenant, Etienne Pierre, 375
Venus, 32
Vérard, Antoine, 19, 20
Vesalius, Andreas, 37, 39, 86
Vesling, Johannes, 98, 141
Victor Amadeus II, Duke of Savoy, 270
Vincent of Beauvais, xxxi
Vines, Sydney Howard, 45, 46
Virgil, 170
Virgilio, Marcello, 31, 42
Vitsen, Nicolaes (Witsen), 140, 151
Vogelin (publisher), 304
Voorst, Adolph van (Vorstius), 88–89
Voorst, Everard van, 88
Vroede, François de, 264

W

Wale, Samuel, 206, 208
Wandelaar, Jan, 268
Warren, John C., 355, 365
Waterhouse, Benjamin, 355, 362
Webster, Joshua, 230–35
Wedel, Georg Wolfgang, 111–114
Wedel, Johann Georg, 111
Weiditz, Hans, 32, 34, 44
Weinmann, Johann Wilhelm, xl, 210–17
Wesley, John, 347
Wesling, Johannes, 200
Wieland, Melchior, *see* Melchiorre Guilandino
Wilhelm IV, Landgrave of Hesse, 246
Wistar, Caspar, 161, 356
Withoos, Alida, 266
Withoos, Mattias, 266
Witsen, Nicolaes, *see* Vitsen, Nicolaes
Witsen, Robert, 151
Wood, Alphonso, 383
Woodville, William, 354, 383
Wotton, Sir Henry, 51

Y

Yates, Francis, 112
Young, William, 147
Zannichelli, Gian Giacomo, 202

Z

Zelada, Francesco Saverio de (Cardinal), 285
Ziegler, Jerg, 40
Zuccari, Federico, xxvii, xxx

PHOTOGRAPHY BY
SUSAN LEOPOLD AND KIMBERLEY FISHER
OAK SPRING GARDEN LIBRARY

PRINTED IN DUOTONE AND PROCESS COLOR BY
FINLAY PRINTING
BLOOMFIELD, CONNECTICUT

BOUND IN JAPANESE CLOTH BY ACME BOOKBINDERY
CHARLESTOWN, MASSACHUSETTS

DESIGNED AND COMPOSED
IN DIGITAL MONOTYPE BEMBO BY
MARK ARGETSINGER
ROCHESTER, NEW YORK